Current Topics in Behavioral Neurosciences

Volume 19

About this Series

Current Topics in Behavioral Neurosciences provides critical and comprehensive discussions of the most significant areas of behavioral neuroscience research, written by leading international authorities. Each volume offers an informative and contemporary account of its subject, making it an unrivalled reference source. Titles in this series are available in both print and electronic formats.

With the development of new methodologies for brain imaging, genetic and genomic analyses, molecular engineering of mutant animals, novel routes for drug delivery, and sophisticated cross-species behavioral assessments, it is now possible to study behavior relevant to psychiatric and neurological diseases and disorders on the physiological level. The *Behavioral Neurosciences* series focuses on "translational medicine" and cutting-edge technologies. Preclinical and clinical trials for the development of new diagnostics and therapeutics as well as prevention efforts are covered whenever possible.

More information about this series at http://www.springer.com/series/7854

Grace Lee · Judy Illes · Frauke Ohl
Editors

Ethical Issues in Behavioral Neuroscience

Springer

Editors
Grace Lee
Judy Illes
Faculty of Medicine
National Core for Neuroethics
University of British Columbia
Vancouver, BC
Canada

Frauke Ohl
Faculty of Veterinary Medicine
University Utrecht
Utrecht
The Netherlands

Co-editor of Part I
Franck Meijboom
Ethics Institute
University Utrecht
Utrecht
The Netherlands

ISSN 1866-3370 ISSN 1866-3389 (electronic)
ISBN 978-3-662-44865-6 ISBN 978-3-662-44866-3 (eBook)
DOI 10.1007/978-3-662-44866-3

Library of Congress Control Number: 2014954602

Springer Heidelberg New York Dordrecht London

Printed on acid-free paper

Springer is part of Springer Science+Business Media (www.springer.com)

Preface

We are pleased to present this volume on ethical aspects of studying behavior in psychiatric and neurological disorders as part of the *Current Topics in Behavioral Neurosciences* (CTBN) series. We have brought together a collection of chapters that provides both critical reviews of current advances in the field and key analyses of related ethics issues. The volume aims to bridge disciplines of neurobiology and psychology to provide a contemporary overview of the literature relevant to understanding neurobehavior and how ethics informs and reflects on neurobehavioral research. There is dual emphasis on ethical challenges in experimental approaches and in clinical research involving human participants. In essence, the central theme is one of Neuroethics, the field formalized in 2002 that is dedicated to interlocking the excitement of advances in basic neuroscience and clinical neurology with human values and the diversity of our societies.

With the range of topics covered, we hope that the volume will appeal to CTBN's readership of all behavioral neuroscientists, animal science researchers, clinical scientists, allied health professionals, applied ethicists, and to scholars in the social sciences alike. We also deeply hope that as neuroscience has an impact on and visibility in the daily lives of people in both resourced and under-resourced parts of the world, the volume will serve as a useful resource for early career scientists and scholars who must actively evaluate their research through an ethics lens today more than ever before.

This book has been a collaborative international effort from start to finish. Professor Frauke Ohl had primary responsibility for the first six chapters of the volume on the ethics of using animal subjects for neurobehavioral research, and was assisted by Dr. Franck Meijboom. Postdoctoral Fellow Grace Lee and Professor Judy Illes took the lead on the ten chapters that engage readers in a discourse on ethical issues for neurobehavioral research using human subjects, with a chapter linking pre-clinical and clinical research.

We gratefully acknowledge the support of all who generously fund the research and knowledge translation activities of both our organizations. At the University of Utrecht in the Netherlands, Drs. Ohl and Meijboom thank the Dutch Ministry of Public Health, the Dutch Ministry of Economic Affairs, Neuroscience and

Cognition Utrecht, and the Netherlands Organization for Scientific Research (NWO) provided direct or indirect support to this work. At the National Core for Neuroethics at the University of British Columbia in Canada, Drs. Lee and Illes thank The Canadian Institutes of Health Research, the National Institutes of Health Research, the Canadian Foundation for Knowledge Innovation, the British Columbia Knowledge Development Fund, GenomeBC, GenomeCanada, the Vancouver Foundation, the Stem Cell Network, NeuroDevNet, Inc., the Vancouver Coastal Health Research Institute, the Foundation for Ethics and Technology, the Dana Foundation, and the North Growth Foundation.

We are grateful to CTBN Editors Mark Geyer, Bart Ellenbroek, and Charles Marsden for the opportunity to create this volume and Susanne Dathe at Springer, for engagement in bringing the final product to you.

Vancouver — Grace Lee, Judy Illes

Utrecht — Frauke Ohl

Contents

Part I
Experimental Animal Research

Ethical Issues Associated with the Use of Animal Experimentation in Behavioral Neuroscience Research

Frauke Ohl and Franck Meijboom

Abstract This chapter briefly explores whether there are distinct characteristics in the field of Behavioral Neuroscience that demand specific ethical reflection. We argue that although the ethical issues in animal-based Behavioral Neuroscience are not necessarily distinct from those in other research disciplines using animal experimentation, this field of endeavor makes a number of specific, ethically relevant, questions more explicit and, as a result, may expose to discussion a series of ethical issues that have relevance beyond this field of science. We suggest that innovative research, by its very definition, demands out-of-the-box thinking. At the same time, standardization of animal models and test procedures for the sake of comparability across experiments inhibits the potential and willingness to leave well-established tracks of thinking, and leaves us wondering how open minded research is and whether it is the researcher's established perspective that drives the research rather than the research that drives the researcher's perspective. The chapter finishes by introducing subsequent chapters of this book volume on Ethical Issues in Behavioral Neuroscience.

Keywords Animal behavior · Translational value · Animal ethics · Animal model

Contents

F. Ohl (✉) · F. Meijboom
Department Animals in Science & Society, Faculty of Veterinary Medicine, University Utrecht, Yalelaan 2, PO Box 80.166, 3508 TD Utrecht, The Netherlands
e-mail: f.ohl@uu.nl

F. Meijboom
Ethics Institute, Utrecht University, Utrecht, The Netherlands

Curr Topics Behav Neurosci (2015) 19: 3–15
DOI: 10.1007/7854_2014_328

Published Online: 15 July 2014

1 Reasons for Reflection?

The first part of this book on Ethics in Behavioral Neuroscience explores the question of whether it is worthwhile, or even necessary, to reflect specifically on animal experimentation in Behavioral Neurosciences in extension of more general considerations on Animal Ethics in the broader sense. Are there distinct characteristics in this field of research that demand specific ethical reflection?

Of course, there is an obligation to reflect on the use of animals as models in Behavioral Neuroscience. But, research on animals has already triggered considerable attention during the last decades, exploring whether it may be justifiable to use animals for experiments at all and, if so, how to weigh the costs of such use against its benefits (e.g. Singer 1975; Van Zutphen et al. 1993; Brom 2002; Nuffield 2005) and these same questions hold for other areas of research and are not unique to the field of Behavioral Neuroscience.

More recently however, Neuroethics has emerged as a distinct field of applied ethics within the philosophy of neuroscience (Stefansson 2007; Illes and Sahakian 2011). Neuroethics deals with a wide range of questions related both to the ethical implications of practical experimentation in neuroscience and the application of the results of such neuroscientific research as well as, in turn, the consequences of neuroscience for ethics (cf. Roskies 2002; Buller 2014). In practice however, it appears that, to date, these discussions have mainly focused on humans—as for example, discussions on the moral rights and wrongs of the enhancement of brain function, or questions related to the concept of free will and moral agency. Thus, although Behavioral Neuroscience does raise specific ethical questions in relation to experimental animal research, the attention of neuroethicists has not, at least to this point, been specifically concerned with this wider context of the ethics of animal experimentation in neuroscience.

Yet there are very specific issues which are raised by the use of animal experiments in this particular area of neuroscience; it is because of those specific aspects, which lie in the interactions between the fields of animal ethics and neuroethics, that we consider it relevant to dedicate a section of the book to the ethical issues of animal-based research in Behavioral Neuroscience. Alongside the more basic questions of animal ethics, a research field that is often dependent on modeling distinct mental capacities and behavioral responses in animals, may have specific implications on considerations on the moral status of animals. Thus, the very criteria that lead us to judge some animal a valid research model in Behavioral Neuroscience are pretty much the same as we would use to grant animals moral consideration for their own sake, which inevitably leads to some conflict in terms of the acceptability of their use for experiments.

Therefore, we argue that although the ethical issues in animal-based Behavioral Neuroscience are not necessarily distinct from those in other research disciplines using animal experimentation, this field of endeavor makes a number of specific, ethically relevant, questions more explicit and, as a result, may expose to discussion a series of ethical issues that have relevance beyond this field of science.

In addition to the conflict which may result from the fact that the most valid animal models may also be those which we might consider, from those same characteristics, as having the highest claim to be worthy of specific moral consideration, other questions may, for example, be related to the predictive power of specific animal models and the degree to which results gained on those models may be truly translated to other systems or species (including humans) (Rollin and Rollin 2014). How should we deal with uncertainties regarding the predictive and construct validity of given (animal) models (cf. Geyer and Markou 1995)? How much research is needed before it is justified to move from work on animals to take the step into (pre)clinical trials? And finally: how can we balance the potential benefit of using animal models that might have higher mental capacities (thus enhancing possible translational value to humans) against the cost that such higher mental capacities may imply greater suffering as the result of experimental manipulations?

This chapter briefly introduces ethical questions raising from animal-based Behavioral Neuroscience, each of which will be developed in more detail in the subsequent chapters of this section.

2 The Moral Status of Animals as a Start of Ethical Concerns About Their Use in Experiments

The use of animals in experimental research in general has raised many concerns over the years. While perhaps earliest concerns about experimentation involving live animals arose in the UK in the nineteenth century (Franco 2013), debate about the moral status of animals is not restricted to Europe, but is nowadays of concern in many countries including the US, Australia and Asian countries (cf. Bovenkerk 2012; Linzey 2014; Nuffield Council 2005). The origin of these discussions lies in the recognition of animals as moral subjects toward which we can have moral duties (Warren 1997). A significant number of ethicists concede that animals have some moral value that is independent of their use by humans. However, there is a diversity of arguments that underlie the recognition of this moral standing of animals. Some start in the recognition of animals as living beings that have a good of their own. This is based on the idea that animals develop, maintain their life, and can adapt successfully to their environment. As a consequence, they have inherent worth as animals (Taylor 1986). Others argue for the moral considerability of animals by virtue of their being able to feel (e.g. Rollin 2011)

It is beyond the scope of this chapter fully to elaborate on the diversity of views that have characterized the debate in the past few decades (Callicott 1980; Carruthers 1992; DeGrazia 1996; Midgley 1983; Korsgaard 2005; Nussbaum 2006; Regan 2004; Rollin 1981; Rowlands 2002; Singer 1995)—and these arguments are rehearsed in greater detail in later by Bovenkerk and Kaldewaij (this volume) and Vieira de Castro and Olsson (this volume). However, both within the field of animal ethics and in formal regulations on the use of animals in research there is a consensus that we have valid and sufficient reasons to consider animals as legitimate objects of our moral concern (cf. De Cock Buning et al. 2009 ; EU 2010).

In a nutshell, such recognition implies that animals should be taken into account in our moral reasoning for their own sake. In animal research the health and welfare of animals is of course taken into account, because compromise of either state may frustrate the research or influence the results in some way. However, speaking about animals as moral subjects implies a further step: if animals are acknowledged to be worthy of consideration and significant entities in their own right, we have direct moral reasons to ensure that our actions take account of their interests as well as our own. How this consideration can be translated into practice is not always immediately clear. Some argue that, as a consequence, any type of animal research is unacceptable (Regan 2004). Others stress that there are also legitimate ethical positions that aim to take the interests or value of animals seriously, yet do not exclude the option that using animals for research can morally be justified (cf. Rollin and Kessel 1990; and see Rollin, this volume; Vieira de Castro and Olsson, this volume). This implies that, on the one hand, using animals is not something that is to be rejected by principle; on the other hand, although animals continue to be used, such use demands a careful consideration.

Frequently, such consideration is based on an analysis of the comparative costs (i.e., harm to individual animals) and benefits (see again Vieira de Castro and Olsson, this volume). Determining the moral justification of animal research in terms of such cost–benefit analysis, in effect gives particular emphasis to two central questions: does the expected result of the experiment or project outweigh the potential suffering of the animals; and is the experiment being performed in the best possible way with regard to the principles of Replacement, Reduction and Refinement (Russel and Burch 1959). Such an evaluation process implies that the ethical justification of animal experiments demands that there shall be specific benefits as a result of any experiment that are considered important enough to outweigh the costs for the animal. In general, the benefit of using animals in experiments is argued in terms of its contribution towards reduction of suffering in humans as an immediate or ultimate aim. This holds equally for experimental animal research in Behavioral Neuroscience.

The majority of such experiments is aimed, if sometimes indirectly, at gaining knowledge about the executive function of the brain. Most commonly, it is the dysfunctioning of particular processes that is of especial interest, because some specific dysfunction of the CNS underlies a variety of disorders that can have a severe impact on (human) quality of life. Since many ethical frameworks stress that we have a duty to take action in the face of human suffering, there is a moral imperative to perform some form of research in this field. Having accepted such duty to care for the health and wellbeing of humans, however, there is no automatic logical presumption that animals have to be used or that use of animals is automatically justified. Therefore, an important aspect of the ethical justification of animal experimentation is discussion both of the need to use animals at all *and* on the relevance of animal models in research (to ensure that animals used genuinely do provide appropriate models for human systems or disorders, rather than simply mimicking symptoms but in an unrelated way). We should, therefore, take a closer look at the validity of the animal models used in this field of research, and their relevance for transference of results to other systems and species.

3 Relevance of Animal Models?

The actual relevance of animal models for a distinct field of research is difficult to assess. One may get some impression of the current [quantitative] importance of animal models in experimental Behavioral Neuroscience by way of a literature research, although, of course, there is virtually no way to assess whether the use of particular animal models employed, has indeed resulted in relevant output. Given such reservations, however, it appears from a rough and explorative online screening for recent literature, that of the 7,500 original research articles that have been published on this topic during the last 5 years (PubMed 2009–2013), more than 40 % of the papers at least make some reference to animal models. More specifically, PubMed reports the following number of articles published in the last 5 years when searching with the key-phrase "behavioral neuroscience" together with [....]:

[humans]: 2400
[either humans or other animals and (computer modeling)]: 56
[either humans or other animals and (in vitro)]: 190
[other animals]: 3665

While such numbers cannot tell us anything about the actual contribution of animal studies to developments, and valid advances, within this field of research, such an overview suggests that studies in humans and animals each contribute almost equally to the overall publication output in neurobehavioral research. Given all the recent technical developments and the range of opportunities now available to perform non-invasive experiments in humans, as well as to model neural processes in vitro, it seems somewhat intriguing that animal-based experiments continue to play such a big role in Behavioral Neuroscience. For this to remain true, the results gained from animal experiments in Behavioral Neuroscience are obviously assessed, at least by the researchers themselves, or the wider research community, as of importance—perhaps because they are thought to contribute as much to the development of the research field as do studies in humans, or perhaps for other reasons. It may, for example, be that animal experiments are considered more ethically acceptable than pre-clinical studies in humans; it is also possible that research, or at least the publication of research, constrains itself by following distinct traditions, such as demanding the validation of novel findings by comparing them to already published animal models and test procedures.

One significant question arising from the continued extensive use of animals is embedded in the broader debate on the possibility of replacement of animal experiments, the first of the 3R-principles (Russel and Burch 1959). Although, the search for animal-free methods is complex in any research field (Doktorova et al. 2012; Hendriksen 2009; Huggins 2003; Manciocco et al. 2009; Penza et al. 2009), in Behavioral Neuroscience in particular, the modeling of complex systems such as executive processes of the brain or the central nervous system (CNS) may indeed limit the possibility of finding alternatives and may thus demand use of animal models; at least at present, available in vitro methods, and computer models seem unable to display the complexity of CNS-generated, behavioral-cognitive processes. It may be of note, however, that the declared goal of one of the current EU flagship programs (the Human Brain Project) is: "to build a completely new ICT infrastructure for neuroscience, and for brain-related research in medicine and computing, catalyzing a global collaborative effort to understand the human brain and its diseases and ultimately to emulate its computational capabilities."[1]

While waiting for the results of such initiatives, the use of methods that avoid the use of live animals is still quite limited. But even if we do accept the need to base parts of research in Behavioral Neuroscience on the use of animals, some ethically relevant questions remain to be considered. And first among these questions, as above, is: what it is that animals are supposed to model and are we choosing the correct models?

If we look in more detail at the specific areas of animal experimentation, a literature search using the term "animal model" in combination with some general topics reveals that use of animal models in many cases is related to research into a variety of human-specific, mental disorders. Such a literature scan, again performed on articles listed by PubMed and over the same time period, picks out the following number of publications with the combined keywords [animal model] and [...]:

[stress]: 13561
[alzheimer]: 2568
[depression]: 2918
[schizophrenia]: 1464
[anxiety]: 2340
[mood disorder]: 982
[hyperactivity]: 924
[addiction]: 868
[post traumatic stress disorder]: 247
[eating disorder]: 219

This simple screening results in the identification of more than 25,000 articles on this (artificial) selection of human mental states/disorders. [For comparison: a search on [animal model] and [cancer] delivers 20,304 hits]. Without going too far in interpreting such a crude literature search, we may feel confident enough to suggest that animal models are still considered important in investigating human mental states and/or functions; indeed this use of animal models in exploration of

---

[1] see https://www.humanbrainproject.eu/.

human mental function comprises the majority of those animal studies uncovered in our initial literature review.

From any consideration of the ethics of animal experimentation, such extensive usage of animals begs the question as to whether the obvious importance of animal models genuinely translates into actual useful and relevant output, since the assumption that animals are relevant models can be seen as a pivotal argument in the moral justification of animal use (Rollin and Rollin 2014). A realistic assessment of the benefits and, thus, actual relevance of animal studies is however, more or less impossible to do in practical terms (as explored in more detail in this volume by Viera de Castro and Olsson). Yet the very assumption that the animals chosen as models *are* valid and, thus, relevant models for human mental problems (such as distinct cognitive and emotional capacities) may indicate that these animals share with us morally relevant characteristics that may make them (more) worthwhile protecting, promoting additional concerns about their use in experimental treatments.

Given the need in Behavioral Neuroscience to model complex systems, and perhaps even integrate executive processes, such as learning and social behavior, it may be argued that the best choice for an animal model is the use of animals with 'higher' cognitive capacities, such as primates or dogs. However, as we have noted already, the scientific argument that these animals serve as relevant models because of the greater physiological or behavioral similarity to humans, as compared to other species like fruitflies or mice for example, is often the basis of public concerns because of exactly these same characteristics. As a result, experiments on primates and dogs often raise stronger societal resistance than experiments on rodents or fish (Hagen et al. 2012). In practice this complicates the discussion on the choice for the best possible animal model for a distinct experiment, as in fact the choice of the 'best possible' animal model becomes an interplay between value and scientific judgments.

In this context, it might be of interest to get some idea on what animal species actually are being used to investigate human mental disorders. Once again, we have used PubMed to search for all articles in PubMed which use again [animal model], but this time with [anxiety]. This search delivers 2,340 hits for publications between 2009 and 2013; repeating the same search with reference to individual species gives the following numbers of publications:

[mice or mouse]: 998 (355 on [C57BL])
[rat]: 970 (369 on [Wistar])
[primate]: 615
[fish]: 59
[dog]: 9
[rabbit]: 2

Although surely not fully representative, these findings are at least indicative of current patterns of research publication based on experiments using different animal species: first, we may note that about 25 % of publications within this specific area of research refer explicitly to primates. This high proportion undoubtedly overp-resents the number of experiments actually done in primates, since the proportional representation in publications reported here does not reflect the distribution of species reported as being used in research (reported for example by the EU in 2010 as: mice 59.3 %; rats 17.6 %; other rodents including guinea pigs and rabbits 5.2 %; ungulates 1.4 %; cats, dogs and other carnivores 0.3 %; and non-human primates 0.08 %; birds, reptiles, amphibians, and fish taken together 15.9 %; see Hagen et al. 2012). Secondly, and perhaps not surprisingly, experiments on mice and rats dominate the report on actual animal use (rats and mice combined, around 70 %) and, in this case are represented to about the same extent of (again combined) roughly 60 % of published articles.

What is interesting though is that more than a third of publications on mice refer specifically to the inbred strain C57BL, and that about the same proportion of rat studies seem to involve the Wistar strain. Further, when we look at methodologies employed in experiments, our literature screening on [animal model] with [anxiety] and now specifying [elevated plus maze, or open field, or dark light box] results in 769 hits (again about one-third of the total of 2,340 hits). Overall, there seems to be at least some indication that animal experiments in anxiety research, as merely one example, is being based to a significant extent on only a small number of test systems and primarily on experiments on one distinct mouse or rat strain, respectively.

Such considerations may be of special relevance when considering future developments in experimental Behavioral Neuroscience research. It is predicted that mood disorders in humans, as for example clinical depression, will become one of the leading causes of disability worldwide (Murray and Lopes 1997; Rodríguez et al. 2012). Such a prediction increases the drive to understand better the development and underlying mechanisms of such disorders in order to develop better prevention and treatment; this, in turn, may increase the requirement or motivation to undertake more research, in all probability based in the same way on the use of animal models. This potential development focuses further a debate on the appropriateness and validity of models currently used.

While we would not want to overstate the implications from this limited survey —a more rigorous analysis would clearly demand a much more extensive literature research—we may at least wonder whether indeed the combination of these test systems and strains is genuinely believed to deliver the best possible results in anxiety research or is simply based on tradition, conservatism and lack of exploration of alternative models—or acceptability to journals and their equally conservative referees. To us it seems important at least to raise the question as to whether animal-based research may be self-perpetuating as the result of unimaginative and conventional thinking regarding the choice of animal models and test systems used, and whether such conventional choices are truly the best possible choices in the search for innovative research findings. Gold standards surely have their use, but we should not forget that such standards are established within the frames of knowledge at their time of establishment. Scientific knowledge however develops rapidly—or so we hope—and it may be reasonable to wonder about the half-life of any gold standard, before it turns into fool's gold.

In a recent review article on the predictive value of animals models McGonigle and Ruggeri (2014) state: "For major mood disorders, such as depression and anxiety, inadequacies in the animal models have helped undermine the confidence of major pharmaceutical companies to the point that several, if not the majority have either withdrawn from this therapeutic area or significantly reduced their internal research activities." Indeed it seems of crucial importance not only to try and optimize procedures of animal-based research as such, but carefully to evaluate how appropriate is the model chosen and, in this way not only optimizing the translational value of studies in animal models, but also allowing for actual, retrospective assessment of such translational value. McGonigle and Ruggeri conclude from their review that "Comparison of models within a given therapeutic area, approaches to models and cross fertilization between therapeutic areas will do much to improve translational research. By thinking outside the box that each therapeutic area has created, improvements will be made to existing models to make these more predictive. These advances will inform both the development of new models and biomarkers that will enhance the translational relevance as well as the predictive utility of pre-clinical animal models of human disease, irrespective of therapeutic area."

4 Thinking Out-of-the-Box

Research claims to be innovative, with the exception of experiments that are being done to confirm previous findings. But innovative research, by very definition, demands out-of-the-boxthinking. At the same time standardization of animal models and test procedures for the sake of comparability across experiments inhibits the potential and willingness to leave well established tracks of thinking. Indeed, as Rob Hutter states: "...today's neuroscience research can be described as 'what happens' research versus 'how to make happen' research. One could argue that the former precedes the latter, but there are perspective issues that drive the type of questions researchers are likely to ask as well as the scope of tasks and behaviors that can be included in rigorous experimental conditions."[2] We may thus wonder how open minded research is and whether it is the researcher's established perspective that drives the research rather than the research that drives the researcher's perspective.

In search of the best possible research results in animal-based Behavioral Neuroscience and, thus, in trying to optimize the benefit of animal experiments, while at the same time minimizing the costs, any innovative perspective will be closely linked to the choice of the animal model used. Is it, for example, necessary for an animal to being able to perceive pain in order to resemble a valid animal

[2] DO.Anything; The Science of Intentional Change, posted by Rob Hutter, January 2013; http://robhutter.com/neuroscience/the-neuroscience-of-behavioral-insight/.

model for pain research? Bernard Rollin (this volume) suggests that "the modification of *telos* by way of combining genetic engineering with behavioral neuroscience as a remedy for practices that cause pain or suffering by violation of *telos* represents a whole new approach to intractable problems of animal welfare that emerge from contemporary animal use" and is supported in this by Adam Shriver (this volume) who argues "that we already have, or are extremely close to having, the capacity to dramatically reduce the amount of suffering caused in biomedical research via genetic modification of the animals used in research." The appropriate selection, or perhaps even creation of animal models thus deserves special attention in relation to options for reducing the potential for animal suffering, in relation to the improvement of animal welfare and the considerations of animal integrity (cf. Van der Staay et al. 2009). Such evaluation processes may, however, also profit from some out-of-the-box thinking and the subsequent chapters in this book are intended to stimulate such out-of-the-box thinking in animal-based Behavioral Neuroscience.

Bernice Bovenkerk and Frederike Kaldewaij make a start by reflecting on the tension between the need for translatability in animal models and the moral status of animals. They invite us critically to think about some justifications for the claim that human beings and more complex animals have superior moral status and argue that contemporary approaches which attribute equal moral status to all beings that are capable of conscious strivings (e.g., avoiding pain and anxiety; aiming to eat and play) are based on more plausible assumptions. They further suggest that, while there might be good reasons to assume that more complex beings would be harmed more by a specific physical or environmental intervention, it may also be possible that higher cognitive capacities result in less harm, because of a better ability to cope.

The ultimate use and validity of animal models would require to prove that indeed their use achieves its objective, that is that the results of a given animal study is a benefit that could not be gained otherwise. Ana Catarina Vieira de Castro and Anna Olsson in their chapter explore how cost-benefit analyses currently are being approached, and they conclude that specific 'costs' of animal experimentations in terms of harms inflicted on the animals, are far easier to assess that their benefits—a problem that actually may not be specific for Behavioral Neuroscience. Still, as outlined above, Behavioral Neuroscience often may affect the emotional and/or cognitive state in animals used, and such harm is difficult to counteract. Olsson and Vieira de Castro however come to the conclusion that effective cost-benefit analysis suffers from a lack of realistic ability to assess the true benefits and provocatively suggest that perhaps the benefit assessment should be discarded from any procedural ethical consideration, which, instead, should focus exclusively on the three Rs and improving animal welfare.

Paula Droege and Victoria Braithwaite continue with "a cross-disciplinary debate about the sort of framework that will best organize the growing body of data on behavior, development and anatomy of fish and other non-human animals in order to assess the capacity for consciousness." Fundamentally, considerations on how to assess consciousness in the first place remind us that a taxonomic classification of 'higher' and 'lower' species may be a poor guideline for the assessment of a species capacity to suffer. Instead, as Droege and Braithwaite state, only "once we have a means of determining what sorts of animals feel conscious pain, we can more effectively think about ways to minimize or eliminate their suffering."

Bernard Rollin then reflects on the question why we would consider it ethically problematic or even unacceptable to eliminate an animals' capacity to suffer by means of genetic manipulation, if we do find it acceptable to cause such suffering in the first place? "In biomedical research, we do indeed inflict major pain, suffering and disease on animals. And genetic engineering seems to augment our ability to create animals to model diseases, particularly the more than 3,000 known human genetic diseases. [...] Perhaps one can use the very genetic engineering which creates this dilemma to ablate consciousness in such animal models, thereby escaping a moral impasse." Underlying Rollin's considerations is the understanding that it is the individual one can wrong, not the *telos*.

In the concluding chapter, Adam Shriver explores how genetic manipulation of animals in order to reduce the animal's capacity to suffer would translate into experimental practice. What would be the benefit wnd what the costs of such manipulation? And would the elemination of the animal's capacity to suffer not be the most logical way to solve ethical dilemmas in experimental animal research?

As Bovenkerk and Kaldewaij state in their conclusions: "We have not attempted to give definitive answers here, but rather to raise some moral issues and to point out normative assumptions made in animal experimentation in general, and neurobehavioral research in particular." Indeed, ethical issues, as opposed to neurobehavioral questions, cannot be answered by way of statistical significance, but demand an ongoing and constructive discussion, to which we hope to contribute with this book.

Acknowledgments The authors wish to thank Rory Putman for helpful comments on the manuscript.

References

Bovenkerk B (2012) The biotechnology debate: democracy in the face of intractable disagreement, library of ethics and applied philosophy, vol 29. Springer, Dordrecht

Brom FWA (2002) Science and society: different bioethical approaches towards animal experimentation. ALTEX 19(2/02):78–82

Buller T (2014) Bridging the gap between science and ethics? cambridge quarterly of healthcare ethics, special section: neuroethics and animals animal minds and neuroimaging, vol 23. Cambridge University Press, pp 173–181. doi:10.1017/S0963180113000704

Callicott JB (1980) Animal liberation: a triangular affair. Environ Ethics 2–4:311–338

Carruthers P (1992) The animals issue: moral theory in practice. Cambridge UP, Cambridge

De Cock Buning Tj, Meijboom FLB, Swart JAA (2009) Ethiek en Dierproeven. In: van Zutphen LFM (ed) Handboek proefdierkunde. Proefdieren, dierproeven, alternatieven en ethiek. Elsevier, pp. 321–334

DeGrazia D (1996) Taking animals seriously. Mental life and moral status. Cambridge UP, Cambridge

Doktorova TY, Pauwels M, Vinken M et al (2012) Opportunities for an alternative integrating testing strategy for carcinogen hazard assessment? Crit Rev Toxicol 42:91−106
EU (2010) Directive 201/63/EU on the protection of animals used for scientific purposes. Official Journal of the European Union L 276/33
Franco NH (2013) Animal experiments in biomedical research: a historical perspective. Animals 3:238–273
Geyer MA, Markou A (1995) Animal models of psychiatric disorders. In: Bloom FE, Kupfer DJ (eds) Psychopharmacology: the fourth generation of progress. Raven, New York, pp 787–798
Hagen K, Schnieke A, Thiele F (eds) (2012) Large animals as biomedical models: ethical, societal, legal and biological aspects. Europäische Akademie, Ahrweiler
Hendriksen CFM (2009) Replacement, reduction and refinement alternatives to animal use in vaccine potency measurement. Expert Rev Vaccines, 8:313−322
Huggins J (2003) Alternatives to animal testing: research, trends, validation, regulatory acceptance. Altex-Alternativen Zu Tierexperimenten 20(Supplement 1):3–61
Illes J, Sahakian BJ (eds) (2011) Oxford handbook of neuroethics. Oxford University Press, Oxford Library of Psychology, England
Korsgaard C (2005) Fellow creatures: Kantian ethics and our duties to animals. Tanner Lect Human Values 25:77–110
Linzey A (ed) (2014) The global guide to animal protection. University of Illinois Press, Illinois
Manciocco A, Chiarotti F, Vitale A et al (2009) The application of Russell and Burch 3R principle in rodent models of neurodegenerative disease: the case of Parkinson's disease. Neurosci Biobehav Rev 33(1):18–32
McGonigle P, Ruggeri B (2014) Animal models of human disease: challenges in enabling translation. Biochem Pharmacol 87(1):162–171. doi:10.1016/j.bcp.2013.08.006
Midgley M (1983) Animals and why they matter: a journey around the species barrier. University of Georgia Press, Athens
Murray CJ, Lopez AD (1997) Alternative projections of mortality and disability by cause 1990–2020: global burden of disease study. Lancet 349:1498–1504
Nuffield Council on Bioethics (2005) The ethics of research involving animals. Latimer Trend & Company Ltd, London. (www.nuffieldbioethics.org)
Nussbaum MC (2006) Frontiers of justice. Disability, nationality, species membership. Harvard University Press, Cambridge
Penza M, Jeremic M, Montani C et al (2009) Alternatives to animal experimentation for hormonal compounds research. Genes Nutr 4:165–172
Regan T (2004) The case for animal rights, Updated with a new Preface, The University Press Group Ltd
Rodríguez MR, Nuevo R, Chatterji S, Ayuso-Mateos JL (2012) Definitions and factors associated with subthreshold depressive conditions: a systematic review. BMC Psychiatry 12:181
Rollin BE, Kessel ML (eds) (1990) The experimental animal in biomedical research. CRC Press/ Taylor & Francis, Boca Raton
Rollin BE (1981) Animal rights & human morality. Prometheus Books, New York
Rollin BE (2011) Animal pain: what it is and why it matters. J Ethics 15:425–437
Rollin MD, Rollin BE (2014) Crazy like a fox: validity and ethics of animal models of human psychiatric. Disease Cambridge quarterly of healthcare ethics, special section: neuroethics and animals animal minds and neuroimaging, vol 23. Cambridge University Press, pp 173–181. doi:10.1017/S0963180113000704
Rowlands M (2002) Animals like us. Verso, London
Roskies A (2002) Neuroethics for the New Millenium. Neuron 35:21–23
Russel W, Burch R (1959) The principles of humane experimental technique
Singer P (1975) animal liberation: towards an end to man's inhumanity to animals. Paladin Books, St Albans
Singer P (1995) Practical ethics, 2nd edn. Cambridge University Press, Cambridge
Stefansson H (2007) The biology of behaviour: scientific and ethical implications. EMBO reports vol 8 (Special Issue)

Taylor P (1986) Respect for nature. Princeton University Press, Princeton

van der Staay FJ, Arndt SS, Nordquist RE (2009) Evaluation of animal models of neurobehavioral disorders. Behav Brain Functions 5:11

Warren MA (1997) moral status: obligations to persons and other living things. Clarendon Press, Oxford

Zutphen LFM, van Baumans V, Beynen AC (eds) (1993) Principles of laboratory animal science: a contribution to the humane use and care of animals and to the quality of experimental results. Elsevier, Amsterdam

The Use of Animal Models in Behavioural Neuroscience Research

Bernice Bovenkerk and Frederike Kaldewaij

Abstract Animal models are used in experiments in the behavioural neurosciences that aim to contribute to the prevention and treatment of cognitive and affective disorders in human beings, such as anxiety and depression. Ironically, those animals that are likely to be the best models for psychopathology are also likely to be considered the ones that are most morally problematic to use, if it seems probable that (and if indeed they are initially selected as models because) they have experiences that are similar to human experiences that we have strong reasons to avoid causing, and indeed aim to alleviate (such as pain, anxiety or sadness). In this paper, against the background of contemporary discussions in animal ethics and the philosophy of animal minds, we discuss the views that it is morally permissible to use animals in these kinds of experiments, and that it is better to use less cognitively complex animals (such as zebrafish) than more complex animals (such as dogs). First, we criticise some justifications for the claim that human beings and more complex animals have higher moral status. We argue that contemporary approaches that attribute equal moral status to all beings that are capable of conscious strivings (e.g. avoiding pain and anxiety; aiming to eat and play) are based on more plausible assumptions. Second, we argue that it is problematic to assume that less cognitively complex animals have a lesser sensory and emotional experience than more complex beings across the board. In specific cases, there might be good reasons to assume that more complex beings would be harmed more by a specific physical or environmental intervention, but it might also be that they sometimes are harmed less because of a better ability to cope. Determining whether a specific experiment is justified is therefore a complex issue. Our aim in this chapter is to stimulate further reflection on these common

B. Bovenkerk (✉)
Philosophy Group (CPT), Wageningen University, Wageningen, The Netherlands
e-mail: bernice.bovenkerk@wur.nl

F. Kaldewaij
Research Institute for Philosophy and Religious Studies, Utrecht University, Utrecht
The Netherlands
e-mail: F.E.Kaldewaij@uu.nl

Curr Topics Behav Neurosci (2015) 19: 17–46

DOI: 10.1007/7854_2014_329

Published Online: 17 July 2014

assumptions behind the use of animal models for psychopathologies. In order to be able to draw more definite conclusions, more research will have to be done on the influence of cognitive complexity on the experience of (human and non-human) animals.

Keywords Animal models · Neurobehavioural research · Moral philosophy · Philosophy of animal minds

Contents

1 Introduction

Much research in behavioural neurosciences is aimed at the prevention and cure of cognitive and affective disorders in human beings. These disorders, such as anxiety, depression, and alcohol addiction, have a severe impact on individuals' quality of life. While virtually anyone would applaud the aim of neurobehavioural science to relieve human suffering, the moral acceptability of the use of non-human animals in reaching this aim is a matter of controversy. It is significant that animal models are used precisely because we consider the use of human beings in such experiments morally impermissible. If the use of animal models is morally justified, there must be a relevant difference between human beings and the animals used in these experiments that justifies the differential treatment. Yet, if these animals are indeed good models for certain psychopathologies, it might be considered likely that they have experiences that are similar to human experiences that we consider to have strong reasons to prevent or cure (such as anxiety or sadness). What, then, justifies the use of animal models? Also, in the practice of animal experimentation we see that it is considered preferable to use animals that are less like or further removed from human beings, e.g. rats rather than apes, and zebrafish rather than rats.[1] Is such a preference morally justified?

To determine whether the use of specific animal models is morally justified, we need, first, to determine the basis of moral status. This is an issue that is widely discussed in moral philosophy. The concept of moral status will be explained in more detail below, but roughly, it involves whether and how much a being should count in our moral considerations. We shall see that moral status is often linked to the possession of specific kinds of capacities, e.g. sentience (being able to have negative and positive physical and psychological experiences) or rationality.

Besides determining the sensory or cognitive capacities required for moral status, we need to investigate whether specific species of animals (rats, zebrafish etc.) have these capacities. This will also help us determine whether their interests differ from ours and vary between different kinds of non-human animals. If non-human animals suffer much less from the experiments performed on them than human beings suffer from the pathologies we aim to cure, this might be considered a reason to regard the use of these animal models justified. To find an answer to the question what capacities specific species of animals have, and what they can be thought to experience, we require empirical evidence on different species of animals, e.g., data on their behaviour and neurophysiological responses in certain situations. However, there is an interpretational gap between data and meaning: between test results and what they actually tell us about what certain animals can do and experience. This is why this is also an issue in what is called "philosophy of mind". Philosophy of mind studies the nature of the mind and consciousness, and its relation with the brain.

We do not intend to give an exhaustive discussion of all positions in animal ethics (or moral philosophy, more generally) and the philosophy of animal minds. These are very rich and complex fields, and we cannot fully do them justice in this chapter. We have more modest aims. First, to bring to the fore some of the more important questions that need to be considered to determine whether using animals in neurobehavioural research is morally acceptable and whether it is more justified to use certain animals than others. Second, we want to show that common assumptions about the moral status or capacities of animals that may lie in the background of the use of animal models in the behavioural neurosciences are not uncontroversial, and indeed, that there is good reason to question them.

We will argue that common defences of the view that human beings have a higher moral status than animals (or even that non-human animals lack moral status altogether) involve implausible assumptions or implications. We will present two very divergent positions in contemporary moral philosophy that nevertheless both defend attributing equal moral status to all beings that consciously strive to attain

[1] See Hagen et al. (2012) and Stafleu (1994).

goals, and point out the comparative merits of these views. Furthermore, we shall also question views that less cognitively complex animals have a somehow lesser sensory and emotional experience than more complex beings across the board (or even lack consciousness altogether). We shall argue that while there are good reasons to assume that there are differences in the way that different kinds of animals are affected by negative sensory or emotional states like pain, anxiety and depression, this does not necessarily mean that less complex animals are not seriously harmed by these states.

From the outset, it is important to note that it is extremely difficult to generalise about the cognitive and sensory capacities of animals; thus, different taxa may have widely different capacities for suffering, or for coping with any suffering which may be experienced: mammals may have totally different experiences in a given situation than fish or insects. While animal ethicists tend to talk rather loosely of animals in general, especially for the purposes of this chapter, it makes a lot of difference what type of animal we are discussing. Where appropriate, we will try to specify what group of animals we are discussing, although there remains the problem in many cases that at present, we do not have perfect knowledge about the emotional and cognitive abilities of those different animal taxa, nor do we have enough knowledge on the influence of cognitive complexity on different kinds of emotional suffering.

In our considerations below, we presuppose that all neurobehavioural experiments involve some kinds of physical and environmental interferences with animals, which are aimed at making them models of specific human psychopathologies. The question is whether specific examples of such interferences are morally problematic.

2 Moral Status

2.1 Introduction

To determine whether it is morally acceptable to use specific kinds of non-human animals in experiments in the neurobehavioural sciences, the first question that we need to answer is whether these animals have moral status. If animals have moral status, this means that we should take them into account in our moral decision-making. There are, however, different ways in which things can figure in our moral decision-making: directly or indirectly. Some people have thought that we only have *indirect* duties regarding animals. One example of such a view is that we should not treat animals cruelly only because this is likely to harden us to suffering and therefore to make it more likely that we will violate our duties to other *human* beings (e.g. Kant 2000, p. 6, 442). Also, it might be thought problematic to harm an animal, because in doing so, we harm the owner of that animal. However, the concept of moral status is generally used to signify that a being counts in its own right. If animals have moral status, we that should treat them in a certain way (e.g. not treat them cruelly) *for their own sake*, rather than for the sake of others, say, human beings. We then do not merely have duties *regarding* animals, but also *to* them.

To determine whether animals have moral status, we need to know what is a necessary and sufficient basis for moral status to be accorded to them. We shall first critically discuss some justifications of attributing *unequal* status to human beings and the other animals, and to animals with different degrees of cognitive complexity. These are based on some general assumptions about the nature and basis of morality that we will argue involve implausible assumptions or implications. Then we shall discuss two different approaches in moral philosophy, that both advocate attributing equal status to all conscious animals. As these two authors also conclude, we will argue that it makes sense to consider moral questions from the perspective of all beings that have an evaluative perspective.

2.2 Unequal Moral Status

One view of the basis of morality is the idea that it is in our mutual self-interest to accept moral constraints in our dealings with one another. It might be thought that animals do not have moral status, as we cannot make a mutually advantageous agreement with them, and expect them to uphold their side of the bargain by reciprocating (e.g. Morris 2011). However, we think that the incapacity of animals to reciprocate does not give us a sufficient basis for denying them moral status. Undoubtedly, a lot of rules in social life and much of the practice of politics centre around the idea of reciprocity, but this does not seem to cover the whole content of even human morality. After all, we take it to be wrong to exploit people who are too weak (or too far removed from us) to reciprocate or take their revenge on us. If we think morality goes beyond the confines of mutual interest through reciprocation, we need to find another basis for such duties.

Another proposal for the basis of (human) morality is social sentiment. Most humans are not only motivated to pursue their self-interest, but are at least to some degree sympathetic to others. The famous 18th century philosopher Hume based morality on sympathy. However, he noted that we have limited sympathies, and that our sympathy is greatest for those closest to us and similar to us (Hume 1978; Cohon 2010). While our sympathies are not limited to human beings,[2] it has been noted that we are generally more emotionally attached to members of our own species (Midgley 1998). Wenz (1988), suggesting a "concentric circles" model of justice: we have the strongest duties to those we are in a closest relationship with, and our duties to others become less strict with distance. We do not want to deny here that human social sentiments and capacity for sympathy may play a very large

[2] Indeed, virtue ethical accounts in animal ethics aim to base duties to animals in our sympathy for them (eg. Walker 2007).

role in morality. We do want to question the view that our basic moral duties vary with how close we feel to the other, or what relationships we have with others, especially duties not to harm others. Hume himself noted that our moral judgments on the characters of those who harm or help others do not vary along with our sympathies for those affected. He proposed that we estimate the effects of people's character from a "common point of view", which abstracts from our own self-interest but rather involves the viewpoints of everyone affected by the action (Hume 1978, T 3.3.1).[3] It might be argued that we have stronger positive duties (duties to assist) those whom we have relationships with, but it seems implausible to hold that negative duties (duties not to interfere) depend on the strength of (affective) bonds. Such a view could justify harmful treatment of those with whom one is or feels less connected, like those with a different ethnic background or those on the other side of the world.

A final way to argue for unequal moral status would be to resort to 'everyday moral judgment' which says that rational beings, such as humans, matter more than merely sentient beings, such as many animals. Balzer et al. (2000), for example, say that it fits better with our considered intuitions to assign a hierarchy of inherent moral standing to different kinds of beings. Similarly, DeGrazia (2008) argues that moral status varies with the capacities of beings, e.g. being conscious, self-aware, moral agency, language, and so on. This does appear to be the common view. However, is this view justified? We need to ask *why* exactly it matters whether a being is capable of language or is a moral agent for how we ought to treat them. Sure, it would be problematic to defend a moral theory that has no connection at all to our views about the content of morality. However, we think that a view being commensensical alone does not suffice to justify moral claims. After all, we now consider views that were once common, such as the view that slavery is morally right, as completely morally unjustified. We think we need to dig a little deeper to determine whether our everyday moral judgments are indeed justifiable.

It is important here to consider what a hierarchy of moral status actually means. It means that different creatures would all have moral standing, but would have so to a varying degree. In other words, if we need to decide how to treat two different creatures, the creature with higher moral status would automatically receive preferential treatment, regardless of the specific interest of the creatures involved in that specific dilemma. So, for example, if we must choose to hurt either a rat or a human being, even if their pain would be equally severe, we should choose to spare the human being, because her/his interests matter more *in principle*. However, this begs the question as to why this human being's interests matter more. It cannot be because she/he experiences more pain, because in this example the pain was equally severe for the rat and the human. Could it then be because the human can use

[3] Hume appears to be describing human nature; explaining what human beings do when they make moral judgments. One can question whether and why we *should* take such a common point of view. We describe a utilitarian and a Kantian argument for a similar idea in the next section.

language or is a moral agent? This raises the question why these differences would be relevant in this context. Again, more than a simple reference to common sense is necessary to explain such a position.

2.3 Equal Moral Status

So far, we have argued that three of the most common arguments for attributing unequal moral status to humans and animals are problematic. What bases could there be for attributing *equal* moral status? In this section, we will discuss the views of the prominent practical philosophers from two very different moral-philosophical backgrounds. Peter Singer is a proponent of the theory of utilitarianism, and a prominent animal ethicist. Christine Korsgaard is a Kantian philosopher, and has in recent years discussed the place of animals in her wider philosophical work. While there are important differences between them, the two authors both think that we have moral duties to others that are not dependent on reciprocity or sympathy for others and both are critical of everyday moral judgments. We will now explain how they justify moral claims.

Singer (1999) takes a basic starting point for the moral point of view to be that one should consider what ought to be done not just from the standpoint of self-interest, but from the interests of all involved. The basis of morality, in Singer's view, is the principle of equal consideration of interests: all comparable interests should be weighed equally. If interests differ, however, then this should be taken into account. For example, all people have an equal interest in mobility, but for disabled people this means getting access to facilities like a wheelchair, while for able-bodied people it doesn't. Equal consideration of interests, then, may lead to dissimilar treatment. Singer suggests that not only human beings, but also certain species of animals may have interests. Singer understands interests in terms of the satisfaction or frustration of preferences. The question then is what animals can have preferences. In Singer's view, a minimal requirement to be able to say that a being can form preferences is that the animal can have positive or negative experiences. Singer appears to regard all negative affective states as forms of suffering which they have a preference to avoid and all positive affective states as forms of joy which they have a preference to strive for.[4] If an animal can suffer negative experiences such as pain, or fear, it will have a positive motivation, a preference, to not suffer. Such animals may

[4] Note that it is our aim here to introduce the philosophical reasoning of Singer, and not to add new insights to the debate about what constitutes animal welfare. More in general, suffering could be described as 'strong, negative affective states such as severe hunger, pain, or fear' (Fraser and Duncan 1998) and can result from 'experiencing a wide range of unpleasant emotional states such as fear, boredom, pain, and hunger' (Dawkins 1990). A discussion is possible about the question whether all negative affective states in fact amount to suffering as such. After all, animals can often adapt their behaviour to short-term negative states, such as hunger or fear, in a way that is rather functional for them. Real suffering may result only from intense or prolonged exposure to negative stimuli combined with a negative stance towards such experiences.

also have preferences for positive states, unconnected simply to the avoidance of suffering, e.g. play or food or being with conspecifics.

Singer is a utilitarian, and that means that he thinks that in determining the right thing to do, we ought to compare, aggregate and maximise the interests of everyone involved. Thus, for example, in choosing whether to help someone with her homework, or bring someone with a serious injury to the hospital, we ought to do the latter, because that is here the more important interest. Singer noticed that in practice, even when human and animal interests are considered comparable, for example when humans and animals are thought to experience the same amount of pain after a specific procedure, the human interest is generally considered more important than the animal interest. He posed critical questions about this, and popularized the term "species-ism", meant to signify discrimination on the basis of biological species, which he considers as unjustified as sexism and racism. Only when different species in fact have different interests, it is justified to treat them differently. For example, dogs cannot benefit from human education, so it would not be speciesist to deny them access to schools. He also attacked the idea that it is specific capacities of human beings that make them especially morally significant, such as rationality or their being moral agents. After all, we also think that human babies' pain matters equally to adult beings' pain, even if they are less rational than adult humans, and we accept that just as we may not harm rational humans, nor should we harm intellectually disabled humans.

As a utilitarian, Singer thinks that we should always maximise the satisfaction of the interests of everyone involved. Traditionally, this approach to morality is most contrasted with the moral views inspired by the 18th century philosopher Immanuel Kant. Kant (1785, and more recently reprinted 1998) thinks that we should not act morally for the sake of an external goal, such as self-interest or even the interests of others, but simply from respect for moral law. He sees the moral law not as legislated by an external authority, such as God, but as a law of our own reason. In acting on the moral law, human beings are autonomous (literally: self-legislating). Kant claims that the capacity of autonomy makes human beings "ends in themselves": we ought to respect them for their own sake, not only use them as means to another end (e.g. our self-interest). Kant thinks we do not have any direct moral duties to animals, as they lack the capacity of autonomy. He does think we ought not be cruel to animals, but that is because it undermines a duty to ourselves: to cultivate those capacities (e.g. sympathy) that enable us to do our moral duty (Kant 2000, p. 6, 442).

Christine Korsgaard, a prominent contemporary Kantian author, has offered an internal criticism of Kant's position.[5] Korsgaard (2011) argues that Kant was wrong in thinking we only have duties to autonomous beings. Like Kant, she takes

[5] An external criticism of this view has been given on the basis of the previously mentioned analogy with humans without rational capacities: if we do not have direct duties to animals because they are not rational, what about human beings with similar lack of rational capacity, such as babies or severely mentally challenged people? Should we only not treat them cruelly because of the implications for other beings? Such an argument (e.g. Singer 1999; also Regan 2004) points to an inconsistency in the way that we treat different kinds of beings with similar capacities.

morality to be based on a law that human beings legislate to themselves. As humans, we cannot simply go along with our impulses, but we need to have reasons for what we do. Insofar as we consider our choices rational, we must think that the objects of our choices are objectively good. Korsgaard emphasises, however, that the *content* of our reasons cannot be given by respect for autonomy itself. Rather, we find reasons in what is naturally good for us (Korsgaard 2011, p. 108). While things can be said to be good or bad for plants, only conscious animals care about their own natural good (Korsgaard 2009a, pp. 34–35).[6] Animals can act purposively, to avoid things that they dislike, and to attain things they want (Korsgaard 2009b, pp. 10–15). When we avoid pain and suffering, we act for a purpose we share with other conscious animals. But even if we value ends that other animals do not share, we still value what is good or bad for the kind of beings that we are. When we, rational beings, act for the sake of an aspect of our own good, we take something's being naturally good for us as objectively good: as a law for ourselves and others (Korsgaard 2011, pp. 107–108).

Korsgaard says that we thereby accord ourselves a certain standing: of an end in itself. Kant thought that we only have to respect ourselves as ends in ourselves insofar as we are rational, or autonomous. Korsgaard explains that Kant conflates two different conceptions of the end in itself: (1) the source of legitimate moral claims that should be recognised by all rational agents, and (2) someone who can give the force of law to his claims, or participate in moral legislation. She notes that a law can protect someone who did not participate in the making of it (2005, p. 21). In legislating a law that what is naturally good or bad for us is objectively good or bad, we confer value on our animal selves. We therefore have to accept duties to all those who have a good that they care about, even if they cannot claim respect for it. Korsgaard argues that on the basis of this reasoning, conscious animals too should be regarded as "ends in themselves" (2011, pp. 108–109). We should respect their good for the sake of the individual animals involved, and not just treat them as means for our own ends.[7]

Utilitarianism and Kantianism are usually understood as very different approaches, and some important differences will come to the fore when we apply these theories to the practise of using animal models in neurobehavioural research (in Sect. 4). Here, we want to point out what these specific variants of these

[6] Note that Korgaard is making a philosophical argument here to the effect that those animals who actually experience pain and pleasure and have positive or negative emotions care about their own good in a way that insensate beings cannot. Of course, her argument does not hold for the group of animals who do not have these experiences. To what group of animals such emotions are restricted is a question that should be answered by use of biological research together with reflection about the philosophy of animal minds.

[7] Other animal ethicists, such as Taylor (2011) or Rollin Smulewicz-Zucker (2012) have also emphasised that animals have moral status because they have a good of their own. Korsgaard's theory differs to Taylor's in the sense that in her view, animals should care about their own good in order to have moral status. She differs from Rollin in the structure of her moral theory. Korsgaard tries to show that, as rational agents, we cannot rationally avoid accepting moral duties to all conscious animals.

approaches have in common. They offer basically the same reason for extending equal moral status to all animals that strive to attain goals on the basis of preferences. They attribute moral status to sentient animals, but not, say, to plants, because we can only consider what should be done from the perspective of beings who have preferences or who care about what happens to them. Cars or plants don't care what happens to them, while sentient animals do. We can put ourselves in the place of animals, because it matters to an animal what happens to it.

We cannot completely defend these views of Singer and Korsgaard here, as this would require much more sophisticated reasoning in moral philosophy. We just want to point out that, if we think that those to whom we attribute moral status is something that is not based on reciprocity, then it seems to make sense to take perspective not just from ourselves but also from the other as an experiential being. What animals actually belong to the class of experiential beings is a matter of discussion, even amongst biologists. For example, biologists disagree about the question whether and if so, which, fish can experience pain, and whether they have capacities such as memory and flexible learning. Regarding insects and crustaceans there are even more unknowns. As we will see, an answer to this question depends on how we interpret consciousness, and this requires reflection in the field of philosophy of mind. While this subject is treated exhaustively in the chapter by Droege and Braithwaite (this volume), we are not able, nor intend, to resolve these complex discussions here, but restrict ourselves to pointing out where more research is needed and how this is relevant for animal ethical considerations.

3 Consciousness in Animals

3.1 Introduction

In the previous section we have seen that certain capacities are taken to be the criterion for moral status. Singer takes preference satisfaction as morally important, and Korsgaard argues that all beings who consciously pursue purposes have moral status. These capacities involve that the being in question is sentient, and that it has positive attitudes towards certain goods—such as food or playing—and negative attitudes towards others—such as threats. These attitudes correlate with affective states. For example, fear does result in aversive behaviour *because* it constitutes an unpleasant feeling that motivates a being to avoid what it is afraid of. According to Singer and Korsgaard, consciousness makes a crucial difference with regard to moral status: if a being is not conscious what we do to it will not matter to it (although, of course, it may matter to us). On the other hand, if a being is conscious, it matters to the being in question whether we frustrate or aid its pursuit of goods.

In this section we will focus on the questions whether and how we can know which animals are conscious, and whether there is a difference in consciousness between humans and other animals. Of course, the list of animals that are deemed to have consciousness depends on how one defines consciousness in the first place and is constantly changing, as more research is done on species that were previously assumed to be unconscious. For example, cephalopod molluscs such as the octopus and squid were previously not considered to be sentient and cognitive beings, but are now being recognised as such. They have even been given the status of 'honorary vertebrates' in legislation on animal experiments in many countries (Kolar 2006).[8] Like much of the literature about animal consciousness, we will focus on the question whether animals have phenomenal consciousness, which refers to the experience of sensing what is around you and the feelings and emotions that this creates; also termed 'raw experience' (Block 1995). We assume that when you are conscious there is 'something it is like' to be you (Nagel 1974). The question then is 'can we say it is *like something* to be an animal'? Another way of describing this type of consciousness is as 'the subjective state of feeling or thinking about objects and events' (Griffin and Speck 2004, p. 6).[9]

As we will explain later, we think there are good reasons to believe that consciousness is not an 'on or off' notion, but rather that it is a matter of degree. If so, it may very well be possible that negative experiences as a result of experimentation also come in degrees. The question whether animal consciousness differs in important ways from human consciousness is important in the context of this chapter because it might be thought that, while animals do have moral status, it is less problematic to experiment on animals, if they experience less negative consequences from these experiments. This view seems to be based on the idea that animals are somehow less conscious of what happens to them. But what reasons do we have to conclude that animals are less conscious of pain and suffering than humans and therefore do not have the same interest in avoiding the negative experiences associated with experimentation as humans? The main difference between humans and animals in this context appears to be humans' greater cognitive complexity. Therefore, we need to address the question what the influence of cognitive complexity is on suffering. In order to do this, we first need to ask if and how we can know whether animals are conscious.

3.2 Can We Know Whether Animals Are Conscious?

Both in the philosophy of mind and in biology, we encounter scepticism about the question whether animals are conscious. One reason for this scepticism is that we simply do not—and in a strict sense cannot—know exactly what animals experience. What animals actually belong to the class of experiential beings is a matter of discussion, even amongst biologists. For example, biologists disagree about the

[8] Thanks to Ruud van den Bos for pointing this out to us.

[9] Note that a distinction is sometimes made between consciousness and awareness. We will use the terms interchangeably.

question whether and if so, which, fish can experience pain, and whether they have capacities such as memory and flexible learning. Regarding insects and crustaceans there are even more unknowns. This has led scientists in the past to ignore the study of animal consciousness. As Griffin and Speck (2004, p. 5) put it, 'many behavioural scientists have been extremely reluctant to consider non-human consciousness on the grounds that it is impossible to obtain objective evidence about subjective experiences'. Therefore, some remain agnostic about animals' consciousness and others simply assume that an animal doesn't have experiences and cannot suffer pain. The obvious problem with this last line of reasoning is that it commits the fallacy of ignorance: lack of knowledge of a certain fact doesn't make the opposite true.

Moreover, as Panksepp (2011) convincingly argues, neuroscience does now give us objective evidence about animal feelings, at least about mammals. As he explains, historically, it was believed that 'emotional feelings are a subset of cognitive processes' and many still believe this to be the case (Panksepp 2011, p. 4). This has meant that without higher cognitive functions, animals were not regarded as being able to experience emotions. However, animals, including humans, that had their brain's cortex removed, still showed emotional responses (Panksepp 2011, p. 6).

Similarly, in the philosophy of mind it has historically been thought that because of their lack of cognitive complexity, animals lack the human characteristics that are necessary for consciousness, namely language or higher-order thought. Language is considered important because in order to be conscious a creature must have something going on in its mind. Purposive action is usually understood to be an interplay between beliefs and desires, e.g. wanting to eat food and believing that the food is in the refrigerator, and therefore going to the refrigerator. To have beliefs and desires a creature must be able to think. But in order to think a creature must have complex concepts and this in turn requires language. We can only say that a dog that chases a cat actually thinks it is chasing a cat, when we assume that this dog has a concept of 'cat'. This requires a higher level of abstraction that can only be reached by creatures with language (Lurz 2009).

This view rests on some problematic assumptions: While we humans need language to be able to use complex and abstract concepts, this does not mean we cannot think about things without language or abstraction. Language is only required to communicate those ideas to others, while rather than necessarily articulating ideas as abstract concepts, it is perfectly possible to visualise such constructs 'in pictures'. Temple Grandin, for example, a well-known animal welfare specialist with autism, reports not having generalised concepts, for example of a tree, but rather seeing many pictures of different trees; she projects that certain animals might similarly 'think in pictures' (Grandin 1995). Further, if we were to accept that language is necessary for consciousness how do we reconcile this with the observation that many humans (most notably babies) cannot use language either? Most people would not conclude that they are therefore not conscious.

Some philosophers of mind claim that most non-human animals are not conscious because second-order or higher order thoughts are necessary for one to be conscious, and animals—perhaps with the exception of some primates and cetaceans—don't have such thoughts. Higher-order thought (HOT) theory of consciousness says 'that what makes a mental state conscious is the presence of a suitable higher-order thought about that state' (Gennaro 2009). Another way of saying this is that only animals that have I-thoughts, or meta-cognition, are conscious. Carruthers (1992) for example argues that all animals (except perhaps primates and dolphins) lack higher-order thoughts and therefore lack phenomenal consciousness. In fact, he goes as far as to claim that only beings that have "theory of mind"—the ability to attribute thoughts to others—are conscious. Gennaro (2009), however, convincingly argues that HOT theory is in fact compatible with ideas of animal consciousness, since a higher-order thought need not itself be conscious. One can make meta-cognitive judgments without being explicitly conscious of them: for example, when a being is afraid, he or she is aware of experiencing fear, but need not know that she/he is aware of experiencing fear (Gennaro 2009, p. 190).

3.3 How Can We Find Out Whether Animals Are Conscious?

In daily life most people have no qualms about attributing consciousness to animals. When we step on a dog's tail and it runs off yelping, we do not find it strange to say that the dog is in pain. When it comes up to us with its leash in its mouth, we have no trouble inferring that it wants to go for a walk. However, some people also tend to talk about plants or computers as if they were conscious, as when we say that a plant is not feeling well, or a computer doesn't feel like working today. How do we discriminate between the latter unwarranted cases of anthropomorphism and warranted attributions of human-like states to non-humans?

We are interested, then, in determining whether there is something "that it is like" to be an animal. When can we take aversive behaviour to be indicative of consciously felt pain, anxiety or sadness? And when can we take appetitve behaviour towards a positive goal as a sign of consciousness? Some biologists base the idea that there are relevant experiential similarities between humans and other animals on the analogy postulate, first proposed by Romanes (1882). The postulate states that:

A greater or lesser degree of similarity in the subjective experience of a certain animal species, and of the human being may be assumed, relative to the degree of similarity between the structure of the sensory nervous system of that animal species with the human sensory nervous system, and relative to the degree of similarity between the reaction shown by the animal to a specific stimulus and the human reaction to the same stimulus' (Verheijen and Buwalda 1988).[10]

---

[10] Note that the analogy postulate would more aptly be named the 'homology postulate', as it is looking at homologous structures and functional homology.

The idea is, simply stated, that if certain animals have similar capacities involved in the experience of pain and suffering as humans and if animals respond with similar behaviour to certain stimuli as humans, we can assume that these animals also experience stimuli similarly to humans. The postulate also leaves open the possibility that some animals have more similar experiences to humans than other animals and this could mean that we find varying degrees of consciousness in the animal kingdom. There are two aspects to the analogy: behaviour and physiology. Many animals, most notably mammals, exhibit similar behaviour to humans when confronted with a stimulus that causes pain in humans. Think of vocalisations, running away, rubbing the damaged spot, and trying to avoid future interaction with the stimulus. Also, the sensory nervous system of many vertebrates shows similarities to that of humans.

Determining an analogy is a difficult matter and it is important to note that considering either behaviour or physiology on their own is not sufficient. Let us discuss the case of pain, as a relatively large amount of research has focused on the question whether we can take pain experience to be conscious. What we need, is a way of distinguishing between a reflex-like response to a noxious stimulus, and an actual experience of pain. Many animals have nociception, which provides the physiological basis of pain. But it is commonly thought that in order for the pain to be experienced, a signal has to be sent from the nociceptors through the spine to the relevant areas in the brain. Flies and sea slugs, for example, have nociceptors, but they lack a central nervous system, and therefore no signal can be sent to their brain. Some argue only on the basis of behavioural responses that certain invertebrates do seem to experience pain. For example, Elwood and Apple (2009) have shown that hermit crabs remember in what type of shell they received an electric shock and tend to avoid such shells in future. Similarly, Sherwin (2001) cites research showing that cockroaches, flies, and slugs have memory and show pain responses. It may be possible, therefore, that pain could be experienced by beings without a conventional (vertebrate) central nervous system and that in organisms with different nervous organisation or neural structuring, pain might still be expressed, simply via alternative physiological systems. Especially if we were to assume that there are varying degrees of consciousness, it may be possible that dissimilar nervous systems might simply result in a different style, or degree of conscious experience between mammals, birds, fish, invertebrates, etc. rather than presence or absence of consciousness as such. However, there remains a lot of discussion about this among biologists themselves and the consensus at this point in time seems to be that only vertebrates experience pain (Braithwaite 2010).[11]

Even when consideration is restricted to vertebrates, there is still discussion about the question whether animals can only sense pain, or can also suffer from it, in the sense that the animals 'mind' the pain. In other words, we can distinguish between a sensory and an affective aspect of pain, and there is discussion about the

[11] New research indicates that crayfish show anxiety-like behavior, which led to an increase in serotonin in the brain and was suppressed by the injection of opiates. See Fossat et al. 2014.

question of whether these two aspects can be completely separated (Sufka et al. 2009). Evidence in favour of such a separation is that people who have been given morphine report feeling pain, but not minding it (Shriver 2006). This observation is supported by the fact that different neural pathways are involved in the sensory and in the affective aspects of pain. Research shows that the anterior cingulate cortex (ACC), which is part of the medial pathway, plays an important role in the affective aspect of pain (Shriver 2006). The medial pathway responds much better to anaesthetics than the lateral pathway, which is associated with the sensory aspect of pain (which, for example, enables us to locate pain in a particular area and assess its intensity). This would explain the disconnect between feeling and minding pain in patients who have been administered morphine (Shriver 2006). While it is therefore at least theoretically possible to sense pain, but not mind it, it is important to note that non-human mammals also have an anterior cingulate cortex, and hence that there is no reason to assume at the outset that only humans, but no non-human animals can both sense and mind pain.[12]

However, only from pointing out *physiological* similarities between humans and animals we cannot conclude that humans and animals must feel the same. A first step towards such a conclusion can be made if we also look at pain *behaviour* in animals. Tests with rats have shown behaviour that could be interpreted as 'feeling pain, but not minding it' (Shriver 2006, p. 437). In short, in these tests rats, who usually prefer to spend more time in dark rather than light chambers, were given shocks while in the dark chamber. When their paws were made very sensitive to noxious stimuli they spent more time in the light chamber, except when their anterior cingulate cortex was lesioned. While they still showed withdrawal reflexes after electric shocks, they stayed in the dark chamber, suggesting that they sensed the pain, but did not mind it (Shriver 2006). In other words, we have reason to believe that rats with an intact ACC not only sense, but also mind pain.

The risk of relying on the analogy postulate is that it is always open to the objection that next to analogies, disanalogies exist between humans and animals (Allen 2011). The question is, therefore, whether we should focus on the similarities or on the differences between humans and animals. Such criticism of the analogy postulate can be dispelled when a theoretical underpinning can be given for why similarities in certain behaviours and physiological states are relevant (Allen 2011). One consideration of such theoretical underpinning could be the role of pain in learning. As Shriver (2006, pp. 438–439) explains, 'Minding a sensation often causes us to desire to avoid it in the future… there appears to be an important relationship between the affective pathway and learning, which may in turn be suggestive of an evolutionary role that the conscious experience of pain could play'. It could still be objected that even lesioned spinal cords can learn to respond to noxious stimuli. However, more complex forms of learning are more likely to require consciousness (Allen 2004).

[12] It is as yet unclear whether animals besides mammals, such as fish, have something functionally similar to an anterior cingulate cortex. More research is needed into this question.

Researchers tend to take flexible learning as a sign of consciousness, as non-learned responses to novel situations cannot be based on a reflex response or innate, 'pre-programmed' behaviour. If neither the animal nor its ancestors have encountered this novel situation before, the animal's response cannot be the result of 'instinct'. If an animal can deal with a new and unpredictable situation this seems to imply that it can combine different informational cues in its mind and can plan at least in the short term, a phenomenon referred to as insight learning (Griffin and Speck 2004). In this context a distinction could be made between goal directed and merely goal oriented behaviour (Saidel and Lurz 2009). Only for the former one needs desires and beliefs. For example, evolution can structure plants to follow the position of the sun, but the plant does not thereby have a desire to turn to the sun. In contrast, organisms that display goal-directed behaviour do so in order to achieve a goal they have; they can overcome obstacles in a flexible manner. They need to have distinct representations of the goal and the means to achieve that goal in order to have goal-directed behaviour. This means that they could stop a particular behaviour and adopt another behaviour while still aiming at the same goal (for instance when a barrier is erected so the previous behaviour will not allow the animal to reach the goal anymore; it then has to respond flexibly). This in turn means that animals that have goal-directed behaviour are able to learn in a way that other animals cannot. They have to learn to form a new association between goal and means to achieve the goal. Conditioned learning, on the other hand, does not depend on having representations (although Griffin and Speck (2004) cite evidence showing that even Pavlovian conditioning in humans cannot be accomplished without awareness, and they suggest that it is unlikely that it would be possible completely without awareness in animals). If animals can respond flexibly to their environment, for example by adopting a new way to get food, this will give us evidence of mental representations. Tool-making in primates is an example of such evidence, and more evidence of this kind exists in, for example crows (Emery and Clayton 2004) and many other animals (Shettleworth 2012).

Besides looking at criteria such as analogies between brain areas displaying behaviour indicative of painful and pleasant experiences, and the ability for flexible or insight learning, other criteria can be put forward to strengthen reasoning by analogy. Elwood and Apple (2009, p. 1243) mention 'showing trade-offs between stimulus avoidance and other motivational requirements; having opioid receptors; and reduction of responses to noxious stimuli by analgesics and local anaesthetics'. In this vein, Sufka et al. (2009) argue that researchers implicitly hold that their animal models display the same symptoms as human patients. If an animal model performs well, and for example responds to the same treatment as human patients, they propose that the emotional states of these patients are accurately simulated in the animal model. Their premise is that 'if the animal simulation is valid—that is, it compares well with its corresponding human neuropsychiatric disorder in terms of aetiology, symptomatology, pathophysiology, and response to treatments—one is entitled to argue that the animal shares an emotion similar in kind to the human counterpart' (Sufka et al. 2009, p. 533).

Of course, the fact that researchers use animals and that their experimental set-up seems to work on its own is not sufficient to conclude that animals experience something similar to humans. However, the fact that medication works in the same way for certain animals and humans certainly is a telling fact. Additionally, as Panksepp (2011) explains 'if our *predictions* about changing internal feelings in humans, derived from the animal data, are supported by human self-reports, as has often been the case, we have additional reasons for confidence that both humans and animals are having similar (albeit not identical) experiences (our italics)'. However, even though we have good reasons to assume that animals with similar behaviour and nervous system as humans can have similar experiences, we should be careful not to conclude that those animals that do *not* exhibit these similarities therefore do not have similar experiences as humans.

We have two reasons to be cautious of drawing hasty conclusions: First, we are so different to some animals that we have difficulty relating to them and we may not be able to interpret their behaviour appropriately. Fish, for example, do not have facial expressions that we can discern, they do not make sounds that we can hear, and the fact that they live in a different medium to us—water—also makes that their behavioural repertoire is by necessity very different to ours. And even in the case of closely related species we have difficulty interpreting facial expressions; we might interpret upturned corners of a gorilla's mouth as a smile, while in reality it could be a sign of aggression. In order to find out whether they do exhibit behaviour indicative of feeling pain, we need to devise tests that are sensitive their way of being (see for example tests carried out by Sneddon et al. 2003). In general we might not be able to discern from animal behaviour that an animal experiences something, but by way of preference tests we can see that the animal avoids particular stimuli. We need to bear in mind, however, that for different animals different behaviour is adaptive. In nature it is often functional not to show pain, because this may attract predators or might show an opponent that one is afraid.

Secondly, the fact that some animals do not have brain structures similar to those of humans does not mean that these animals do not have different brain structures that have the same function. In other words, having a different neuro-sensory organisation does not necessarily mean that an animal cannot feel or experience in an analogous way. For example, the eyes of octopus have a completely different structure to those of humans, but they also see in colour and may have an even greater visual acuity. Similarly, birds do not have a prefrontal cortex, which is a part of the human brain associated with complex cognition. However, certain birds, such as corvids, do exhibit complex cognition; there is evidence of causal reasoning, flexibility, imagination and prospection in these birds. In their brain the 'prefrontal' functions can be found in the parts that are analogous in function to the mammalian cortex, namely the mesopallium and the nidopallium (Seed et al. 2009). In the same way, while fish do not have a prefrontal cortex, some of the functions of this brain area appear to exist in fish brains as well, in the telencephalon (Chandroo et al. 2004). We should therefore not simply look at similar brain structures, but at *functional* analogies. In general, it is important to stress that the elements of the postulate should not be examined in isolation. Just looking at animal vocalisations after painful stimuli is not enough, for example. Tests with rats and pigs show increased vocalisations after a stimulus even if this stimulus has led to less sensitivity or when the animal was actually anaesthetized (Allen and Bekoff 2007).

3.4 Why Caution Requires Attributing Consciousness to Certain Animals

Many reasons have been put forward, then, to argue that at least certain animals are conscious: they share behavioural repertoires and physiological make-up with humans, they show complex and flexible learning, they can be used as valid models for human disorders, and they respond similarly to anaesthetics. In addition, it seems unlikely that consciousness evolved all of a sudden in humans and it is more reasonable to anticipate that both rudimentary as well as more evolved levels of consciousness can be found throughout the animal kingdom. However, these arguments may not convince sceptics about animal consciousness. For any piece of empirical evidence cited, it will never convincingly prove that an animal is conscious. In theory, it is possible that animals, like complex machines, can display very sophisticated behaviour without actually experiencing anything. The difficulty is that we can never have access to another being's experiences. This goes for other humans as well; I can have a look at brain scans and at behaviour of other humans and I can listen to what a person tells me about her experiences. On the basis of this I can try to put myself in this person's place, but I can never know for certain that there is something that it is actually like to be that person, nor what it feels like exactly. With animals it is even more difficult to determine what it is like to be them, because they cannot tell us.

The question is where lies the burden of proof. From a sceptic's point of view, it might be though most legitimate to assume that animals are unconscious, until proven otherwise, but this is actually not at all obvious. From an evolutionary perspective it seems plausible to think that creatures that were confronted with similar environmental and social pressures as our ancestors would have been selected similarly as those ancestors and therefore consciousness would have evolved at least in animals of similar sociality. As Jamieson (2009, p. 17) eloquently puts it: 'it would be surprising, perhaps even the biological equivalent of the Immaculate Conception, if we were nature's only minded creatures'. As discussed above, there are many similarities between humans and animals' behaviour and physiology that suggest continuity in their experiences.

There are also moral reasons not to start with the assumption that animals are unconscious. As Shriver (2006) points out, for the purpose of making a moral judgment about our treatment of animals, we do not require a knock-down argument for or absolute proof of animal consciousness. As long as it is reasonable to hold that at least certain animals are conscious—and we believe it should be clear from the above that it is reasonable—this is sufficient to call for caution in our dealings with these animals. After all, if we erroneously treat animals as if they were unconscious when they in fact are, the moral costs are very high (Shriver 2006). According to the precautionary principle, if there is a reasonable projection of harm, we should not wait for absolute proof before we act to remedy this harm.

3.5 Degrees of Consciousness, Pain and Suffering

When we look at empirical evidence based on the analogy postulate and additional criteria mentioned above, it seems likely that all vertebrate species do experience pain (Braithwaite 2010). Moreover, some invertebrates with more complexly organised brains, in particular cephalopods and some crustaceans, show behaviour that could be interpreted as pain behaviour (Sherwin 2001; Elwood and Appel 2009). And if we are willing to look beyond the analogy postulate and consider that differently organised species may have evolved different ways of structuring pain sensations, other animals that can experience pain may be identified in the future. We should bear in mind, however, that different species may experience sensations differently: concluding that animals can experience emotions does not entail concluding that these emotions are identical to human emotions.

As mentioned before, we think we should regard consciousness not as an 'on- off' notion; it is not as if you are either conscious or you are not, but rather, some animals have higher, or more complex, levels of consciousness than others. How should we understand these different levels of consciousness? What consequences does the view that consciousness comes in degrees have for the question whether it is more justified to use certain animals than others for neurobehavioural research? In order to answer this question, we need to consider whether more complex cognition makes suffering worse. As Yeates (2011) explains, it is far from self-evident that more complexity leads to more pain. For example, more cognitively complex animals can in some cases cope better with pain, if the pain is short and the animals realise the pain will be over quickly. On the other hand, if they realise the pain is chronic, they could cope less well, because they know the pain will go on.

No clear picture emerges when we take these kinds of considerations into account. In some cases animals with simpler cognitive capacities are likely to experience pain more frequently and more intensely, for example because they cannot distract their attention and tend to focus on the most biologically important stimulus (namely pain). In other cases, animals with more complex cognitive capacities will experience more pain, for example when they suffer from anxiety, which tends to increase pain experience. In any case, Yeates (2011) raises doubts about the question whether these cognitive influences make so much difference on pain experience that this is meaningful for the animals in question.

Of course, pain experience is only one aspect of suffering. Suffering also—perhaps even more urgently—has to do with the frustration of one's goals and there are forms of suffering that are not so much experienced as pain sensations, but rather as adverse mental states. These kinds of suffering are typically involved in research into cognitive and affective disorders. A relevant question is then what it is like for animals to be, for example, afraid, anxious, or depressed. Because humans have more cognitive complexity, we will have different, and perhaps more varied, experiences than animals, but does this mean it is worse for us to have negative experiences than it is for animals? We think that, like in the case of pain, it is not self-evident that the answer is yes; rather, a more varied picture emerges if we consider how cognition influences mental and emotional suffering. While we have not done empirical research into this question, we want to at least raise some questions and make the point that the answers are not self-evident.

In experiments into neurobehavioural disorders the emotional states that are examined are for example fear, anxiety, depression, and paranoia. If we assume that animals live more in the present moment than humans do, could we say that they therefore experience these emotions to a lesser degree than humans? At first sight, we may say this is the case. After all, the 'subject' or 'content' of the depression, paranoia or anxiety is more complex in humans. However, we might also come to the opposite conclusion: these experiences might be completely overwhelming for them exactly because they live in the present moment. They have no way of telling themselves that their suffering is relative, or will be over soon. They can take no distance from their own suffering, but can only undergo it.

There are types of fear, anxiety or paranoia that animals likely do not experience, because they require abstract concepts. For example, animals are not likely to have existential anxiety as they have no concept of death and do not wonder what the meaning of their life is. Neither will they suffer from thoughts of conspiracies mounted against them. Having abstract concepts, such as a concept of death, does make humans and animals different in a morally relevant way. It may mean that killing a human being is worse when she/he can anticipate his/her death, because it creates supplementary suffering of anticipation that the animal does not experience, but having such abstract concepts does not automatically make all fear experiences worse. Fear, after all, is a very primordial experience, which animals share with humans.

The next question we need to raise is what consequences the presence of higher-order thought has for the interests that a being has. For example, human beings can experience a tension between 'the person they want to be' and 'the person they are' (see also Korsgaard 2009c, pp. 13–20). They may regret the impact of their cognitive or affective disorders on their lives, for example because it interferes with their achieving their larger goals in life. Also, human beings may be aware that their condition influences their social status. This in turn may impact their self-esteem. As a consequence, for human beings, conditions like chronic anxiety disorders, or depression may completely undermine their ability to live the lives they wish to live. It is difficult to determine whether the feelings of anxiety or sadness caused to animals in experiments have a similar dimension for animals. On the one hand, it seems that animals have less at stake: a big picture of who they want to be and the lives they wish to lead. On the other hand, as Korsgaard (2009b, pp. 21–30) also points out, while it matters *to humans* to live out their lives as a project at which they can succeed or fail, this does not matter to animals (and, we might add, not to all human beings either). Animals may care as much about their own purposes as we do about ours. Emotions like anxiety, or the effects of alcohol dependency, are very unpleasant on their own, and will also undermine animals' ability to effectively pursue their other purposes.

It is difficult to determine the comparative meaning of cognitive or affective disorders for different types of animals. On the one hand, it seems reasonable to assume that such disorders are much worse for human beings, as they are acutely aware of how these disorders impact their lives. On the other hand, primitive feelings such as sadness or anxiety might be just as vividly experienced by animals like mice and zebrafish, and perhaps even more so, as these animals cannot learn ways of dealing with their emotions, put them into perspective, or even give meaning to their suffering. In certain cases it is quite possible that cognitive complexity makes suffering (or in general: the frustration of interests) as a result of neurobehavioural disorders worse (for example when a human being has time to reflect on the terrible experiences that lie ahead of her), but other cases are imaginable where the opposite is true (for example when an animal is very afraid and has no way to deal with the feeling, and no idea that it may pass).

There is a reason why these considerations are all rather speculative. As Panksepp (2011) notes, while earlier brain functions that generate primary-process emotions form a common basis for the experience of, at least, all mammals, there is more diversification in the animal kingdom when it comes to later brain functions, in particular those involved in higher cognition. It is a lot more difficult, therefore, to use a principle like the analogy postulate to make inferences about similarity in experiences on the cognitive level than regarding pain experience. At the same time, this raises the question whether animal models in neurobehavioural research can really give us answers regarding our own situation. While the evolutionary sources of human emotion can be studied with the use of animal models, as these stem from 'the fundamental level of brain organisation upon which the rest of the mental apparatus relies… studying the primordial sources of emotional feelings, important as they are, cannot clarify the whole emotional story' (Panksepp 2011, p. 8). As Panksepp (2011, p. 9) argues, 'such thoroughly cognitivized tertiary-processes thought-related emotions, engendered culturally by social learning are, at present, next to impossible to study rigorously at causal levels and most certainly not in animal models'.

4 Should We Use Animals for Neurobehavioural Research?

4.1 Introduction

In this section, we will apply the considerations above to the practice of neurobehavioural research. The preliminary question is whether we should use animals at all in experiments to develop cures and find ways to prevent cognitive and affective disorders for human beings. We have argued that it is not self-evident that humans are morally more important than other animals, and that there seem to be good reasons to attribute moral status to all beings that are sentient and pursue purposes. We have also argued that it makes sense to attribute consciousness, which is required for these capacities, to at least certain kinds of animals, who are, e.g. capable of flexible learning.

We shall see what this means for the acceptability of this kind of research. First, what moral considerations count against using animals in such research? Second, can the benefits that the outcomes of these experiments may give to human beings justify such research? We shall conclude that there can be different views about this from utilitarian and Kantian positions. The discussed Kantian approach will categorically oppose such experiments, while utilitarians may think that, in limited cases where the benefits outweigh the harms, such experiments are justified.

If we do think that experiments on animals are in principle justified, then is it worse to use cognitively more complex animals than less complex animals? We have argued above that there are differences between animals and human beings and between different kinds of animals, related to their different cognitive capacities, and different interests. However, the relevance of this will depend on the specific experiment, and how the beings involved are affected by it.

4.2 Moral Considerations Against the Use of Animal Models

What follows from the general approaches of Singer's utilitarianism and Korsgaard's Kantianism for the acceptability of using animals in neurobehavioural research? To know that, we have to consider what specific moral duties follow from their general frameworks. In general, on the basis of these approaches, we should not harm others, understood as treating them in a way that they have a negative attitude to. This generates duties not to cause others pain, but also other types of suffering, e.g. anxiety, sadness, extreme boredom. If, then, specific kinds of animals experience these kinds of suffering, this is prima facie morally problematic. We have suggested that at least vertebrates, mammals, birds and fish, and the "honorary vertebrates" (cephalopods) at least seem to consciously experience pain. Whether they actually experience the other types of suffering is more difficult to determine.

To take anxiety as an example, animal models are used in some protocols (e.g. zebrafish) that show fear responses to certain stimuli (e.g. Gerlai 2012). Braithwaite and Droege in this volume cite many forms of evidence that indicates considerable flexibility in fish behaviour, which together with physical structures in the fish, seem to suggest consciousness (Sneddon et al. 2003). They explicitly do not claim not to have the final word on whether and how fish are conscious, but based on precautionary reasoning suggested above, it may be right to assume that these fish at least experience a primitive type of conscious fear. There may, however, be important differences between human and zebrafish anxiety. This may make the use of human beings (or more cognitively complex animals) more problematic, but it does not mean that using zebrafish would be morally *un*problematic. If they experience some primitive type of fear (which may be quite intensely experienced), this is a moral reason against using them in this type of research.

It is generally agreed that we should not cause animals unnecessary pain. Killing animals painlessly, on the other hand, is often regarded as unproblematic. The question whether painless killing is indeed morally unproblematic is very relevant for the issue of experimentation, as it routinely involves "sacrificing" the animals involved after the experiment. The issue of the wrongness of killing is a rather vexing one, also in the case of human beings. There is a philosophical problem about death: it is not a bad state to be in, so how can it be thought to harm someone when we kill them? Hence sometimes it is said that it is only bad for those left behind. Still, we commonly think that death is a serious harm. We might think death is bad because we do not want to die. But then the value of life reduces to this single desire. Sometimes people are suicidal, but we still consider death bad for them, for example if there is a possibility for them to be cured from a depression. Another view is that death is bad because it is a deprivation of future goods (Nagel 1993). Life is not in itself valuable (e.g. living in a permanent state of coma does not seem preferable to death), but life is valuable as a precondition for doing whatever we want to do. In death we lose the possibility of future activities and experiences that we would have valued had we not been killed. This goes for animals too (Kaldewaij 2008; Bovenkerk and Braithwaite, forthcoming).

This could provide an additional reason why experiments on animals, or at least on certain types of animals are morally problematic. The animals involved are not just caused suffering, they are also deprived of all possibility of future experience. If killing animals is indeed morally problematic, this is also relevant when we are faced between a choice of using more animals but causing them less discomfort in an experiment, or using less animals with more discomfort. It is by no means clear that we should use more animals, while discounting for the death of these animals. It should also be pointed out here that an animal interest in life sheds an alarming light on the fact that many animals are bred for research purposes but killed without ever being used, for a variety of reasons, such as the creation of a specific line, or the necessity of only using males of one line. In the Netherlands alone this can amount to 530,000 bred but unused animals annually.[13]

If we follow either of the two philosophical approaches selected for discussion, we have prima facie moral duties not to harm, including not to kill, animals. We speak of "Prima facie moral duties", because our overall duties may be thought to depend on the duties in favour of these experiments. These are negative duties, duties not to interfere with them in certain ways. These same philosophical approaches also give a basis for positive duties to others: to assist others in specific ways. Singer's view leads to quite extensive duties to others, as he holds that we ought to further the preference-satisfaction of everyone affected by our actions,

---

[13] http://www.google.nl/-workshop-gefokt-maar-niet-gebruikt-bred-but-not-used.pdf.

including animals who have preferences. Positive duties also seem to follow from Korsgaard's account, notably from her idea that we must assign normative value to the natural good of a being who cares about its own good. If so, it seems that we ought to further the ends of others when we can (when this does not conflict with other, more stringent, moral duties that we have). However, Korsgaard says that, while she also recognises that it would be good if we had a world in which every animal is happy and comfortable, for the Kantian, morality is not about bringing about a world that is as best as possible, but about how we should interact and relate to others. She, therefore, focuses on treating those with whom we interact in a beneficial and fair way (Korsgaard 2011, p. 111). What may be thought problematic about animal experiments, on this basis, is that they not only make animals suffer, but also deprive them of living a good life, while we should help them to attain this.

4.3 Do the Benefits to Human Beings Justify the Harms to Animals?

There are, then, strong moral reasons against using animals as models in experiments in the neurobehavioural sciences: when these experiments involve causing animals suffering, when they are not able to live a good life as an animal of its kind, and when they are killed far before the end of their natural life span. Adding the extra responsibility we have for beings for which we are responsible, and which are completely dependent on our care, this seems to amount to a strong case against animal experiments. However, the experiments are undertaken to provide benefits to human beings. They are aimed at finding cures and ways to prevent debilitating human diseases involving a lot of suffering, such as anxiety disorders, depression, alcohol addiction, and so on. Can the benefits to humans justify using animals in these experiments? This is a question that utilitarians, like Peter Singer, and Kantians, like Christine Korsgaard, would answer in very different ways.

As explained, Singer thinks ethics is about weighing the interests of different individuals equally, and then trying to maximise the total amount of satisfaction of interests. This is actually quite similar to common thought behind public policy about animal experiments and the procedures used in ethics committees. In both cases, human and animal interests are weighed against each other, and the aim is to find the route to the optimal outcome. However, in practice, in determining public policy and in ethics committees, human interests are given much more weight, even when they are considered comparable to the animal interests in question. Singer would object to this, and say that the interests have to be given equal weight when they are comparable (e.g., when the same amount of pain or other types of suffering is involved).

For Singer, then, the question is exactly how much suffering these experiments would be able to prevent, and weighing this against how much suffering they cause to the animals involved. This is a very difficult calculation to make, as we are not quite sure what the animals involved experience. There may be important differences between the exact interests at stake for different beings. Again, let us, using precautionary principles, assume that some experiments at least involve some pain and other kinds of suffering, like nausea, fear or sadness, which will probably be quite intensely experienced. In addition, killing the animals involves taking away future well-being or preference-satisfaction. If many animals are used, then this is a very large cost to be outweighed. We know that not all experiments lead to positive results for human beings immediately. However, when such positive results are found, they might lead to a large improvement in well-being for many people suffering from anxiety disorders, depression, and so on. We also have to take into account future generations that might benefit from the cumulative benefits of these experiments. It is still a real question whether this would outweigh the harms to the animals involved. More empirical research on the actual benefits to be expected from these experiments and the animals used, and the harms involved, would be needed to find a satisfactory answer to this question. Moreover, when we also take into consideration the (possible) development of alternatives that lead to less animal suffering, the calculation might lead to a rejection of animal experimentation.

Kantians do not agree with the view of moral decision-making as weighing benefits against harms. In Korsgaard's view, it is a mistake to aggregate the interests of everyone involved and try to maximise total well-being. There is no one who benefits from an increase in overall well-being. Rather, we should respect individuals as "ends in themselves". Respecting an individual involves respecting its good (2009b, p. 25). Korsgaard notes that animals cannot give informed and uncoerced consent to how we treat them. She suggests that we treat animals in ways to which we think they would consent if they could, that is, in ways that are mutually beneficial and fair. Using animals in ways that cause them suffering for scientific information, she says, is something they would not consent to (Korsgaard 2011, p. 110). In general, in Kantian views, we may not violate a strict duty to an individual, e.g. not to kill them or harm them in a serious way, for the sake of providing benefits to other people. This makes experiments in which subjects are harmed and killed morally impermissible. It involves using a being as a means for the satisfaction of another being's preferences.

One might wonder what the Kantian view implies for the practical assessments of animals experiments. For example, EU guidelines require us to make a harm-benefit analysis for each experiment. The guidelines are based on the idea that the use of animals in experiments should be justified. There is a tendency to translate this into a simple cost-benefit analysis, but from a Kantian perspective we have reasons to take a wider perspective on such justification. While even a Kantian has to take into account the interests of all those involved, it is important to realise that not everything is reducible to a weighing of minuses and plusses.

It may be thought that that using animals in experiments is 'a necessary evil' or 'the lesser of two evils', suggesting that we see ourselves faced with an unavoidable choice between harming an animal or allowing human beings to be in a harmed state. While there is good reason to assume that when we are in an emergency situation, such as in a lifeboat, it is more justified to throw a dog overboard than a human being, if one of them has to go (Regan 2004), we reject the comparison between animal experiments and life-boat situations. In a life-boat situation immediate action has to be taken and there is no alternative. In the practise of animal research there are often alternatives and even if there are not, it is unlikely that one specific experiment will directly lead to saving human lives. Also, and even more importantly, an animal experiment is not a case in which everyone will die, unless one of them is sacrificed. In the case of an experiment to find cures for others, individuals are actively brought to the laboratory and used as means for the sake of saving others.

There is, then, no consensus among those attributing moral status to animals on whether we may use them in experiments to benefit other beings. Utilitarians think we may, if all similar interests involved are weighed equally, while Kantians think we may never use a being as a means to further the well-being of others.

4.4 When We Do Decide to Use Animals in Research, Which Animals?

We started this chapter by noting a tension between, on the one hand, using animals in research because they are in the relevant respects similar to human beings, but on the other hand, assuming we are justified in using them because their interests matter less. If they are so similar to humans, should they not also have similar moral status? As we have argued, we cannot simply assume that it is better to use less complex animals because they may not suffer less. What could be good reasons to choose one animal model over another?

In practise, the choice for a particular species as a model tends to be based on practical considerations, such as ease of handling, space needed, reproduction rate, and costs. However, if we would decide purely on a normative basis what species to use as a model in scientific or medical research, only two considerations should be relevant: first, the impact that the experiment has on animal interests, the discomfort and suffering the animals are likely to experience; and secondly, the quality of the research design. When we are talking about suffering in the case of neurobehavioural research, we refer not only to pain, but also to emotional suffering such as fear and anxiety, and also to the fact that positive experiences may be taken away from the animals due to the research setting. The quality of the research design is important, of course, because if research has been poorly designed it is not likely to yield useful results and animals will have been sacrificed for nothing. But it can also be a relevant criterion for animal choice, because if we ensure that for a given disease the animal model that has most predictive value for the human situation is chosen, the number of animals that need to be used may be reduced.

Still, we can see in the practise of animal experimentation that different animal species are treated differently. Researchers are more reluctant to use large animal models and three species of animal in particular have a privileged status in legislation: cats, dogs, and non-human primates. Cats and dogs are given privileged status for reasons of public concern (Hagen et al. 2012). We can wonder whether public concern is in itself morally relevant, however. To the extent that people were to suffer from the knowledge that cats and dogs are used in research, according to utilitarianism this could be a reason to restrict the use of these animals. But one may wonder whether there is not as much reasons for public concern about mice and zebrafish as there are about cats and dogs, if they are similarly affected by these experiments.

Moreover, EU Directive 2010/63/EU stipulates that non-human primates may only be used when no other animals can be used to study a specific research question (Hagen et al. 2012). A recent lawsuit in the United States aims to grant chimpanzees the right to freedom, by requesting that these animals be given the status of persons.[14] Three reasons have been mentioned for this: (1) public concern; (2) highest similarity to humans; (3) more complex cognitive abilities and advanced social and behavioural skills (Hagen et al. 2012). Are these arguments sound? Neurobiological research makes much use of non-human primates, especially because their cognitive capacities are so similar to those of humans (Brukamp 2012). Their ability to form social bonds and their communication skills make them interesting animal models, for example because they respond similarly to social isolation as humans. At the same time these characteristics may give them a higher moral status according to some theories. Non-human primates are often granted a special status based on their resemblance to humans. Brukamp (2012, p. 66) for example states that 'primates possess extraordinary cognitive, social, and moral capabilities, which closely relate them to humans, and therefore, primates may need to be ascribed a higher status as moral subjects in comparison to other animals'. But is this justified? Certainly resemblance to humans on its own is not a good reason to grant an animal higher moral status. The question is whether this resemblance is morally relevant. This comes back again to the question of whether these specific resemblances make the animals suffer more from experiments. Only if animals with more complex cognitive capacities suffer more from experiments than less complex animals, then the former should be replaced by the latter.

Is it possible to study such disorders without also creating negative experiences in these animals? And can we know with certainty that they don't experience these states similarly to us, or whether they are even more overwhelmed by them than us? Of course, at least in theory it could be possible to examine only part of the mechanism of depression without creating a whole depressed animal model, but the question then rises whether this still counts as a good model for human depression. On the other hand, if animals can really be depressed, then this raises questions about the moral acceptability of these experiments.

---

[14] This lawsuit was filed on December 2, 2013, by the Non-human Rights Group with the help of animal rights lawyer Steven Wise. See http://www.nonhumanrightsproject.org/2013/12/02/lawsuit-filed-today-on-behalf-of-chimpanzee-seeking-legal-personhood/.

5 Conclusion

When deciding whether to use animals in experiments, we should only take into account the importance of the purpose of the experiment, the quality of the research setup, including a consideration of which animals offer the best translation to the human situation, and the effects on the animals' interests. If animals are indeed good models of human psychopathologies like anxiety and depression, they are likely to have experiences that are similar to human experiences that we consider to have good reason to avoid and cure. More research is needed linking cognitive complexity to suffering. However, this type of research is by definition very difficult, for due to the layered organisation of the brain, there is more diversity between animals on the cognitive 'tertiary level'. Consequently, it is a lot more difficult to use a principle like the analogy postulate to make inferences about similarity in experiences on the cognitive level than regarding pain experience. Even though it is reasonable to assume that consciousness comes in degrees, and more conscious animals may in many cases have richer experiences, in our view it is not self-evident that cognitive complexity will always make suffering worse, although there are definitely forms of mental suffering that animals will not experience. In the latter case, the question should be raised how well the animal model can be translated to the human situation under study. Of course, there is also a more fundamental question that we have to consider: whether moral acceptability is indeed a matter of cost-benefit analysis, or whether it is in principle impermissible to use some beings in seriously harmful ways for the benefit of others.

Acknowledgments We would like to thank Franck Meijboom, Frauke Ohl, and several anonymous reviewers for their very helpful comments. Work for this chapter was supported by the Netherlands Organisation for Scientific Research (NWO) under grant number 275-20-038.

References

Allen C (2004) Animal Pain. Nous 38:617–643

Allen C (2011) Animal consciousness. Stanford Encyclopedia of Philosophy. http://plato.stanford.edu/archives/win2011/entries/consciousness-animal/

Allen Colin, Bekoff Marc (2007) Animal minds, cognitive ethology, and ethics. J Ethics 11(3):299–317

Balzer P, Rippe KP, Schaber P (2000) Two concepts of dignity for humans and non-human organisms in the context of genetic engineering. J Agric Environ Ethics 13(1):7–27

Block Ned (1995) On a confusion about a function of consciousness. Behav Brain Sci 18:227–287

Bovenkerk B, Braithwaite V (Forthcoming) Beneath the surface. killing of fish as a moral problem. In: The end of animal life. ethical and societal considerations on killing animals, (ed.) Franck Meijboom and Elsbeth Stassen. Wageningen Academic Publishers, Wageningen

Braithwaite V (2010) Do fish feel pain? Oxford University Press, London

Brukamp K (2012) Research involving non-human primates: treatment guidelines and ethical frameworks. In: Hagen K, Schnieke A, Thiele F (ed) Large animals as biomedical models: ethical, societal, legal and biological aspects. Europäische Akademie, Berlin
Chandroo KP, Duncan JH, Moccia RD (2004) Can fish suffer? perspectives on sentience, pain, fear and stress. Appl Anim Behav Sci 86(3):225–250
Cohon R (2010) Hume's moral philosophy. In: Zalta EN (ed) The stanford encyclopedia of philosophy. (Fall 2010) http://plato.stanford.edu/archives/fall2010/entries/hume-moral/
Dawkins MS (1990) From an animal's point of view: motivation, fitness, and animal welfare. Behav Brain Sci 13(01):1–9
DeGrazia D (2008) Moral status as a matter of degree? South J Philos 46(2):181–198
Elwood RW, Appel M (2009) Pain experience in hermit crabs? Anim Behav 77(5):1243–1246
Emery NJ, Clayton NS (2004) The mentality of crows: convergent evolution of intelligence in corvids and apes. Science 306(5703):1903–1907
Fraser D, Duncan JH (1998) 'Pleasures', 'pains' and animal welfare: toward a natural history of affect. Anim Welf 7(4):383–396
Fossat P, Bacque-Cazenave J, De Deurwaerdere P, Delbecque J-P, Cattaert D (2014) Anxiety-like behavior in crayfish is controlled by serotonin. Science 344:1293
Gennaro R (2009) The philosophy of animal minds: animals, consciousness, and i-thoughts. Cambridge University Press, Cambridge, pp 184–200
Gerlai R (2012) Using zebrafish to unravel the genetics of complex brain disorders. In: Behavioral neurogenetics. Springer, Berlin, pp 3–24
Grandin T (1995) How people with autism think. In: Learning and cognition in autism. Springer, New York, pp 137–156
Griffin DR, Speck GB (2004) New evidence of animal consciousness. Anim Cogn 7(1):5–18
Hagen K, Angelike S, Thiele F (2012) Large animals as biomedical models: ethical, societal, legal and biological aspects. Europäische Akademie, Berlin
Hume D (1978) A treatise of human nature. Clarendon Press; Oxford University Press, Oxford (Selby-Bigge LA, Nidditch PH, ed)
Jamieson D (2009) What do animals think? In: Lurz RW (ed) The Philosophy of Animals Minds. Cambridge University Press, Cambridge, pp 15–34
Kaldewaij F (2008) Animals and the harm of death. In: Susan A, Richard GB (ed) The animal ethics reader, 2nd ed. Routledge, New York, pp 59–62
Kant I (1998) Groundwork of the metaphysics of morals. Cambridge University Press, Cambridge
Kant I (2000) The metaphysics of morals. Cambridge University Press, Cambridge
Kolar R (2006) Animal experimentation. Sci Eng Ethics 12(1):111–122
Korsgaard C (2005) Fellow creatures: kantian ethics and our duties to animals. Tanner Lect Hum Values 25:77–110
Korsgaard C (2009a) Animal nature and the good. http://www.people.fas.harvard.edu/~korsgaar/Essays.htm
Korsgaard C (2009b) Human nature and the right. http://www.people.fas.harvard.edu/~korsgaar/Essays.htm
Korsgaard C (2009c) Human beings and the other animals. http://www.people.fas.harvard.edu/~korsgaar/Essays.htm
Korsgaard C (2011) Interacting with animals: a kantian account. In: Beauchamp T, Frey RG (ed) The oxford handbook of animal ethics. Oxford University Press, New York
Lurz RW (2009) The philosophy of animal minds. Cambridge University Press, Cambridge
Midgley M (1998) Animals and why they matter. University of Georgia Press, Athens
Morris CW (2011) The idea of moral standing. In: Beauchamp T, Frey RG (eds) The Oxford Handbook of Animal Ethics. Oxford University Press, New York, pp 255–275
Nagel T (1974) What is it like to be a bat? Philos Rev 83(4):435–450
Nagel T (1993) Death. In: Martin Fischer J (ed) The metaphysics of death. Stanford University Press, Stanford, pp 59–69

Panksepp Jaak (2011) Cross-species affective neuroscience decoding of the primal affective experiences of humans and related animals. PLoS One 6(9):e21236
Regan T (2004) The case for animal rights: updated with a new preface. University of California Press, Berkeley
Romanes G (1882) Animal intelligence. Kegan Paul, London
Saidel E, Lurz RW (2009) Attributing mental representations to animals: the philosophy of animal minds. Cambridge University press, Cambridge, pp 35–51
Seed A, Emery N, Clayton N (2009) Intelligence in corvids and apes: a case of convergent evolution? Ethology 115(5):401–420
Sherwin CM (2001) Can invertebrates suffer? or, how robust is argument-by-analogy? Anim Welf 10(Supplement 1):103–118
Shettleworth SJ (2012) Do animals have insight, and what is insight anyway? Can J Exp Psychol Rev Can Psychol Exp 66(4):217
Shriver A (2006) Minding mammals. Philos Psychol 19(4):433–442
Singer P (1999) Practical ethics, 2 edn. Cambridge University Press, Cambridge (reprinted)
Smulewicz Z, Gregory R (ed) (2012) Strangers to nature: animal lives and human ethics. Lexington Books, La Vergne
Sneddon LU, Braithwaite VA, Gentle MJ (2003) Novel object test: examining nociception and fear in the rainbow trout. J Pain 4(8):431–440
Stafleu FR (1994) The ethical acceptability of animal experiments as judged researchers. Universiteit Utrecht, Faculteit Diergeneeskunde, Utrecht
Sufka KJ, Weldon M, Allen C (2009) The case for animal emotions: modeling neuropsychiatric disorders. http://philpapers.org/rec/SUFTCF
Taylor PW (2011) Respect for nature: a theory of environmental ethics. Princeton University Press, Princeton
Verheijen FJ, Buwalda RJA (1988) Doen pijn en angst een gehaakte en gedrilde karper lijden? = do pain and fear make a hooked carp in play suffer?. RU Utrecht, Utrecht
Walker R (2007) He good life for non-human animals: what virtue requires of humans. In: Walker R, Ivanhoe PJ (ed) Working virtue: virtue ethics and contemporary moral problems. Oxford University Press, Oxford
Wenz PS (1988) Environmental justice. SUNY Press, New York
Yeates JW (2011) Brain-pain: do animals with higher cognitive capacities feel more pain? Insights for species selection in scientific experiments. Large animals as biomedical models: ethical, societal, legal and biological aspects. In: Hagen K, Schnieke A, Thiele F (eds).Large animals as biomedical models: Ethical, social, legal and biological aspects. Europaische Akademie, Bad-Neuenahr-Ahrweiler

Does the Goal Justify the Methods? Harm and Benefit in Neuroscience Research Using Animals

Ana Catarina Vieira de Castro and I. Anna S. Olsson

Abstract The goal of the present chapter is to open up for discussion some of the major ethical issues involved in animal-based neuroscience research. We begin by approaching the question of the moral acceptability of the use of animals in research at all, exploring the implications of three different ethical theories: contractarianism, utilitarianism, and animal rights. In the rest of this chapter, we discuss more specific issues of neuroscience research within what we argue is the mainstream framework for research animal ethics, namely one based on harm–benefit analysis. We explore issues of harms and benefits and how to balance them as well as how to reduce harm and increase benefit within neuroscience research.

Keywords Animal model · Benefits · Harms · Harm-benefit analysis · Ethics · Neuroscience

Contents

A. C. Vieira de Castro · I. A. S. Olsson (✉)
Laboratory Animal Science Group, IBMC–Instituto de Biologia Molecular e Celular, Universidade do Porto, Porto, Portugal
e-mail: Olsson@ibmc.up.pt

Curr Topics Behav Neurosci (2015) 19: 47–78

DOI: 10.1007/7854_2014_319

Published Online: 22 May 2014

1 Introduction

1.1 The Ethical Dilemma of Animal Research

Studies on live animals play an important role in neuroscience research. In *basic* neuroscience research, animals are studied to understand the functioning of the nervous system and the mechanisms involved in the diseases that affect it. In *applied* neuroscience research, animals are used to develop and test therapies for such diseases. The ultimate aim of both lines of research is commonly to extrapolate results to the human case. When animals are used as models of human diseases—which constitutes the bulk of animal-based neuroscience research—the object is to induce in them conditions which, at least in some aspects, mimic the conditions that researchers aim to understand in humans and for which they wish to develop appropriate treatment. Disease-oriented research in neuroscience includes the study of both psychiatric and neurological disorders. The former consists of disorders of mood and thought associated with either no apparent signs or at most only minor physical signs in the motor and sensory systems, and includes diseases such as schizophrenia (SCZ), depression or anxiety. The latter refers to nervous system disorders that also present somatic signs and include neurodegenerative diseases, such as Alzheimer's, Parkinson's, Huntington's disease, Amyotrophic Lateral Sclerosis, or stroke, and pain (Baker et al. 2002). Animal models are used in research in both psychiatric and neurological disorders.

The use of animals for research remains a controversial issue. Most experimental procedures are likely to inflict at least some harm on the animals that are studied. During experimentation, animals may be in relatively limiting conditions, and deprivation of food and water often forms part of behavior testing schedules. Varying degrees of physical or psychological harm can result from the procedures used to induce in animals conditions mimicking the human diseases under study, as well as from the conditions themselves. Distressing or painful interventions may be part of experimental protocols and, not least, most animals are killed at the end of experimental trials. Unlike humans participating as subjects in research, however, animals cannot consent to their own participation. Moreover, since most research is intended to benefit humans and not animals, there is no benefit for the research subjects themselves. While perhaps few researchers would question the desirability of discovering new ways to prevent, alleviate, or cure human diseases, the question remains: Are we, as human beings, morally justified in using animals as tools for research?

The answer to this question, of course, depends on one's ethical framework. In society[1], we find a range of opinions: while some people are outraged by the idea of inflicting suffering in innocent animals, others consider it acceptable to do research on animals, if there are no alternatives available and if the outcome of the research is valuable; at the extreme, some see no problem in animal experimentation at all (e.g., Nuffield Council 2005). In a pluralist society, there is a room for such a variety of opinions to coexist, but there is also the need for finding a working compromise which can be accepted by a majority, for example, in order to draft policy and legislation.

The main driver for regulating the use of animals in experiments is the demand from society to protect these animals. Public concern for the well-being of animals has a long history and the protection given to animals has gradually been formalized, to the extent that animal welfare today is mentioned in several constitutional documents—including those of the European Union. Legislation protecting animals used in research was first introduced in the second half of the nineteenth century with the enactment of the UK Cruelty to Animals Act (1876), the first legislation to place "conditions on the way in which (experiments on animals) were done, the reasons for which they were done, and the qualifications of persons responsible for performing them" (Bayne et al. 2011). Today, the use of animals in research is commonly regarded as requiring explicit ethical justification. Such justification has become an integral part of the review process for research proposals involving animals and is prescribed in legislation worldwide. Usually, a research proposal is understood to being justifiable only if the benefits of the study can be demonstrated to outweigh the expected harms to the animals used.

Basing ethical justifications for a given procedure on some evaluation of the relative magnitudes of harm and benefits is currently the central requirement in legislation worldwide, as apparent in official guidelines and policy documents for ethical review. For example, the Australian Animal Welfare Act 1999 states that the key principle underlying the application approval process is that "the use of animals in research, testing, and teaching is confined to cases in which there is good reason to believe the benefits (…) are not outweighed by the likely harm to the animals." More recently, the new European Directive 63/2010/EU also makes explicit that "(…) an impartial project evaluation, independent of those involved in the study, should be carried out (…) The project evaluation shall consist in: (…) (d) a *harm-benefit analysis* of the project, to assess whether the harm to the animals in terms of suffering, pain and distress is justified by the expected outcome taking into account ethical considerations, and may ultimately benefit human beings, animals or the environment."

---

[1] In this chapter, we use the word 'society' to refer to a wider public in the industrialized countries.

Before further expanding on the mechanics of this process of weighing harms and benefits of particular research programs, however, we will first look briefly at the ethics underlying the wider issue of the overall acceptability of any form of animal experimentation.

1.2 Ethical Theories

Assessment of harms and benefits has become the mainstream, approach in examining the potential justification for individual, animal-based, research protocols. However, this approach presupposes from the outset that animal experimentation is, in principle, acceptable to society, within defined constraints. Hence, assessment of the harms and benefits of a particular piece of research does not address the more fundamental ethical issue of animal experimentation in general. There are various ethical theories regarding human duties to animals, which offer different perspectives on the acceptability of the use of animals in research. Here, we will expand on three of them: contractarianism, utilitarianism, and animal rights. These three ethical perspectives, representing a significant proportion of the debate over animal research, reflect three relevant but markedly different views on the issue.

According to the contractarian view (e.g., Cohen 1997), animals are morally insignificant or lack moral status. Contractarians regard morality as a system of hypothetical contracts that rational individuals tacitly enter into with one another. Animals cannot be direct parties to such contracts, or agreements, because they lack the linguistic and intellectual skills to do so. Hence, animals are not bearers of duties or rights. However, for those who adopt the contractarian view, the way animals are treated is not by definition irrelevant. To the extent that people care about animals, the tacit contracts that constitute morality will contain clauses affording some protection to them. If some people are emotionally attached to certain types of animals and disapprove of their use in experiments, this becomes an ethical concern. For the contractarian, public concern is the overarching issue determining the ethical acceptability of animal experiments. Harm–benefit analysis is only relevant to the extent that it affects public acceptance of research—it is not relevant in itself.

The animal rights view (e.g., Regan 1983), as the name indicates, defends that some animals (the *experiencing subjects of a life*—usually understood as conscious animals) have rights, at least the basic right always to be respected as an end in themselves and never to be used merely as a means to an end. From this perspective, animal experimentation is considered to be one of the many examples of human use of animals in which this right is disrespected—it implies using animals as instruments for reaching human purposes. Thus, the animal rights view has an abolitionist position as regards animal experimentation. It does not matter that an experiment will cause only minor suffering to the animals or that it will have an extraordinary contribution to humanity: experimentation with sentient animals is in itself unacceptable. Hence, the animal rights view rejects harm–benefit weighing as a measure of the acceptability of animal research.

According to the utilitarian view (e.g., Singer 1975), morality has one basic rule: always act so as to maximize the well-being of those affected by your actions or, in other words, to create the greatest amount possible of good. In the utilitarian approach to animal ethics, the good to be maximized is defined in terms of pleasure and the absence of suffering. Moral consideration is therefore entitled to every sentient creature. In the utilitarian approach, then, ethical decisions require us to strike the most favorable balance of benefits and harms for all sentient individuals affected by what we do. Utilitarians support the use of animals in research if and only if the harm to animals is outweighed by the benefits of research. In fact, harm–benefit analyses are the core of the utilitarian approach.

The question raised earlier about whether humans are morally justified to use animals for experimental purposes has, thus, no single answer. There are a number of different ethical perspectives, of which we have here advanced three. These three views cover the most common arguments heard in the public debate about the use of animals in experiments. These arguments, however, focus on the fundamental principle of whether or not it is ethically legitimate to undertake experiments on animals at all. In the rest of the chapter, we will work within a practical reality where such experimentation is legal and accepted by the majority of society, if within certain constraints. Within such a framework, we will explore the more applied aspects of animal research ethics; that is, how can animal ethics influence and affect actual animal experiments, their regulation, and how they are carried out.

In this more practical application of how ethical considerations may be applied to actual research protocols, issues of harm versus benefit play a central role, as well as considerations of how these two factors may be weighed against each other. It is important to stress that we are not favoring one particular ethical perspective. Rather, we attempt to explore specific issues of animal research ethics in neurosciences within the utilitarian-based, mainstream, position. Although the harm-benefit weighing has a strong component of utilitarianism, this does not mean that considerations of harm and benefit are only relevant from the utilitarian perspective. Neither contractarians nor animal rights defenders would be against the minimization of the harms caused to the animals used in research and at least contractarians would support the maximization of the benefits to be gained from it.

2 Harms

In this chapter, we follow the standard approach used in ethical evaluation of animal research and consider harms as negative impacts, actually or potentially, caused to the welfare of animals. That includes adverse effects on health, as well as all the adverse subjective experiences animals might undergo, such as pain, fear, or anxiety. We also include killing as a harm, based on the understanding that killing rarely is in the animal's own best interest and it prevents it from all potential future positive experiences (see Yeates 2010).

2.1 What Are the Harms of Neuroscience Research?

If we follow the approach outlined in legislation, the ethical acceptability of animal research is determined by balancing its benefits and harms. Harms to animals used in neuroscience research can potentially result from inappropriate transport, housing, handling, and care, from the procedures used to inflict the disorders to be studied, from the experimental techniques to administer treatments and monitor parameters, and from badly conducted euthanasia. Using concrete examples, we discuss specific harms to animals associated with neuroscience research. The cases discussed have been selected to illustrate important issues, with no claim for the list to be exhaustive.

As our first illustration: the $SOD1^{G93A}$ mouse model of Amyotrophic Lateral Schelerosis (ALS) is an example of a genetically modified neuroscience model. In this model, the onset of this neurological disease is characterized by weakness and tremors of hind limbs. Disease progression then leads to paralysis of hind limbs, accompanied by increased difficulty to eat, drink, and swallow, terminating in complete paralysis (Lever et al. 2009, 2010). Mice die of respiratory failure due to paralysis of diaphragm (Solomon et al. 2011). Discomfort is unavoidable at least in the more advanced stages of the disease in this animal model due to the progressive loss of motor capacity which characterizes the disease. In some studies, the animals reach an advanced stage in which they have difficulty in reaching food and water provided at the cage top and performing behaviors as simple as chewing. Animals kept beyond this stage will die from the disease. The pain that patients might experience has been a neglected aspect of ALS. In a recent review, Handy et al. (2011) raised this concern. Although not generally associated with ALS, pain has been reported to occur in nearly 70 % of ALS patients at some point in time. This raises an additional concern regarding the welfare of animals used as models of ALS. They may, like humans, feel pain during disease progression.

Whereas the $SOD1^{G93A}$ mouse phenotype is well known by now, special attention must be given to genetically altered animals whose phenotype is not (yet) thoroughly characterized, as unpredictable changes in physiology or behavior may also occur (Morton and Hau 2011).

Different welfare concerns can be found in the next illustrative case, rats with neonatal hippocampal lesion (NHL), used as a model of SCZ. In the NHL model, the disorder is surgically induced, in that, the hippocampus of these rats is lesioned few days after birth by way of an injection. The animals are then returned to their mothers and weaned normally. This model involves a number of potential stressors for the animal: maternal separation before surgery, the surgery itself, postoperative recovery, and presumably postoperative pain and discomfort. At this point, pain and stress appear as the major animal welfare issues. As adults, these animals present several behavioral deficits, such as increased response to stress, deficient prepulse inhibition and latent inhibition, impaired social behaviors, and working memory problems (e.g., Lipska 2004). Although these impairments do not generate physical suffering, they reflect a loss of capacities which may have psychological repercussions for the animals. For example, it can be assumed that handling and external disturbances are more stressful for animals with hyper-responsiveness to stress or with memory impairment, which may result in a diminished capacity for behavioral habituation. Also, these rats may undergo stress by being housed in groups due to their impaired social behavior.

Another important area of neuroscience research with its own welfare issues is pain research in general, and neuropathic pain studies in particular. Neuropathic pain is defined as a form of chronic pain that results from damage or abnormal function of the central or peripheral nervous system (Abdi et al. 2004; Woolf 2004). Patients with neuropathic pain frequently report sensory abnormalities including burning sensations, exaggerated responses to noxious stimuli (hyperalgesia), pain sensations resulting from innocuous stimuli (allodynia), and spontaneous pain episodes (dysesthesia) (Gilron et al. 2006). Due to its severity, chronicity, and resistance to some classical analgesics (Gilron et al. 2006), it has received much attention in research.

A widely used animal model of neuropathic pain involves the surgical placement of tubing cuffs around the main branch of the sciatic nerve in rats. These animals are expected to undergo substantial pain in the course of the experiments. In this type of animal research, pain is the major welfare concern and the fact that pain is also the characteristic under study makes it more challenging to diminish it than in many other research situations. Besides the direct experience of pain, it has been demonstrated that long-term neuropathic pain can cause anxiety (after 4 weeks) and depression-related behaviors (after 6–8 weeks) (Yalcin et al. 2011). Such harms should be avoided where possible because, unlike pain, they are not a central feature of most studies on neuropathic pain. Although the results presented by Yalcin et al. (2011) show that mood disorders and neuropathic pain may be connected, it is not a necessary condition for the study of pain that the animals are left anxious or depressed.

Harms such as those presented in this section affect the well-being of the animals, but they also have important implications for the quality of experimental data. Animals under stress may, for example, have altered physiological parameters which can interfere with the conclusions drawn from the studies. Hence, there are two major reasons for reducing harm to animals in research: to preserve animal welfare and to guarantee the cleanness of experimental data (Russell and Burch 1959; Smaje et al. 1998).

2.2 How Can We Assess Harms?

In the previous section, we gave some examples of harm inflicted to animals in experimental research in the neurosciences. These harms include effects on animal health but also adverse subjective experiences resulting from the experimental interventions. Health and subjective experiences both form part of what is commonly understood to represent "animal welfare," and considerable research is directed toward understanding how animal welfare is affected by internal and external factors, as well as toward developing methods for assessing welfare status or changes to welfare. In the context of neuroscience, animal welfare science is particularly relevant when it comes to understanding and measuring subjective experiences, including pain. Here, the two disciplines overlap considerably and there is potential for methods as well as ideas to flow both ways. Two relevant examples are the use of cognitive bias to assess animal welfare and the development of a face expression scale to assess pain in mice.

Based on the idea that the way information is processed by humans is affected by their mood and that thus, those suffering from a mood disorder are likely to make biased judgments (e.g., anxious and depressed people tend to make negative judgments about events and to interpret ambiguous stimuli unfavorably), Mendl and collaborators tested whether laboratory rats also show such cognitive bias. Rats were trained on a discrimination task in which one tone predicted the arrival of food (positive event) and the other the onset of white noise (negative event). When the rats had reliably learned to discriminate the tones, they were presented with ambiguous stimuli, that is, tones that were intermediate between those signaling positive and those signaling negative events. Rats having experienced chronic mild stress (according to a protocol often used in behavioral pharmacology as a model of depression, e.g., Willner 1997) were more likely to respond to ambiguous cues as if they were predicting a negative event—that is they showed a negative cognitive bias (Harding et al. 2004). This research has received considerable attention in animal welfare science as a promising method to assess subjective experiences; similar results have been demonstrated for a range of other mammal and also bird species (e.g., Mendl et al. 2009) and the approach may thus have considerable potential to help to quantify harm in animal experiments.

When Mogil and collaborators tested whether facial expression in mice reflects the level of painful stimuli to which they are exposed, their aim was to improve the relevance of animal-based pain research for human benefit. The subjective component is an essential part of the human pain experience, which is not presently available in nonverbal animals, and as the authors argue "[t]his measure of spontaneously emitted pain may provide insight into the subjective pain experience of mice" (Langford et al. 2010). However, it may also be used to develop measures to recognize pain in mice which can underlie decisions to apply analgesia or apply other measures of refinement, though such applications still have to be established.

By increasing our understanding of subjective experiences in animals, measures such as those outlined above may complement clinical measures of animal health when the welfare of animals is to be formally assessed. Such practical assessment is likely to play a larger role in the management of ethical issues in animal research, as formal assessment of actual harm resulting from a given protocol is being given greater emphasis. One example of this is the retrospective assessment of the effects of experimental procedures which is now being implemented at the European level (see Sect. 2.3.3).

Currently, however, assessment of probable harm is primarily an exercise of prediction within ethical evaluation for a proposed procedure. Toward this, guidelines and policy documents for the evaluation of animal experiments have suggested lists of criteria to be assessed. These criteria usually include the quality of the facilities, the experience of the personnel caring for and carrying out the procedures on animals, the number of animals, the animal species, the husbandry and housing conditions, the scientific procedures themselves (including the killing method), the duration and the intensity of the pain or distress likely to be inflicted on the animal, the fate of the animals at the end of the experiments, and the endpoints to be applied (e.g., APC 2003; FELASA 2005; Smith and Boyd 1991).

Some regulatory systems further ask for classification of the severity of procedures or experiments, that is, the degree of pain or suffering likely to be experienced by animals. In assessing the negative impact on the animal, the duration and frequency of a procedure is considered. Table 1 presents, as examples, the severity classifications adopted by the European Union in Directive 2010/63/EU and by the Canadian Council on Animal Care (CCAC). The new European Directive requires that experiments are attributed a severity classification (non-recovery, mild, moderate, or severe) determined by the degree of pain, suffering, distress, or lasting harm expected to be experienced by an individual animal during the course of the procedure. Following the European Directive, in assigning the category of severity, one shall take into account any intervention or manipulation of an animal within a defined procedure, namely (1) the type of manipulation, handling, (2) the nature of pain, suffering, distress, or lasting harm caused by (all elements of) the procedure, and its intensity, the duration, frequency, and -multiplicity of techniques employed, (3) the cumulative suffering within a procedure, and (4) the prevention from expressing natural behavior including restrictions on the housing, husbandry, and care standards. For more than 20 years, evaluation under the Canadian Council on Animal Care (CCAC) has applied a similar scale of "Categories of Invasiveness in Animal Experiments" ranging from experiments on most invertebrates or on live isolates (A) to procedures in vertebrates and some invertebrates which may cause severe pain (E) (CCAC 1993). Protocols must be submitted to an appropriate review committee for all studies which involve the use of vertebrates and some invertebrates in categories B through E. Both the European Directive and the CCAC guidelines give potential examples of experimental procedures which are considered to be representative of each category. Although neurosciences are poorly represented in these documents, we advanced a few examples in Table 1.

Table 1 Severity classifications of animal experiments from the new European Directive 63/2010/EU (*left column*) and the Canadian Council on Animal Care (CCAC 1993, *middle column*) and examples from neurosciences illustrating the different categories (*right column*)

EU directive	CCAC guidelines	Potential examples in neurosciences
Non-recovery Procedures which are performed entirely under general anesthesia from which the animal shall not recover consciousness		Some studies of neuroconductivity
	A Experiments on most invertebrates or on live isolates	Studies of neurobiology in *C. elegans*
	B Experiments which cause little or no discomfort or stress	Studies of cognitive bias in companion dogs without food or water deprivation
Mild Procedures on animals as a result of which the animals are likely to experience short-term mild pain, suffering, or distress, as well as procedures with no significant impairment of the well-being or general condition of the animals	**C** Experiments which cause minor stress or pain of short duration	Mouse model of human repetitive mild traumatic brain injury
Moderate Procedures on animals as a result of which the animals are likely to experience short-term moderate pain, suffering, or distress, or long-lasting mild pain, suffering, or distress as well as procedures that are likely to cause moderate impairment of the well-being or general condition of the animals	**D** Experiments which cause moderate to severe distress or discomfort	NHL model of schizophrenia
Severe Procedures on animals as a result of which the animals are likely to experience severe pain, suffering, or distress, or long-lasting moderate pain, suffering, or distress as well as procedures, that are likely to cause severe impairment of the well-being or general condition of the animals		Huntington's disease ALS Learned helplessness
	E Procedures which cause severe pain near, at, or above the pain tolerance threshold of unanesthetized conscious animals	

2.3 How to Reduce Harms?

According to the mainstream approach on which this chapter is based, for an animal experiment to be ethically acceptable, the expected benefits not only must outweigh potential harms, but the harms caused to the animals must be reduced to a minimum as well, or in other words, animals shall not undergo unnecessary suffering. The three Rs (replacement, reduction, refinement), proposed by Russell and Burch (1959), are widely recognized principles in the attempt to minimize harms to animals and, hence, to perform ethically acceptable research. Whereas the Replacement and Reduction principles reduce harm by avoiding animal use, the Refinement principle addresses the welfare of individual animals which are actually used in experiments. We will discuss each of the three Rs in the context of neuroscience research.

2.3.1 Replacement

Replacement is the first of the three Rs, for several reasons:

> Replacement enjoys a particular standing among the three Rs. It was the first of the Rs to be introduced by Russell and Burch (1959), reflecting the intended order in which the Rs were to be considered. Questions about Reduction and Refinement are only relevant if Replacement has first been considered and excluded. The goal of Replacement also has received widespread support, in part because it is the only goal that is fully compatible with the animal rights perspective that animal use solely for human benefit should not be permitted (Olsson et al. 2011).

The main point of this principle of Replacement is that the use of animals should be replaced by nonanimal methods whenever this is possible without compromising the research objective. Replacement methods can be divided into four main types: in vitro (e.g., cell lines), ex vivo (e.g., tissue cultures), in silico methods (e.g., bioinformatics), and research with human volunteers. The idea that studies in human volunteers would be an ethical alternative to the use of animals in research may seem provocative, and it is of course a sine qua noncondition that such a study meets the ethical standards for research with human subjects. That said, in the neurosciences there seems actually to be real potential for this approach where replacement not only spares animals but also increases the relevance of the research itself. Focusing specifically on pain research, a workshop which brought together neuroscientists with proponents of nonanimal research methods came up with a number of suggestions for situations in which studies with human volunteers could replace animal studies. They are all based on the use of low-risk minimally invasive techniques (e.g., functional imaging and microdialysis) in humans, and draw on the fact that it is much easier to evaluate subjective experience—a key aspect of pain—in verbal humans (Langley et al. 2008).

Another example of a replacement strategy in neuroscience research was recently described by Barnett and collaborators (Sorensen et al. 2008; Boomkamp et al. 2012). These authors proposed an in vitro method for research on spinal cord injury, a disorder that has depended mainly on animal research. Spinal cord injury is a complex injury, caused by traumatic accidents. Traumatic injury disrupts spinal white matter tracts, resulting in loss of sensory and motor function. This loss of function is generally permanent because the central nervous system has a restricted regenerative capacity (Fawcett and Asher 1999; Rudge and Silver 1990). After the initial injury, which results from direct mechanical disruption of spinal cord integrity, glial scars are formed, which inhibit central nervous system repair by creating both physical and biochemical barriers to axonal growth (Boomkamp et al. 2012).

An example of an animal model of spinal cord lesion is a wire knife lesion, generated by inserting the knife into the dorsal column and pulling up a piece of tissue. The method results in a cavity and glial scarring that mimics human spinal cord injury. Disadvantages to rat models of spinal cord injury include the need for large numbers of animals, the severity of the procedure for the animals, the long time frame for results, and the high expensive of the experiments (International Animal Research Regulations 2012).

In the nonanimal model proposed by Barnett and collaborators, embryonic spinal cord cells from rats are layered on top of an astrocyte monolayer derived from embryonic tissue. Growth in culture over time leads to complex axonal/glial interactions resulting in myelinated neurons. This system allows for the study of contact between astrocytes and how they communicate with the axons, which is necessary for understanding the problems in spinal cord injury. The researchers also have induced lesions in the cell culture by cutting with a scalpel to studying axon density and myelination adjacent to the lesion and cell growth into the damaged area (International Animal Research Regulations, Impact on Neuroscience Research, Workshop Summary, Institute of Medicine (US); National Research Council (US) 2012).

Overall, the greater the role of nonanimal replacement in research, the fewer animals will be needed in total for research purposes. In this way, replacement is also directly related to the second R, Reduction.

2.3.2 Reduction

The aim with the principle of Reduction is to use the smallest possible number of animals to obtain valid information. Its main ethical purpose is to reduce collective animal harm, understood as the number of animals on which harm is inflicted. One important measure is to use correct and careful statistics, namely by carrying out appropriate power analysis prior to study commencement. Sample sizes can also be decreased by controlling variance associated with different environmental and genetic conditions, as for example, by using uniform housing conditions and inbred animals.

Reduction is probably the most controversial of the three Rs. There is a great political value in bringing down numbers of animals used in experimental procedures as a whole, as the number of animals reported in annual statistics is a very visible and easily understood aspect of research animal ethics. This also holds for replacement—performing fewer experiments is also immediately recognizable in the statistics. However, the problem with reduction is that, as detailed analyses have repeatedly shown, in actual research the number of animals used in an individual experiment is often too small for results to be reliable. This of course has important implications for the validity of the research results. Within a larger review of methods in neuroscience, Button and collaborators (2013) examined the statistical power of animal experiments investigating sex differences in water maze and radial maze performance. The effect (i.e., how large a difference is between male and female animals) was calculated through a meta-analysis, and the authors then established how many animals a single study would need to detect effects of this magnitude with different levels of statistical power. To achieve 80 % power (a common standard), 134 animals would be needed for a water maze experiment and 68 for a radial maze, whereas the average sample sizes were 22 and 24 animals, respectively. The authors commented on the ethical consequences of underpowered studies:

> There is ongoing debate regarding the appropriate balance to strike between using as few animals as possible in experiments and the need to obtain robust, reliable findings. We argue that it is important to appreciate the waste associated with an underpowered study—even a study that achieves only 80 % power still presents a 20 % possibility that the animals have been sacrificed without the study detecting the underlying true effect. If the average power in neuroscience animal model studies is between 20–30 %, as we observed in our analysis above, the ethical implications are clear.
>
> Low power therefore has an ethical dimension—unreliable research is inefficient and wasteful. This applies to both human and animal research. The principles of the 'three Rs' in animal research (reduce, refine and replace) require appropriate experimental design and statistics—both too many and too few animals present an issue as they reduce the value of research outputs.

Based on this, it does not seem appropriate to apply reduction through uncritically decreasing sample sizes in individual experiments. Additional approaches in experimental design are needed if the aim is to bring down animal numbers. This could include the use of imaging techniques allowing the study of disease progress in the same animals rather than in separate groups for separate time points, or greater use of nonanimal approaches before moving to an animal model.

2.3.3 Refinement

Whereas the Replacement and Reduction principles reduce harm by avoiding the use of animals, the Refinement principle addresses the welfare of individual animals which are actually used in experiments. This principle states that all experimental procedures shall be adjusted to minimize any pain or discomfort they may cause to the animals. Experiments can be refined in several ways, from the use of anesthesia and analgesia, to housing adaptations and the establishment of human endpoints. Appropriate measures need to be defined for each individual study, taking into account the nature of the harms which need to be mitigated. The scheme for welfare assessment recently proposed by a European working group allows refinement measures to be integrated into the assessment. Table 2 displays examples of refinement measures that can be applied in neuroscience studies whose harms were presented in Sect. 2.2 (and are now summarized in the scheme).

The principle of the three Rs is already present in much legislation. For example, the new European Directive states that "To ensure that the way in which animals are bred, cared for and used in procedures within the Union is in line with that of the other international and national standards applicable outside the Union, the principles of replacement, reduction and refinement should be considered systematically when implementing this Directive." Although the three Rs principle was not explicitly referred to in previous European legislation, researchers were asked to use animals only when necessary, to use as few animals as possible and to use procedures having as little impact as possible.

Unfortunately, systematic reviews of the implementation of refinement measures in biomedical research indicate that the present situation is far from ideal. For example, between 2000 and 2002 pain relief was administered in only around 20 % of studies subjecting rodents to potentially painful procedures (Richardson and Flecknell 2005). In 2009, humane endpoints were only reported in about 20 % of studies of mice models of the neurodegenerative disorder Huntington's Disease, with no significant increase in the reporting of this refinement measure during the preceding 10-year period (Franco and Olsson 2012). There is thus considerable potential for improvement in the application of refinement.

2.3.4 Is Species Choice a Way to Reduce Harm?

In this final section regarding harm, we will address an idea that is recurrent in the discussion of ethical evaluation and regulation of research: that research will be more or less harmful depending on the animal species chosen (see also Chapter "Would the Elimination of the Capacity to Suffer Solve Ethical Dilemmas in Experimental Animal Research?" of this book). There are sometimes obvious physical justifications having to do with the size of the animal in relation to the minimum amount of tissue needed for analysis or the minimum size of lesion determined by human dexterity and instruments used—in such cases the smaller the animal the larger the proportional impact will be.

But there is also the widespread idea that animals of different species vary in their capacity for subjective experience. This idea is put forward in the European Directive which requires that if several methods are available one shall choose those that "involve animals with the lowest capacity to experience pain, suffering, distress or lasting harm" (Directive 2010/63/EU, Article 13). This seems to indicate that animals can be different on their capacity to suffer—but no guidelines are given for how to assess this capacity. Smith and Boyd (1991) proposed a

Table 2 Schematic approach for assessing severity proposed by the European Commission Expert Working Group

Example	What does this study involve doing to the animals?	What will the animals experience? How much suffering might it cause? What might make it worse?	How will suffering be reduced to a minimum?	
		Adverse effects	Methodology and interventions	Endpoints
Genetically-modified $SOD1^{G93A}$ mouse model of ALS		Discomfort associated with motor capacity loss and difficulties to eat, drink and swallow	Housing adaptations (e.g., placing mashed food at a low level and adjusting of bedding to facilitate movement)	
		Pain	Use of painkillers	
		Animals may reach complete paralysis		Euthanasia of the animals as soon as possible to avoid unnecessary suffering
NHL rat model of schizophrenia	Maternal separation	Anxiety associated with maternal separation	Reduce duration of separation up to a minimum	
	Surgery for hippocampus lesion	Pain and discomfort associated with surgery	Appropriate anesthesia and analgesia	
		Potential stress resulting from handling and external disturbances	Handling and external disturbances avoided up to a minimum	
		Potential stress resulting from group housing	Rats housed in groups should be monitored for anxiety behaviors related to social contact and maybe housed individually	
Rat model of neuropathic pain	Placement, through surgery, of tubing cuffs around the main branch of the sciatic nerve	Pain	Adequate anaesthesia and analgesia during and immediately after surgery	
		Possible anxiety and depression-related behaviors		Early endpoints to avoid the development of anxiety and depression-related behaviors

Adapted from the document at http://ec.europa.eu/environment/chemicals/lab_animals/pdf/examples.pdf

systematic method of assessment consisting of a checklist of neuroanatomical/physiological and behavioral criteria to determine whether a nonhuman animal has the capacity for pain, stress, and anxiety. On the neuroanatomical side the criteria include (1) the possession of receptors sensitive to noxious stimuli (nociceptors), (2) the possession of higher brain centers (especially a structure analogous to the human cerebral cortex), (3) the possession of nociceptors connected to these higher brain structures, and (4) the possession of opioid-type receptors. On the behavioral side the criteria include (5) responses to painful stimuli modified by analgesics, (6) avoidance or escape responses to painful stimuli, (7) responses to noxious stimuli that persist, and (8) the capacity to associate neutral with noxious stimuli. However, looking at how taxonomically distinct animals used in research fare in this assessment it becomes clear that (1) if we complement the information available in the original 1991 analysis with contemporary knowledge about fish, at least all vertebrate animals meet the criteria for pain and (2) information about nonvertebrate sentience is too limited to allow species to be identified as less sentient with reasonably reliability. That is, it is highly unclear what animal researchers are to choose to ensure "lowest capacity to experience". Colin Allen (2004) proposed to use learning abilities as indicators of capacity to suffer. This would include operant learning—which appears to require a brain (Grau 2002), and certain kinds of classical conditioning (e.g., trace conditioning). However, this also does not help much to draw the distinction between species. There are no significant differences in the learning abilities between mammal species, and most likely all vertebrates and even some invertebrates would still fall within the same category in terms of their "capacity to experience".[2]

On the other side of the sentience coin, we find the concern that some species might have a *higher* capacity to experience. In a position paper, a European Science Foundation working group argued that nonhuman primates (NHPs) have a greater potential for suffering since they are : "distinguished by the very advanced nature of their social, cognitive, sensory and motor functions" (ESF 2009). A reasonable interpretation of this is that NHPs will be more harmed by research than other laboratory animals. But in which way?

We have analyzed this question in some detail elsewhere (Olsson and Sandøe 2010). In summary, in terms of capacity for sentience, it is unclear how most NHPs are different from other mammals which also share the capacity for experiencing pain and distress. Capacity for self-awareness may affect potential for suffering, but reasonable evidence to attribute this capacity only exists for great apes. The biological difference with clearest welfare relevance between NHPs and other mammal species used in research seems to be that primate species are not fully domesticated, making it more challenging to meet their needs in captive housing. On the other hand, there are also aspects in which primates may be better off in research than, for example, rodents: primates are usually trained to collaborate rather than restrained,

[2] See Chapter "Telos, Conservation of Welfare, and Ethical Issues in Genetic Engineering of Animals" of this book for an extensive discussion on animal consciousness.

their greater similarity to human beings facilitates the recognition of signs of poor welfare and higher concern for their welfare might encourage scientists to be more careful in how primates are treated.

Very recently, working on an analogy with pediatric research ethics, Fenton (2014) advocated that the cognitive capacities of chimpanzees may allow them to dissent from participating in research. It is not consensual whether chimps fulfill the conditions set, including, for example, whether they are capable of planning the future—which is far from being consensual (e.g., Shettleworth 2010; Suddendorf et al. 2009). Furthermore, it remains to be seen how to allow such dissent to be expressed in practice in a meaningful way. However, if these hurdles were overcome, this may be an interesting approach to develop a research ethics for nonhuman primates which not only respects but actually relies on their cognitive capacities.

In summary, with present knowledge there is little support for establishing differences which can motivate species choice to be a useful measure to reduce animal harm. Instead, the differences that society and the research community tend to make between less and more ethically problematic species are best understood in the light of the socio-zoological scale. This scale rates animals in terms of how greatly they are valued by humans, and places companion animal species and nonhuman primates at the top and rodents, fish, and invertebrates quite further down (Arluke and Sanders 1996). That the socio-zoological scale is based on what humans think about animals rather than on the characteristics of the animals themselves does not mean that it is ethically irrelevant. But, in our opinion, the difference between using a rhesus macaque or a fish in a given experiment is better described as more or less harmful to public sensitivity than more or less harmful to the animal.

3 Benefits

3.1 What Are the Benefits of Neuroscience Research?

We now turn to the other side of the harm–benefit equation, the benefits. Overall, research in neuroscience aims to deliver benefits for scientific knowledge and for human health and welfare, that is, benefits for humans. On one hand, basic research is conducted with the aim of understanding the functioning of the nervous system and the mechanisms underlying the diseases that affect it. On the other hand, applied research is carried out to develop treatments for such disorders.

3.2 How Can We Quantify Benefits?

3.2.1 Assessment of Potential Benefits

It is very difficult or almost impossible to predict accurately whether a research project will improve our understanding of important mechanisms or lead to the development of therapeutics. Science has a considerable element of unpredictability; even when armed with well-defined hypotheses and carefully executed experiments, it is impossible to guarantee that a research project delivers its intended benefits, in particular, if these are defined on the level of the practical impact, the study will have in the scientific field or in society. Especially with basic research it is difficult to anticipate the direction of the findings (i.e., whether they will support the researchers' hypotheses) and the long-term impact of such results for human health and the society. Nevertheless, assessing benefits is fundamental if we are to weigh them against harms in order to justify animal experimentation. Also, using animals for research with no clear or intended benefit would be unethical virtually for every ethical position.

Official documents provide some, although little, guidance as to how to evaluate benefits. An expert working group set up by the Federation of European Laboratory Animal Science Associations (FELASA 2005) described and explored a set of principles for how to conduct ethical reviews of laboratory animal use and proposed an outline scheme for the assessment of benefits and harms in scientific projects involving animals. On the benefit side, the questions to be answered included:

- How will the results add to existing knowledge? What practical applications, if any, are envisaged at this stage?
- What is the potential value of these insights and/or applications?
- Are the objectives of the project original, timely, and realistic?
- How does the present proposal relate to what was done before? What progress was made in previous studies, and what scientific or other benefits have resulted?
- What is the relevance of this project to other studies in this field of research and what might be the implications for other areas of research, if any?

Similar questions are proposed by other policy documents and reports. Both The Canadian Council on Animal Care guidelines (CCAC 1997) and the recent Working document on Project Evaluation and Retrospective Assessment, for example, claim for clear statements of the scientific objectives and potential value of the study in terms of originality and importance of the new information, as well as the need for the experimental project.

Many of these issues are challenging to evaluate, to say the least. There are practical challenges having to do with the difficulty of predicting outcomes, but also ethical challenges in terms of judging whether a certain scientific objective is more valuable than another. For which purpose animals are used plays a role in determining acceptability of research in society (see Lund et al. 2012), but it is unclear to what extent this is also reflected in practical decision-making, and official documents guiding ethical review focus more on the assessment of the likelihood that the proposed benefits will be achieved.

3.2.2 Assessment of Likelihood That Potential Benefits Will Be Achieved

At least as long as benefit is understood in terms of knowledge gains, these questions are of a more technical nature and, thus, are easier to evaluate and can be assessed more objectively. They include evaluating (1) the appropriateness of the animal model and the scientific approach, (2) the validity of the experimental design, (3) the staff competence, (4) the appropriateness and quality of facilities, and (5) the communication of results (e.g., APC 2003; FELASA 2005; Smith and Boyd 1991). This kind of evaluation will tell whether a proposed study will be able to provide reliable answers to the questions it poses, without making any judgment on the relevance of these questions.

The choice of an appropriate animal model is crucial for a research project to succeed. However, critical analysis of what characterizes a good animal model is curiously rare in the scientific literature. Most review papers on animal models limit themselves to an overview of the existing models in a field and discussion of the results obtained in studies using them. In neuroscience, it is widely recognized that existing animal models are insufficient (e.g., Micale et al. 2013). The animal models of psychiatric diseases have been especially criticized (e.g., Schapiro 1998; Rollin and Rollin 2014), mainly due to the impossibility for any animal model to accurately mimic all aspects of mental illnesses, some of which are unique to humans (American Psychiatric Association 2000). These criticisms have led to greater attention to questions of model validity in neurosciences than in, for example, infectious disease research. In 2012, National Institutes of Health (NIH) workshop aimed to increase the utility and translation of animal models in neuroscience research, experts highlighted the need for improved animal models and better matching between animal model research and the human disease phenotype. Several other joint efforts in the scientific community indicate that there is a heightened awareness of the present shortcomings of animal models in this field. It remains to be seen how well this translates into better practice. An important consequence of using animal models of unreliable validity is that scientific progress may be retarded and, as a consequence, animals are used unnecessarily.[3]

Regarding the second issue, the experimental design of animal experiments, systematic reviews have recurrently shown significant shortcomings in how studies

[3] In this context, it is interesting to note that NIH recently announced it will only fund clinical trials in psychiatry which address underlying biological mechanisms (Reardon 2014). Although it is too early to tell whether this will affect how and which animal models are used in this field, this measure may contribute further to raising the validity requirements in this field of research.

are designed and carried out in neurosciences. Most of this knowledge comes from extensive systematic reviews and meta-analyses of experimental animal research in acute ischemic attack (stroke). In this field, a wide number of neuroprotective drugs which had proven effective in animal models later failed to work in clinical trials on humans (van der Worp et al. 2005). In many of these experiments, the efficacy of the treatment was probably overestimated as a result of design bias. Often animals were not randomly allocated to treatments, and researchers who were not blinded when they administered the treatment or assessed the outcome may unconsciously have influenced the measurements (van der Worp et al. 2005; Crossley et al. 2008). Additionally, there were obvious significant methodological differences between pre-clinical and clinical trials, in that the animals used were generally young and healthy before the experimentally induced stroke, while human patients were often elderly and hypertense (Macleod and Sandercock 2005). The same sort of shortcomings was also identified in studies on intracerebral hemorrhage (MacLellan et al. 2012).

The likelihood of the proposed benefits being delivered also depends on the competence of the people involved in the project, namely on the experience the researchers have in conducting research in the field or in using the proposed animal model, on their general scientific capacity, and on their level of training in laboratory animal science. The fact that the experiments are conducted in facilities with the necessary and appropriate conditions and equipments is also essential to guarantee that the experiments produce reliable results.

Finally, if the intended benefits of research are to be achieved in practice, the results of the experiments must be made public; hence communication is central. Publication in peer reviewed journals is a central feature of modern academic research and, as is well known, the performance of today's researchers is measured largely on the basis of the number of publications they have in influential journals. However, it is generally difficult to get negative results (no effect of treatment) published. As a direct consequence of this, publications are likely to reflect only part of the research that has been carried out in a field—the research in which differences were found between treatment groups. This has wide-ranging ethical consequences. Importantly, it affects the number of animals used in research. Also, poor concepts, hypotheses, and models may survive, notwithstanding a vast amount of contradictory data, all of this merely because these data are not made available to the scientific community. Publication bias—that positive results are more likely to be published, has indeed been pointed out as a major problem in neuroscience research (Sena et al. 2010).

3.3 How Can We Improve Benefits?

In order to overcome such shortcomings and improve the benefits delivered by animal research, some issues are particularly pertinent.

First of all, more rigorous preclinical study methodology is clearly necessary. A relevant handicap of animal-based neuroscience research has been the difficulty in translating apparent therapeutic successes from in vivo preclinical studies into human clinical trials. Given this remarkably low success rate, some attempts have been made to bring researchers together to discuss how to do preclinical research in a given field in the best way. An example is the European ALS/MND group, which organized two workshops, one in 2006 and one in 2009 (Ludolph et al. 2007, 2010). Another example is the 2011 meeting of the National Institute of Neurological Disorders and Stroke (NINDS), the Eunice Kennedy Shriver National Institute of Child Health and Human Development (NICHD), the International Rett Syndrome Foundation (IRSF), the Rett Syndrome Research Trust (RSRT), the National Institutes of Health (NIH), the US Food and Drug Administration (FDA), the pharmaceutical industry, and private foundations in the USA (Katz et al. 2012). These meetings aimed to identify the reasons for the translational failure in ALS and Rett's syndrome research, respectively, and to establish guidelines to standardize and improve preclinical research in these fields.

The major barriers to translational success identified in these workshops were a lack of rigorous standards and transparency in reporting preclinical studies, similar to what was reported by der Worp (2005) in ischemia stroke research and by MacLellan et al. (2012) in intracerebral hemorrhage research. The subsequent published guidelines from these groups outline some of the principles and standards of good study design and report when conducting preclinical trials of candidate therapeutics—e.g., allocation concealment, blinded assessment of outcome, random allocation of subjects to experimental groups, and other methods designed to minimize bias and Type 1 ("false positive") errors. By raising standards and awareness, these initiatives strive to increase the reliability, reproducibility, and predictive value of preclinical research, and ultimately improve the likelihood of success on clinical translation.

Research on spinal cord injury has also encountered a low translational rate. Results of many treatment strategies showed some beneficial effects in animals on the anatomical/histological and/or the functional level, but subsequent pilot clinical trials failed or were abandoned. However, in this case, the problem does not appear to result from experimental design shortcomings. A major reason for such translational failure, which has been less discussed, may derive from the high control of variability in the animal experiments (Fili and Schwab 2012). Animal studies are designed to minimize all variables within an experiment (age, sex, and strain of animals; extent and location of the lesion; application of medication, etc.), which is in contrast to the clinical situation, where large heterogeneities exist at all levels and thus may hamper the detection of potential treatment effects. This implies that beneficial effects of a given treatment in animal models have to be robust, functionally meaningful, and reproducible in independent laboratories. Minor functional improvements as a consequence of a novel therapy or effects at only the anatomical/histological level are unlikely to materialize as detectable functional benefits in clinical trials, where the patient population is more heterogeneous and the assessment of functional outcome is even more difficult (Fili and

Schwab 2012). In summary, improving the standards for considering therapeutic treatments as ready for clinical trials would be likely to make the use of animals more profitable and increase the translational relevance of animal models.

Second, we believe that more rigorous and extensive evaluations of animal models are necessary. As mentioned above, there are fields where it is widely recognized that existing animal models are insufficient. As stated by Micale et al. (2013) regarding neuropsychiatry, "The field desperately needs better animal models of depression and SCZ because of the partial efficacy of present pharmacological treatment. Without improved models of human disease, we cannot know whether particular molecular and cellular findings in animals are relevant to the clinical situations." We propose that, for improved ethical animal experimentation, animal research in these situations should be avoided until improved models are developed. It is not clear that in these cases the benefits of research justify the harms caused to the animals.

Lastly, publication of negative findings from well-conceived and performed studies should be encouraged. It can help investigators to evaluate and ultimately abandon the development of invalid and irrelevant animal models, which may be hampering the progress of neuroscience. Importantly, this will result in efforts for the development of good animal models and their validation, which will improve the likelihood that the benefits of research are delivered.

4 Weighing Harms and Benefits

Harm–benefit assessment is widely regarded as the standard approach for assessing the ethical justification of experiments within the framework of regulations. It is advanced in several guidelines and recommendations for ethical review as the way to determine the acceptability of animal experiments (e.g., APC 2003; CCAC 1997; FELASA 2005) and it is also explicitly required in legislation in many countries (e. g., EU through Directive 2010/63 and Australia through Australian Animal Welfare Act (1999)).

4.1 Committees

The decision to authorize a research project with animals is usually based on some form of assessment of the relative harms and benefits, with that evaluation, in most countries carried out by a review committee. Such review may take place at an institutional, regional, or national level, with different countries using different organizations. Animal Ethics Committees (AECs)—or Institutional Animal Care and Use Committees (IACUCs) as they are called in the USA and Canada, thus have the major responsibility for the performance of these harm–benefit assessments. These committees may have an advisory (e.g., UK) or regulatory role (e.g., Canada).

Their function is to ensure that animal experiments are only performed when they are ethically justified, i.e., that the impact on animal welfare is minimized and that the proposed benefits of the research are likely to be achieved.

Typically, review committees are composed of a number of people representing different competences, including scientists, animal technicians/caretakers, veterinarians, nonanimal researchers, animal welfare and/or animal ethics experts, and lay people. Such composition is, in many countries, effectively mandated by law, including the USA, Australia, and European countries such as Sweden, Germany, Denmark, and Switzerland. The majority of members belong to the scientific community and in the case of institutional committees they also typically belong to the institution. Noninstitutional members and lay people constitute a minor percentage of the committees' composition. The largest representation seems to be in Sweden, where half of the committee members are nonscientists (lay persons or representatives of animal protection NGOs). The Australian Code currently states that at least one-third of committee members should be lay persons or representatives of animal welfare groups. In the USA, according to the legislation, "... at least one member... shall not be affiliated with (the research) facility other than as a member of the Committee." Moreover, the unaffiliated member may not be a close relative of a person affiliated with the facility. The idea of such composition is, by having members that are external to the institution, to avoid potential conflicts of interests in project evaluations and, by having lay people, to provide representation for general community interests, and to ensure that institutional committees go beyond a purely scientific analysis in making decisions on laboratory animal care and use (Dresser 1999).

There has been some research into how this committee review system works. Dresser (1989) surveyed the responses of 32 IACUCs when asked to review four hypothetical protocols. Results suggested that that committees showed little consensus in their approach to assessing the justification for animal use, although there was broad agreement on the need to refine particular procedures. Forsmann (1993) analyzed the written reports from the regional committees in Sweden between 1979 and 1989 and found that they acted primarily as technical committees, with refinement of procedures being the predominant focus of deliberations. Hagelin et al. (2003) found that between 1989 and 2000 the practice had not changed in Sweden: the majority of the modifications continued to relate to the goals of refinement. However, the authors concluded that the AECs were being effective in improving animal welfare. Houde et al. (2003) observed three Canadian IACUCs over 1 year period and the data revealed that most comments were of a technical nature, with 16 % related to what had been defined as the "explicit ethical categories" of the three Rs. In interviews, with members of Swedish AECs, Ideland (2009) found that they focused on methodological improvements instead of on weighing research aims against animal suffering. More recently, Schuppli (2011) conducted an ethnographic study involving participant observation and in-depth interviews with members of university AECs in Canada and also found that the major focus of protocol review was reducing harms to animals, with some members stating that assessing ethical justification was the role of scientific peer review. In summary, these studies showed that the focus of committees was mainly reducing harms to animals, with less importance given to the justification of research and evaluation of potential benefits, despite this being stressed in policy as a goal of AECs.

Another focus of research on ethics committees has been the degree of agreement in the decision-making process both between and within committees. In her 1989 study, in which 32 IACUCs evaluated four animal research protocols, Dresser found low levels of inter-committee agreement. More recently, Plous and Herzog (2001) evaluated both inter and intra-committee agreement. The authors randomly selected IACUCs from US universities and colleges. Next, each IACUC was asked to submit its three most recently reviewed protocols involving animal behavior. Each protocol was randomly assigned to be reviewed a second time by a participant of another IACUC. Once they had received the reviews from individual committee members, the IACUCs were asked to meet as a group and render a final evaluation for each of the three protocols. The authors found a low level of agreement both between and within committees. Although strong conclusions and generalizations cannot be drawn only from these two qualitative studies, these results are important indicators that the review process performed by ethics committees may be working with low levels of agreement.

4.2 The Challenge of Harm–Benefit Weighing

The studies referred to above indicate two major problems in the process of ethical review. On one hand, it appears that committees are neglecting the assessment of benefits and their weighing against harms and that they are focusing mainly on reducing harms. On the other hand, the reliability of the ethical review process appears questionable, as studies pointed to a poor agreement in protocol evaluation both by the members of the same committee and by different committees. Despite this general low agreement, some aspects appear to gather consensus, namely the evaluation of animal suffering (Plous and Herzog 2001) and the evaluation and improvement of technical and methodological issues (Ideland 2009). Hence, there is some suggestion that problems in the reliability of the ethical review process appear to result mainly from the lack of agreement on benefit evaluation and harm–benefit weighing.

One explanation for the quasi-absence and the lack of agreement on benefit assessment and harm–benefit weighing may be the fact that guidelines for ethical review are very general, especially as regards how to evaluate benefit, or how to weigh it against harms (e.g., Directive 63/2010/EU, IACUC Guidebook, CCAC Guidelines). Hence, they are not very helpful to these respects. Stafleu (1994) further suggested that ethics committee members have probably little notion of how they are to balance the significance of research against the interests of the animals. In support of this view, more recently, Schuppli (2011) found a large variation in how harm–benefit analysis was applied by ethics committee members.

It seems probable that this is due to a general absence of concrete guidance in the guidelines as to how to evaluate benefit and how to weigh it against harm. Of the two elements in the harm–benefit assessment, the question of harm (or more precisely, how to reduce it) has been central to research and teaching in laboratory animal science worldwide over the last couple of decades and has thus had a longer history of research and greater contextual framework underpinning its assessment. In fact, harms may be easier to identify and define than the benefits (Voipio 2004). What we currently know about the mechanisms of pain in vertebrates and of their physiological and behavioral needs allow us to make relatively objective predictions of the impact of research on them. Guidelines and examples are commonly provided to help bring some uniformity to harm assessment (Orlans 1996). Classification systems for ranking the degree of animal pain and distress have been elaborated for ethical reviews (e.g., Directive 63/2010/EU; CCAC 1997). The fact that these are more immediate and explicitly visible may also make it easier in any evaluation to assess harms, rather than benefits which are more general and further away in time.

In the meantime, a few ethical schemes or scoring systems have been proposed to help in the ethical review process (e.g., Bateson 1986; Delpire et al. 1999; Smith and Boyd 1991; Porter 1992; Stafleu et al. 1999). These systems are meant to support the process of assessing and weighing the various considerations that determine the ethical acceptability of an animal experiment and to help people to do so in a systematic way. In an attempt to make them as objective as possible, some of these schemes are based on mathematical calculations (e.g., Stafleu et al. 1999; Porter 1992). These have a series of categories to which scores are attributed. These scores are then incorporated in a computational rule, whose result, when compared to a cut-off value, determines whether an experiment is or is not ethically acceptable. In Stafleu's scheme, for example, scores are calculated for the human interests, the relevance of the animal experiment and the harms caused to the animals. To calculate a score for each one of these, several points are considered. For example, for calculating a score for the human interest, assessors are advised to consider whether the benefits are for human health, for scientific knowledge or for economy, with each one deserving a different weight. For calculating a score for the relevance of the animal experiment, issues such as the methodological quality, the quality of the research group, and the necessity of the experiment are scored. For the harm score, the duration and intensity of discomfort, the number of animals involved and their psychological complexity are considered. In the end, the score for human interests is multiplied by the score for the relevance of the animal experiment and this value is compared with the harm score for animals. This will determine the acceptability of the experiment. If the former is higher than the latter, the experiment is acceptable, otherwise it is unacceptable.

However, not everyone supports the use of schemes based on detailed algorithms. The argument is that they try to imply accurate measures where there can be none—harms and benefits are not quantifiable in this way. This idea was widely defended in 2009, at a workshop organized by the Cooperation Group for

Laboratory Animal Sciences within the Finnish Ministry of Education. This workshop gathered participants representing the scientific community, animal welfare organizations, and regulators from Nordic and Baltic countries and The Netherlands. A classification of harms and benefits into three degrees (low, medium, and high; e.g., Bateson 1986) was preferred over the scoring systems (Voipio 2004). In Bateson's model, the probability of benefits, the quality of research and the animal suffering are attributed a classification of low, medium, or high. If these three dimensions are then considered as the three dimensions of a cube, there will be a zone where the probability of benefits and the quality of research will be high and the animal suffering will be low—if a research proposal falls into this zone it is ethically acceptable, and there will be a zone where the probability of benefits and the quality of research will be low and the animal suffering will be high—if a research proposal falls into this zone, it is ethically unacceptable and should not be approved. A revised version of the Bateson model is proposed in the most recent guidelines prepared by the European Expert Working Group for Project Evaluation for how to evaluate projects under Directive 2010/63/EU.

Finally, the fact that AECs are not assessing benefits may have to do with some confusion over the relation between AEC review and scientific peer-review by granting agencies—some committee members believe that ethical justification is decided by scientific peer-review (e.g., Graham 2002; Schuppli 2011).

Overall, there seem to be a discrepancy between the intended content of an ethics review and what is currently being evaluated in actual practice by animal ethics/animal care and use committees. Committees spend much effort on evaluating issues related to animal harm, and on considerations of how to reduce said harm (by way of applying the three Rs). Benefit, on the other hand, is not always addressed and when it is, there is little agreement between committees and between individual members. We will end this section by reflecting on what can be done to change this discrepancy.

There are several possible ways to improve how benefit is evaluated and weighed against harm. A first possibility for improvement would be the development of more detailed guidelines for ethical review. Specifically, more detailed classification systems for benefits and harms might help. A good example is the report recently produced by the European Commission Expert Working Group. The fact that Plous and Herzog (2001) found a relatively high intra-committee agreement when the committee members used a pain scale to rate animals' pain or stress, suggests that when committee members are given detailed classification criteria they may achieve a relatively high degree of agreement. Elsewhere it was already advanced that the use of detailed classification systems is essential for careful ethical analysis (Orlans 1997). A second way ethical review effectiveness could be improved would be by clarifying the roles of AECs and scientific peer-review. As referred above, AECs may not be assessing benefits and weighing them against harms because they believe it is not their role. Finally, giving the committee members enhanced training in ethics and/or philosophy could also help, as suggested by Houde et al. (2009) and Schuppli and Fraser (2005).

However, another possibility would be to change the aims of ethics review. Based on the variety of problems resulting from the evaluation of benefit and from weighing benefit against harm, we wonder whether benefit assessment should not be openly (and expressly) discarded and whether ethical evaluation should not focus on applying the three Rs and thus upon improving animal welfare only. For all intents, this largely seems to be what is currently happening anyway. An interesting comparison can be made with farm animals: While broad societal concern exists around the intensive rearing of animals for human consumption, regulation quite strictly focuses on the improvement of animal welfare or, in other words, the reduction of harms caused to the animals, and the harms for the animal are not weighed against the benefits. Is there a significant reason for demanding a different approach in laboratory animals?

Another possibility would be to restrict full harm–benefit analyses to experiments likely to cause severe harm to the animals, though this implies that in all other situations the benefit would be considered to outweigh the harm without explicit reflection. Still, since it is such a difficult exercise, we could reserve such explicit reflection for those situations where animal suffering appears so great that a strong justification is needed for the research to be ethically acceptable.

5 Summary

- Studies on live animals play a crucial role in neuroscience research, but their use as research subjects is a controversial issue.
- Different ethical theories have different positions as regards the morality of using animals in research. Contractarianism, utilitarianism, and animal rights are three ethical perspectives that represent a great proportion of the debate.
- Harm–benefit weighing, a utilitarian-based perspective, is currently the mainstream approach for determining the acceptability of animal experiments in practice and it is prescribed in legislation worldwide: If the benefits outweigh the harms, the research is considered justified.
- Harms caused to animals in neuroscience research include effects on animal health but also adverse subjective experiences. These harms can range, among others, from the loss of capacities to behavioral deficits, to pain and anxiety.
- Harm assessment is one part of the ethics evaluation of animal experiments. Guidelines and policy documents have advanced lists of criteria to be assessed. Usually, these criteria include the quality of the facilities, the experience of the personnel, the number of animals, the animal species, the husbandry and housing conditions, the scientific procedures themselves (including the killing method), the fate of the animals at the end of the experiments, and the end-points to be applied.
- The harms inflicted to research animals have implications not only for the well-being of the animals, but also for the quality of experimental data. There are thus two major reasons for reducing harm to animals in research: to preserve animal welfare and to guarantee the cleanness of experimental data.

- The three Rs (replacement—replacing animals with nonanimal alternatives, reduction—reducing the number of animals used, refinement—adjusting experimental procedures to minimize any pain or distress) are widely recognized principles for minimizing harms to animals.
- A recurrent idea when considering the harms to animals is that research will be more or less harmful depending on the animal species chosen, which is based on the idea that animals of different species vary in their capacity for subjective experience. We argue that this claim is of limited validity in distinguishing between different vertebrate species, and that the moral difference between using different animals is better described as an issue of public sensitivity than as an issue of animal harm.
- Overall, the benefits of neuroscience research encompass increased scientific knowledge and the improvement of human health and welfare.
- Official documents provide some, although little, guidance as to how to evaluate benefits. Usually it includes statements of the scientific objectives and potential value of the study in terms of originality and importance of the new information, timeliness and the need for the experimental project. However, it is very difficult or almost impossible to assess what will be the benefits of a given research project.
- Besides harm and benefit assessment, ethical reviews also involve the assessment of the likelihood that the potential benefits will be achieved. This involves evaluating the appropriateness of the animal model and the scientific approach, the validity of the experimental design, the staff competence, the appropriateness, and quality of facilities and the communication of results.
- Benefits can be improved in neuroscience research through more rigorous preclinical study methodology, more rigorous extensive evaluations of animal models, and publication of negative results.
- There seems to be a discrepancy between what the ethics review is said to consist in and what AECs actually evaluate. Committees spend much effort on evaluating issues having to do with animal harm and how to reduce it (three Rs), but benefit is not always addressed and when it is, there is little agreement between committees and between individual members.
- Explanations for this discrepancy may include the fact that guidelines for ethical review are very general, especially as regards benefit assessment and harm–benefit weighing, and the fact that there is some confusion over who bears responsibility for benefit assessment—AECs or scientific peer-review.
- Improvement of ethical review may be achieved by the development of more detailed guidelines, by clarifying the roles of AECs and scientific peer-review and giving the committee members enhanced training in ethics and/or philosophy. Another possibility would be to change the aims of ethics review. Benefit assessment could be discarded and ethical evaluation could focus only on applying the three Rs and improving animal welfare, or AECs could be required to evaluate benefits and perform harm–benefit analysis only when the proposed experiments appear to be highly severe for the animals.

Acknowledgments We would like to thank the editors of this book for the challenge to write this chapter, the opportunity to publish it and for constructive comments on previous versions of the text. This work was funded by FEDER funds through the Operational Competitiveness Programme—COMPETE and by National Funds through FCT—Fundação para a Ciência e a Tecnologia under the projects FCOMP-01-0124-FEDER-037277 (PEst-C/SAU/LA0002/2013), FCOMP-01-0124-FEDER-029527 (PTDC/MHC-ETI/4890/2012), and Projeto "NORTE-07-0124-FEDER-000001—Neurodegenerative disorders" cofinanciado pelo Programa Operacional Regional do Norte (ON.2—O Novo Norte), ao abrigo do Quadro de Referência Estratégico Nacional (QREN), através do Fundo Europeu de Desenvolvimento Regional (FEDER) e por fundos nacionais através da FCT—Fundação para a Ciência e Tecnologia.

References

Abdi S, Haruo A, Bloomstone J (2004) Electroconvulsive therapy for neuropathic pain: a case report and literature review. Pain Physician 7:261–263

Allen C (2004) Animal pain. Noûs 38:617–643

American Psychiatric Association APA (2000) Diagnostic and statistical manual of mental disorders, 4th edn. American Psychiatric Association, Washington

APC (2003) Review of cost-benefit assessment in the use of animals in research. Report of the cost-benefit working group of the animal procedures committee. Home Office, Communication Directorate, London

Arluke A, Sanders CR (1996) Regarding animals. Temple University Press, Philadelphia

Australian Animal Welfare Act (1999) http://www.legislation.govt.nz/act/public/1999/0142/latest/DLM49664.html)

Baker M, Kale R, Menken M (2002) The wall between neurology and psychiatry. BMJ 324:1468–1469

Bateson P (1986) When to experiment on animals. New Scientist 109:30–32

Bayne K, Howard BR, Kurosawa TM, Nájera MEA (2011) An overview of global legislation, regulation and policies. In: Hau J, Schapiro SJ (eds) Handbook of laboratory animal science: essential principles and practices, vol 1, 3rd edn. Boca Raton, pp 39–64

Boomkamp SD, Riehle MO, Wood J et al (2012) The development of a rat in vitro model of spinal cord injury demonstrating the additive effects of Rho and ROCK inhibitors on neurite outgrowth and myelination. GLIA 60:441–456

Button KS, Ioannidis JPA, Mokrysz C et al (2013) Power failure: why small sample size undermines the reliability of neuroscience. Nat Rev Neurosci 14:365–376

CCAC (1993) Guide to the care and use of experimental animals. Canadian Council on Animal Care, Ottawa

CCAC (1997) CCAC guide to protocol review. Canadian Council on Animal Care, Ottawa

Cohen C (1997) Do animals have rights? Ethics and behavior. Harvard University, Cambridge

Crossley NA, Sena E, Goehler J et al (2008) Empirical evidence of bias in the design of experimental stroke studies—a metaepidemiologic approach. Stroke 39(3):929–934

Delpire VC, Mepham TB, Balls M (1999) A proposal for a new ethical scheme addressing the use of laboratory animals for biomedical purposes. Altern Lab Anim 27:869–881

Dresser R (1989) Developing standards in animal research review. J Am Vet Med Assoc 194:1184–1191

Dresser R (1999) Community representatives and nonscientists on the IACUC: what difference should it make? ILAR J 40(1):29–33

ESF-EMRC (2009) Position on the proposal for a directive on the protection of animals used for scientific purposes, 2nd edn, March 2009. http://www.esf.org/fileadmin/Public_documents/Publications/AnimalsProtection_2ndEd.pdf. Accessed Feb 2014

European Commission Expert Working Group. Examples to illustrate the process of severity classification, day-to-day assessment and actual severity assessment. http://ec.europa.eu/environment/chemicals/lab_animals/pdf/examples.pdf. Accessed Jan 2014
Fawcett JW, Asher RA (1999) The glial scar and central nervous system repair. Brain Res Bull 49:377–391
FELASA (2005) Principles and practice in ethical review of animal experiments across Europe: a report prepared by the FELASA working group on ethical evaluation of animal experiments
Fenton A (2014) Can a chimp say "no"? Reenvisioning chimpanzee dissent in harmful research. Camb Q Healthc Ethics 23:130–139
Fili L, Schwab ME (2012) The rocky road to translation in spinal cord repair. Ann Neurol 72:491–501
Forsman B (1993) Research ethics in practice: the animal ethics committees in Sweden 1979–1989. In: The Royal Society of Arts and Sciences in Gothenburg (eds) Studies in research ethics no. 4. Centre for Research Ethics, Goteborg
Franco NH, Olsson IAS (2012) How sick must your mouse be? An analysis of Huntington's disease research using animal models. ATLA Altern Lab Anim 40:271–283
Gilron I, Watson CP, Cahill CM, Moulin DE (2006) Neuropathic pain: a practical guide for the clinician. CMAJ 175:265–275
Graham K (2002) A study of three IACUCs and their views of scientific merit and alternatives. J Appl Anim Welfare Sci 5:75–81
Grau J (2002) Learning and memory without a brain. In: Bekoff M, Allen C, Burghardt GM (eds) The cognitive animal. MIT Press, Cambridge, pp 325–334
Hagelin J, Hau J, Carlsson HE (2003) The refining influence of ethics committees on animal experimentation in Sweden. Lab Anim 37:10–18
Handy CR, Krudy C, Boulis N, Federici T (2011) Pain in amyotrophic lateral sclerosis: a neglected aspect of disease. Neurol Res Int 2011:403808
Harding EJ, Paul ES, Mendl M (2004) Animal behaviour: cognitive bias and affective state. Nature 427:312
Houde L, Dumas C, Leroux T (2009) Ethics: views from IACUC members. ATLA 37:291–296
Ideland M (2009) Different views on ethics: how animal ethics is situated in a committee culture. J Med Ethics 35:258–261
International Animal Research Regulations, Impact on Neuroscience Research, Workshop Summary, Institute of Medicine (US); National Research Council (US) (2012) National Academies Press (US), Washington (DC)
Katz DM, Berger-Sweeney JE, Eubanks JH et al (2012) Preclinical research in Rett syndrome: setting the foundation for translational success. Dis Models Mech 5:733–745
Langford DJ, Bailey AL, Chanda ML et al (2010) Coding of facial expressions of pain in the laboratory mouse. Nat Methods 7:447–449
Langley CK, Aziz Q, Bountra C et al (2008) Volunteer studies in pain research–opportunities and challenges to replace animal experiments: the report and recommendations of a focus on alternatives workshop. Neuroimage 42(2):467–473
Lever TE, Gorsek A, Cox KT et al (2009) An animal model of oral dysphagia in amyotrophic lateral sclerosis. Dysphagia 24:180–195
Lever TE, Simon E, Cox KT et al (2010) A mouse model of pharyngeal dysphagia in amyotrophic lateral sclerosis. Dysphagia 25:112–126
Lipska B (2004) Using animal models to test a neuro developmental hypothesis of schizophrenia. J Psychiatry Neurosci 29(4):282–286
Ludolph AC, Bendotti C, Blaugrund E et al (2007) Guidelines for the preclinical in vivo evaluation of pharmacological active drugs for als/mnd: Report on the 142nd enmc international workshop. Amyotroph Lateral Scler 8:217–223
Ludolph AC, Bendotti C, Blaugrund E et al (2010) Guidelines for preclinical animal research in als/mnd: a consensus meeting. Amyotroph Lateral Scler 11:38–45

Lund TB, Mørkbak MR, Lassen J et al (2012) Painful dilemmas: a study of the way the public's assessment of animal research balances costs to animals against human benefits. Public Underst Sci 0:1–17
MacLellan CL, Paquette R, Colbourne F (2012) A critical appraisal of experimental intracerebral hemorrhage research. J Cereb Blood Flow Metab 32:612–627
Macleod M, Sandercock P (2005) Systematic reviews improve clinical research design—can they help improve animal experimental work? RDS News Winter, pp 8–10
Mendl M, Burman OHP, Parker RMA, Paul ES (2009) Cognitive bias as an indicator of animal emotion and welfare: emerging evidence and underlying mechanisms. Appl Anim Behav Sci 118:161–181
Micale V, Kucerova J, Sulcova A (2013) Leading compounds for the validation of animal models of psychopathology. Cell Tissue Res 354:309–330
Morton DB, Hau J (2011) Welfare assessment and humane endpoints. In: Hau J and Schapiro SJ (eds) Handbook of laboratory animal science: essential principles and practices, vol 1, 3rd edn. Boca Raton, FL, pp 535–572
Nuffield Council on Bioethics (2005) The ethics of research involving animals. Nuffield Council on Bioethics, London
Olsson IAS, Franco NH, Weary DM, Sandøe P (2011) The 3Rs principle: mind the ethical gap! Altex 29, Special Issue. In: Proceedings of WC8. (http://www.altex.ch/en/index.html?id=90)
Olsson IAS, Sandøe P (2010) "What's wrong with my monkey?" An ethical perspective on germline transgenesis in marmosets. Transgenic Res 19:181–186
Orlans FB (1996) Invasiveness scales for animal pain and distress. Lab Animal 25(6):23–25
Orlans FB (1997) Ethical decision making about animal experiments. Ethics Behav 7:163–171
Plous S, Herzog H (2001) Reliability of protocol reviews for animal research. Science 293:608–609
Porter DG (1992) Ethical scores for animal experiments. Nature 356:101–102
Reardon S (2014) NIH rethinks psychiatric trials. Nature 507:288
Regan T (1983) The case for animal rights. University of California Press, California
Richardson CA, Flecknell PA (2005) Anaesthesia and post-operative analgesia following experimental surgery in laboratory rodents: are we making progress? Altern Lab Anim 33(2):119–127
Rollin MDH, Rollin BE (2014) Crazy like a fox: validity and ethics of animal models of human psychiatric disease. Camb Q Healthc Ethics 23:140–151
Rudge JS, Silver J (1990) Inhibition of neurite outgrowth on astroglial scars in vitro. J Neurosci 10:594–3603
Russell W, Burch R (1959) The principles of humane experimental technique. Methuen, London
Schapiro KJ (1998) Animal models of human psychology: Critique of science, ethics and policy. Hogrefe & Huber, Seattle
Schuppli CA (2011) Decisions about the use of animals in research: ethical reflection by Animal Ethics Committee Members. Anthrozoos 24(4):409–425
Schuppli CA, Fraser D (2005) Factors influencing the effectiveness of research ethics committees. J Med Ethics 33(5):294–301
Sena ES, van der Worp HB, Bath PMW et al (2010) Publication bias in reports of animal stroke studies leads to major overstatement of efficacy. PLoS Biol 8(3):1–8
Shettleworth SJ (2010) Clever animals and killjoy explanations in comparative psychology. Trends Cogn Sci 14(11):477–481
Singer P (1975) Animal liberation. Avon Books, New York
Smaje LH, Smith JA, Ewbank R et al (1998) Advancing refinement of laboratory animal use. Lab Anim 32:137–142
Smith JA, Boyd KM (1991) Lives in balance. The ethics of using animals in biomedical research. Oxford University Press, Oxford
Solomon JA, Tarnopolsky MA, Hamadeh MJ (2011) One universal common endpoint in mouse models of amyotrophic lateral sclerosis. PLoS One 6:e20582

Sorensen A, Moffat K, Thomson C, Barnett SC (2008) Astrocytes, but not olfactory ensheathing cells or schwann cells, promote myelination of CNS axons in vitro. GLIA 56:750–763
Stafleu FR, Tramper R, Vorstenbosch J, Joles JA (1999) The ethical acceptability of animal experiments: a proposal for a system to support decision-making. Lab Anim 33:295–303
Stafleu FR (1994) The ethical acceptability of animal experiments as judged by researchers. PhD Thesis, Utrecht University
Suddendorf T, Corballis MC, Collier-Baker E (2009) How great is great ape foresight? Anim Cogn 12(5):751–754
Van der Worp HB, de Haan P, Morrema E et al (2005) Methodological quality of animal studies on neuroprotection in focal cerebral ischaemia. J Neurol 252:1108–1114
Voipio H-M, Kaliste E, Hirsjärvi P et al (2004) Nordic-European workshop on ethical evaluation of animal experiments. Scand J Lab Anim Sci 31:251–267
Woolf CJ (2004) Dissecting out mechanisms responsible for peripheral neuropathic pain: implications for diagnosis and therapy. Life Sci 74:2605–2610
Willner P (1997) Validity, reliability and utility of the chronic mild stress model of depression: a 10-year review and evaluation. Psychopharmacology 134:319–329
Yalcin I, Bohren Y, Waltisperger E et al (2011) A time-dependent history of mood disorders in a murine model of neuropathic pain. Biol Psychiatry 70:946–953
Yeates JW (2010) Death is a welfare issue. J Agric Environ Ethics 23:229–241

A Framework for Investigating Animal Consciousness

Paula Droege and Victoria A. Braithwaite

Abstract An assessment of consciousness in nonverbal animals requires a framework for research that extends testing methods beyond subjective report. This chapter proposes a working definition of consciousness in terms of temporal representation that provides the critical link between internal phenomenology and external behavior and neural structure. Our claim is that consciousness represents the present moment as distinct from the past and the future in order to flexibly respond to stimuli. We discuss behavioral and neural evidence that indicates the capacity for both flexible response and temporal representation, and we illustrate these capacities in fish, a taxonomic group that challenges human intuitions about consciousness.

Keywords Animals · Behavior · Consciousness · Fish · Flexibility · Goal-directed · Pain · Temporal representation · Telencephalon · Tool use

Contents

P. Droege (✉)
Philosophy Department, Pennsylvania State University, 244 Sparks Building,
University Park, PA 16802, USA
e-mail: pdroege@psu.edu
URL: http://www.personal.psu.edu/pud10/

V. A. Braithwaite
Center for Brain, Behavior and Cognition,
Departments of Ecosystem Science and Management, and Biology,
Pennsylvania State University, 410 Forest Resources Building,
University Park, PA 16802, USA
e-mail: v.braithwaite@psu.edu
URL: http://bio.psu.edu/directory/vab12

Curr Topics Behav Neurosci (2015) 19: 79–98
DOI: 10.1007/7854_2014_278

Published Online: 9 February 2014

Highlights

- consciousness defined in terms of teleosemantic theory of temporal representation,
- temporal representation links phenomenology, behavior, and neural structure,
- four forms of evidence indicate temporal representation and flexible response,
- three neural structures essential,
- capacities of fish considered as test of framework.

The question is not, Can they reason, nor Can they talk, but Can they suffer? Bentham (1789).

This oft-quoted statement succinctly captures the central insight of utilitarian ethics: the capacity to feel pain ought to be the primary marker of moral concern rather than the capacity for abstract reasoning.[1] Rationality may be necessary to be a moral *agent,* but it is not required to be a moral *subject.* The force of this insight goes well beyond utilitarianism and strikes at the heart of an ethical relationship to the world. A crying infant compels care, an injured pet elicits sympathy. In humans, the evolutionarily adaptive capacity for empathy is so well developed that viewers of animated geometric figures will interpret them as agents feeling fear, love, and anger (Heider and Simmel 1944). Of course, animated circles do not suffer, which raises the problem of gauging appropriate ethical response. On one hand, concern for suffering requires that we minimize pain in all cases, but on the other hand, the tendency to overextend this empathetic concern means we need some measures other than our own moral feelings to determine which animals do in fact suffer.

To begin with, we can distinguish three elements in human suffering: sensory responsiveness to noxious stimuli, conscious hurtfulness, and self-conscious understanding of the pain's significance.[2] While each of these elements raises difficult research issues, this chapter will focus on conscious hurtfulness as the *sin qua non* of suffering. Conscious pain is the element in human suffering that we attribute to infants and pets and circles; it is the element that calls for an ethical response.

To determine which animals suffer, then, we need to determine which animals are conscious. In the following, we propose a working definition of consciousness in terms of temporal representation that links internal phenomenology with behavioral function and neural structure. It is worth emphasizing that the proposal offers a *working* definition of consciousness. Temporal representation is a necessary feature of consciousness, we claim, and is sufficient to distinguish conscious

[1] This is not to say that the capacity to feel pain is the only marker for moral concern. Suffering is one very important marker but should figure within a broader ethical theory.

[2] A fourth element could, and perhaps should, be added: the emotional response. The conscious feeling of dislike rivals conscious hurtfulness as the most important feature of suffering. As with pain, conscious rather than unconscious emotions are associated with suffering, so the investigation of consciousness is appropriate in both cases.

from unconscious representations. For a complete account, more must be said about what conditions are necessary for something to be a "representation," among other things.[3] As a working definition, the test of its value is its usefulness in providing an explanation that accounts for the available data. Phenomenological experience forms the foundational data; it is the data that characterizes consciousness from the subjective perspective. Comparative behavioral and neuropsychological evidence characterize consciousness from the objective perspective. A working definition of consciousness is useful to the extent that it brings these two perspectives into alignment and offers testable hypotheses for the future research. We are particularly interested in the research potential of the present proposal and hope the debate about animal consciousness sparks refinement of the theory as well as methodological recommendations for effective tests. If we have more effective means of knowing when animals are suffering, we can more effectively decide what moral response is appropriate. Ideally, the science of consciousness can provide a useful guide for our moral intuitions.

1 Defining Consciousness

The first, and probably most difficult task in this project is to set out a workable operational definition of consciousness. The point here is not to set out necessary and sufficient conditions of the sort proposed in conceptual analysis. Our goal is to identify a tractable target for scientific research, to map *the way a creature experiences the world* to *the world it is experiencing*. For as many years as philosophers have been writing about this topic, the best description of the way a creature experiences the environment around it has been the unhelpful locution "what it's like" to sense red or feel pain (Nagel 1974; Block 1995; Chalmers 1997; Lycan 1996). Though vague, the phrase resonates because it points to the elusive quality that distinguishes an awake state that is responsive to sensory stimuli from fully unconscious states such as dreamless sleep. When unconscious, there is clearly nothing it is like to sense red or feel pain. Similarly, there is nothing it is like to be an object incapable of consciousness, like a table or a laptop computer. For us, by contrast, waking life is full of sights and sounds and tastes and smells which combine and shift to form a vibrant, changing world, and there is something it is like to experience that world. Later we will say more to refine this rough contrast into a sufficient working definition. Even in us, awake responses to stimuli are often unconscious. Consider, for example, your current body position. To maintain balance and prevent injury, you need to shift position periodically in response to proprioceptive signals. Until your attention was drawn to your body,

---

[3] Though the account is compatible with various theories of representation, we favor a teleofunctional view. For more, see footnote 6.

chances are that these shifts in position were accomplished unconsciously. There was "nothing it was like" for you to shift your body position.

But "what it's like" merely points at the target phenomenon without adequately identifying it. One tempting refinement of "what it's like" would be to assume it is a case of self-consciousness. Of course, the instant we begin to think about the nature of consciousness, we become self-conscious by means of introspection. Consciousness cannot be caught unawares, as it were, and this fact raises several methodological difficulties in designing experiments to test human consciousness. Whenever a researcher asks a subject to report on her conscious experience, she must introspect in order to do so. Consequently the capacity for self-consciousness and consciousness appear inseparable. In the investigation of non-human animal consciousness, however, the separability of consciousness and self-consciousness is more clearly apparent. To be conscious of one's self requires an ability to distinguish one's self from other selves.[4] More than the simple recognition that one's body is unique with regard to sensation and action, self-consciousness involves the recognition that one has a unique *mind*. The representation of a unique mind involves, at minimum, a theory of mind, including an understanding of deception, perceptual processes, and empathetic imagination (Perner 1991; Gopnik 1993; Tomasello 1999; Goldman 2006). Given the sophisticated representational abilities required for self-consciousness, it seems likely that consciousness evolved prior to self-consciousness, and so we should expect there to be creatures capable of consciousness that are not capable of self-consciousness. There may be reasons to claim that consciousness entails self-consciousness, but this necessary link must be argued and not assumed.

So far then, we can say that we are looking for a phenomenon that occurs during awake states that are responsive to sensory stimuli, and there is something it is like to be in those states. This definition is insufficient for a science of consciousness. The problem is that it is unclear how to supplement it without begging the question about the nature of consciousness. If we propose that consciousness is a global workspace (Baars 1988, 1997; Baars and Newnan 1994; Dehaene and Naccache 2001; Dehaene and Changeux 2011), this assumes a particular integration function is constitutive of consciousness. If we suggest that consciousness involves being conscious of our mental states, the assumption is that some form of higher-order state is necessary for consciousness.[5] Nonetheless, some starting

---

[4] At least, this is what we mean by 'self-consciousness' here: conscious states about one's own mental states. Conscious bodily states form a 'minimal phenomenal self' (Metzinger 2009) or a core self (Damasio 1999) but they are not self-conscious on our use of the term.

[5] According to higher-order theories of consciousness, a mental state is conscious when there is another mental state that is conscious of it. The appropriate higher-order state might be a thought or a sensation, it might be extrinsic or intrinsic to the lower-order state, and it might be occurrent or dispositional. The theory is higher-order because the relation is a consciousness of consciousness; two levels of consciousness are required (Lycan 1996, 2004; Carruthers 2000, 2004; Rosenthal 2005; Kriegel 2006, 2009; Gennaro 2006, 2012). Theories that involve cognitively higher-order capacities such as conceptual or inferential abilities are not higher-order theories in the sense invoked by higher-order theories of consciousness.

assumption about the nature of consciousness is necessary to articulate a clear target for scientific explanation.

We believe that the phenomenologist Husserl (1905) offered a useful suggestion when he noted the essential role of time in the structure of consciousness. According to Husserl, consciousness represents the world from the perspective of a now-point. We take the items presented to us by consciousness as before us now. Yet this moment is not instantaneous, it includes a brief span of time from just past (retention) to just ahead (protention). A span is necessary to account for the experience of duration, such as the hum of cicadas or the waving of leaves in the breeze. In other words, an essential aspect of what it is like to sense the world is to sense it as now enduring (Droege 2003, 2009).

The representation of time in consciousness differs crucially from the way representations may vary in relation to time. A creature that can vary its action in accord with the seasons or other natural temporal patterns does not need to represent time, it simply needs to respond to the appropriate temporal cues. Even a perception–action sequence that includes a temporal element utilizes time without representing it. The marine Palolo worm *Eunice viridis,* for example, times its reproductive cycle to a 2-h period in late fall. The precise synchronization of thousands of these organisms is attributable to a combination of biologically based oscillations (daily, annually, lunar, and tidal).[6] As long as action is strictly coded to some sort of fixed cycle, no representation *of* time is required. Only when a creature might decide between actions must it be able to assess the environment *as it is now* in order to determine how to proceed in the light of its goals. As we will argue, the evolutionary development of flexible behavior demands the ability to represent time. This psychological connection between flexibility and temporal representation along with the proposed definitional connection between temporal representation and consciousness form the links between private experience and public action necessary for assessing the capacity for consciousness in non-human animals.

This claim is controversial, and thus forms a substantive commitment about the nature of consciousness, just as the claim that consciousness is a global workspace or a higher-order state (Baars 1997, 2003; Rosenthal 2005; Dehaene and Changeux 2011). Nonetheless, this description of consciousness has several advantages over other proposed claims. First, unlike the global workspace, our sense of presence is phenomenologically apparent. The world appears before us as *right here, right*

[6] Gallistel (1990, p. 236). Later Gallistel notes that "the ability to detect and make behavioral use of temporal intervals of arbitrary duration is distinct from the ability to record the time at which something happens" (240). Early on in the book, he also offers a useful definition of 'representation' as "a functioning isomorphism between an aspect of the environment and a brain process that adapts an animal's behavior to it" (3). For a fully developed theory of the biological functional value of representation, see Millikan (1984, 1993, 1998, 2004). Because biological function is necessary to representation on the view we are proposing, a time code cannot simply be added to a computer to make it conscious. A prior question must be answered: whether and how a computer or robot might be capable of genuine representation rather than derived representation (Searle 1980).

now. What it is like to be conscious includes the temporal immediacy of experience. Note that the coordination of sensations and thoughts into a representation of presence is consistent with the integration function proposed by global workspace theorists and other neuroscientific theories of consciousness (Edelman and Tononi 2000; Gulick 2004; Tononi 2004, 2008; Tononi and Koch 2008; Dehaene and Changeux 2011). The present proposal puts the integration function into an evolutionary and developmental context to explain why creatures come to integrate information in the particular form that is conscious representation. A second, related advantage is that the identification in nonhuman animals of the ability to represent the present moment offers an intermediate evolutionary step between simple sensory response and more sophisticated higher-order representation. As noted above, self-consciousness requires additional evolutionary pressure to develop beyond the capacity for consciousness.

In the remainder of this chapter, we will consider the behavioral and physical evidence for temporal representation, and we have chosen to focus on fish as a particularly hard case for demonstrating consciousness (Braithwaite 2010). Fundamental differences between fish and humans stretch the argument from analogy to its breaking point, and this relative weakness places more weight on functional and theoretical argument. If we can make a convincing case that fish are conscious, despite our lack of intuitions about what it might be like to be them, then we can be sure that the framework of explanation is supporting the theory rather than mere anthropocentric inference.

In sum, conscious states are those it is like something to have. When we are awake and responsive to sensory stimuli, there is something it is like to experience (some of) those sensations. Furthermore, our conscious sensory experience presents the world as appearing before us *now*, which includes a brief temporal span of events indicating an enduring environment. While temporal representation is not (yet) universally accepted as a criterion for consciousness, we aim to show that it forms a vital part of a research program to establish a framework for consciousness in non-human animals.

2 Behavioral Function

Behavior is certainly the primary and most potent indicator of mental activity, even if it cannot provide a complete explanation of the mind and its capacities. Dogs, cats, horses, and rabbits behave in ways similar to the ways we behave when excited, fearful, happy, or in pain (Fraser 2009). As a result, it is fairly easy to believe that these creatures have similar mental states. When they are awake and responsive to sensory stimuli, it seems reasonable to assume there is something it is like for them to experience their sensations, i.e., they have conscious states.

To reinforce the earlier claim that consciousness and self-consciousness are separate capacities, note that there is no compelling reason to attribute self-consciousness to these creatures. When the squirrels look to be enjoying their game of chase in the tree, it is difficult to imagine that there is nothing it is like for them to *consciously feel* the joy of their game. However, it is not at all difficult to imagine that they lack the reflective ability to think of themselves *as* squirrels or *as* joyful. They may very well be joyful without reflecting self-consciously on their joy. Our ability to imagine that a creature is conscious without being self-conscious does not prove these capacities are separable, but it does show that the claim of necessary connection must be argued.

The limits of this strategy of imagining what it is like to be a creature become apparent in the borderline cases. Fish exhibit some similar behavior: their avoidance of electric shocks looks like fear, and their brisk interactions with one another looks like playful fun. But is it? How could we be sure these are not simply automated responses to stimuli in the way a thermostat responds to changes in temperature, or the way our knee reflex responds to the tap of a doctor's rubber hammer?

As the question suggests, we need a way to distinguish automated behavior from conscious response. The ability to flexibly adapt to novel environmental conditions is often taken to signal consciousness (Griffin and Speck 2004; Edelman et al. 2005; Seth et al. 2005; Merker 2005, 2007; Edelman and Seth 2009), but the question is, why? Why should consciousness be connected to flexibility, or indeed to any other behavioral or physiological indicator (Dawkins 2006, 2012)? Here is where the definition of consciousness in terms of the representation of presence can help connect behavior to phenomenology. In order to flexibly respond to a stimulus, a creature must be able to represent the present stimulus environment in relation to its goals. The environment appears in a certain way *now*, and this differs in various ways from the desired environment. There is a predator on the horizon, or no food is available, or scratchy bristles are all around. Without an ability to distinguish how things are now from how things might otherwise be, there is no way to consider alternative paths from here to there. An animal might flee or approach as a simple response to a stimulus, but the ability to consider an alternative possibility involves an assessment of the best action in pursuit of a goal, and this more sophisticated alternative depends on a representation of the present moment. In other words, a representation of *now* is essential to flexible behavior as well as being essential to the structure of consciousness from the first-person perspective.[7]

[7] Body illusions, such as autoscopic hallucinations and out-of-body experiences, confirm the role of first-person experience in conscious experiences. Even in the strange case of heautoscopic hallucinations where a person experiences her own body and an illusory body from the point of view of both bodies *simultaneously*, the experience is nonetheless from the first-person perspective in the weak sense of an egocentric spatial frame of reference. How there could be two simultaneous egocentric frames of reference is puzzling, as if Escher constructed one of his visually impossible drawings inside the mind. The sense that subjects are 'two selves' suggests these hallucinations form of consciousness (Blanke and Metzinger 2009).

The methodological value of flexibility as behavioral evidence for consciousness lies in the many ways to demonstrate it: (1) differential response to the environment, (2) adaptation to novel situations, (3) manipulation of the environment to accomplish goals, and (4) explicit representation of absent objects.[8] We will consider each form of evidence in terms of how it shows both flexibility and temporal representation. Because all of this evidence demonstrates the same basic capacity, several examples could be used to illustrate more than one form.

As noted earlier, fish are a particularly useful group for this sort of investigation, because they are at the edge of human intuitions about consciousness (Allen 2011). If we can get a framework to guide our thinking and research with fish, then we can make a compelling argument about consciousness in other vertebrates as well, and possibly even invertebrates.

Another reason in favor of studying fish is the wide variety of species and the diverse evolutionary pressures on differential species development. Very closely related species exhibit strikingly different capabilities depending on factors such as the complexity of their environment, density of predators, and availability of resources (Braithwaite 2005). This diversity presents the possibility of tracing correlations in neurobiology, behavior, and environment in much richer ways than are possible in the study of mammals (Kotrschal et al. 1998; Gonzalez-Voyer and Kolm 2010).

Differential response calibrated to relevant differences in a situation is the minimum requirement for flexible behavior; all three of the remaining ways to demonstrate flexibility also show the capacity for differential response. While simple single-celled organisms have the basic ability to react to variations in stimuli, flexibility depends on using information learned in the past to select among different possible actions in the present situation.

Dennett (1995, pp. 373–378) provides a helpful schematic of various ways an organism might be designed to respond successfully to its environment, which he calls the *Tower of Generate-and-Test*. At the base of the tower are *Darwinian creatures*. Gene recombination and mutation provides these creatures a selective advantage, leading to the reproduction of the advantageous genetic traits. The level of *Skinnerian creatures* introduces the element of phenotypic plasticity. The behavior of these creatures can be modified by positive and negative reinforcement. This design system works effectively when a specific response is appropriate given a specific stimulus. When a creature develops the capacity to respond in different ways to a set of stimuli, it becomes a *Popperian creature*. At this level,

---

[8] In a recent workshop, Investigating Animal Pain and Consciousness, participants debated the appropriate cognitive marker for consciousness. 'Flexibility' may be too vague to adequately characterize the difference between fixed action patterns and more cognitively complex, variable behavior. Other candidates were explicit knowledge, goal-directed behavior, and decision making. This four-pronged strategy for identifying flexibility further specifies the targeted cognitive ability and incorporates the other candidate suggestions. Adaptation to novel situations involves goal-directed decision making. Manipulation of the environment and explicit representation of absent objects involve explicit knowledge.

a creature no longer simply acts based on past associations; it generalizes on past learning conditions to anticipate which current action will best advance its goals. Associative learning remains the foundation for the generalizations made by Popperian creatures. The critical development is that the associations are no longer one-to-one stimulus–response pairings but become a complex weighting of many-to-many relations. Learning, memory, and crucially, integration of information are necessary to differential response. A creature must be able to determine appropriate behavior based on the assessment of a number of variable factors present in the situation (Merker 2007). Popperian creatures behave flexibly, and so on the proposed account they are conscious.

Consider, for example, the complex symbiotic relationship between the cleaner wrasse *Labroides dimidiatus* and the client fish that need parasites removed. Cleaners establish territorial stations where they feed on the parasites of clients, interacting with clients over 2,000 times a day (Bshary and Grutter 2005). While clients with large home ranges have a choice of several cleaners, they tend to return to the same cleaners, as much or more than 100 times a day (Bshary and Würth 2001). It is in the interest of cleaners, therefore, to provide a valuable service, so clients will continue to return to their station. The mutual benefit of this relationship is complicated by the preference of cleaner fish to feed on the mucus of client fish instead of the parasites. This they do by literally biting a chunk out of the client's flesh, an action that causes the client to visibly shudder. When bitten by a cleaner, clients will respond by chasing the cleaner or by leaving the station (Bshary and Grutter 2005). Another complication is that some of the client fish are predators, so they too have interests that both favor and counter cooperation.

The variety of factors in cleaner–client interactions has resulted in surprisingly sophisticated behavior by the cleaner wrasse, and to a certain extent in cleaner gobies that also perform these services (Bshary and Würth 2001; Bshary and Grutter 2005; Danisman et al. 2010). First, clients may punish cleaners by chasing them or leaving the station. This reaction causes cleaners to limit their cheating in order to avoid these penalties (Bshary and Grutter 2005). Second, cleaners offer incentives to clients in the form of tactile stimulation: a cleaner hovers above the client, touching the client's dorsal fin with its pectoral and pelvic fins (Bshary and Würth 2001). Importantly, cleaners offer tactile stimulation in three sorts of situation: (1) in order to induce clients to stop at the station for a cleaning, (2) differentially to predators over non-predator client fish in order to prevent conflict, and (3) after the cleaner has bitten the client, to prevent the client from chasing or leaving.

A second way to demonstrate flexible behavior is the ability to respond to novel situations by inferring future conditions based on information gathered from the past. This description is loaded with mentally sophisticated terms: inference, past/future conditions, information. To warrant attribution of such cognitively rich abilities requires distinguishing them from the simple weighting of past associations. While past associations necessarily figure in the evaluation of a novel situation, the solution must depend on a combination of features never before experienced simultaneously.

According to the temporal representation theory of consciousness, a representation of the present moment allows the creature to assess the current situation in order to determine which action is the best means toward its goals. If a situation is novel, no merely backward-looking algorithm is sufficient to calculate appropriate behavior. Instead the features of the new situation must be combined. In other words, the creature needs a representation of both feature A and feature B as components of the world *now* in order to know that they need to be combined. The result of this combination specifies the best action in that situation.

The ability of the male cichlids to respond appropriately to a novel set of competitors provides just the sort of cognitively rich inference needed to demonstrate flexible response. Because the species is aggressively territorial, it is useful for them to gain as much information about potential opponents as possible. Watching aggressive interactions is one source of information utilized by cichlids to determine which opponents are more threatening. In an experiment by Grosenick et al. (2007), cichlids were allowed to watch neighbor fish fight with one another in order to determine how much information they were able to gather from observation. Subject fish were shown fights in the following order: A beat B, B beat C, C beat D, and D beat E. In the critical trials, observer fish were placed between super-winner A and super-loser E and between marginally stronger B and marginally weaker D to see how the cichlid would react to these novel pairs. In both cases, the observer cichlid moves toward the weaker fish to assert its dominance.

Because the observer fish has never seen A and E or B and D fight one another, the response cannot be simple association based on past experience. Information about relative fighting strength needs to be assessed given the current opponents in order to determine the best action. This result is particularly striking when the fish identifies D as weaker than B. In this case, both B and D have lost one fight and won one fight, so neither is distinguishable simply as "winner" or "loser." Only by recognizing B as, in some sense, "winner over C" and D as "loser to C" can the observer act effectively. In other words, the observer fish utilizes memory to identify the relative strength of the particular individuals in the specific current situation. Representation of the features in the world *now* is needed to respond appropriately.

A third behavioral indicator of flexibility is the ability to manipulate the environment in pursuit of a goal. Differential response and assessment of novel situations are both reactive forms of behavior, showing a highly adaptive but not necessarily forward-thinking creature. Proactive, problem-solving skills, in contrast, require the ability to represent how a change in the current environment might bring about the desired goal. In other words, a creature must be able to represent the difference between the current situation and the goal situation in order to determine what action would bring about the appropriate change.

The best example of manipulating the environment is tool manufacture. To fashion a tool requires an understanding of how a goal can be achieved by means of unrelated items. By twisting or bending or combining objects, an animal can produce a tool specifically designed for a particular task. This shows an understanding of the relation between the tool and the task *prior* to the execution of the task. Tool manufacture depends on the ability to plan.

Since fish have no limbs and live in a viscous environment with little gravitational force, they are not likely to demonstrate the capacity to manufacture tools. Twisting and bending just are not in their ecological repertoire. Nonetheless, some features of tool use in fish suggest control over the environment rather than simple reactivity. It is important that the object used as a tool is not itself interesting in any way; its value is only in relation to the final objective. To be tool use, this connection between tool and goal must be clearly evident. Moreover, the creature must refrain from acting on the desired object directly in order to make use of the indirect advantage provided by the tool. An animal changes the relations of things in its environment to accomplish its goal, even though the change in itself is not valued.

Recent evidence suggests that water and stones may be used by fish as tools. Archerfish (*Toxotes* sp.) squirt water at insects sitting on surface plants or flying overhead, and can modify the trajectory and quantity of water proportional to the motion and size of the prey (Schuster et al. 2006; Brown 2012). In the most remarkable and controversial display, a six bar wrasse (*Thalassoma hardwicke*) carries a large food pellet about 75 cm in order to smash it against a specially selected rock to break up the pellets into smaller, digestible pieces (Paśko 2010). This last behavior in particular involves several important steps in problem-solving behavior. First, there is a search for a solution to the problem of breaking down large items of food. Next, the potential solution of using a rock as an anvil is tested until the preferred tool is found. Finally, the selected rock is remembered and reused unless another object proves more effective or convenient.

Caution is advisable in assessing this evidence, since reports are anecdotal, and the behavior has not yet been rigorously investigated.[9] If confirmed, the flexibility in tool use is obvious; no routinized associative stimulus–response pattern could account for the use of a tool to accomplish an otherwise unrelated task. Even if an animal hits upon the value of the tool accidentally, it must be able to remember when and how to use the tool as a means toward its goal. Less obvious is the temporal representation necessary to exhibit this behavior. Problem solving requires that time stops, in a sense. Goal-directed action must cease in order to determine the route forward.[10] Attention to the collection of items in the current environment is geared toward identifying the object or relation of objects needed to achieve the desired end. A creature incapable of stopping time in this way would not be able to solve a problem through the kind of trial and error process exhibited by tool use. Instead, the goal would simply be abandoned and a search for more tractable goal undertaken.

---

[9] The definition of "tool use" is also a question. As Brown (2012) points out, most definitions of tool use rule out the use of the substrate as a tool. It seems arbitrary to say that smashing an oyster with a rock is tool use, yet smashing a rock with the oyster is not. The relevant question is whether the action demonstrates controlled manipulation of the environment toward a goal. The way the tool is used indirectly as a means toward the goal is another feature that indicates cognitive flexibility rather than simple conditioned response, whether or not we call it "tool use."

[10] This does not mean that the creature literally freezes in its tracks. The point is that its general, exploratory behavior is not a direct means to its goal. It is a way of determining a means to its goal.

The fourth and last indicator of flexibility is the ability to represent both present features of the situation and features that are absent in the immediate stimulus environment. Any representation of what is present entails the corollary representation of absence. This may seem counterintuitive or even fallacious. There is no reason to assume that my representation of an apple as present before me entails the representation of something else, say a pear, as absent. The point is not that every representation of presence entails the representation of something *in particular* as absent. Rather, the possible contrast case of absence is necessary to make presence meaningful. If I am not capable of representing this apple as *either* present *or* absent, the content "present" adds nothing to the content "apple." In other words, a creature without the capacity for temporal representation of presence *and* absence could represent an apple or a pear or whatever else might be in its environment, but it could not represent the apple *as present*, or *as absent* in the form of a desired but not yet actual goal.

Though this way of putting things may sound too theoretical to be naturalistically viable, the ability to represent goals requires the representation of states of affairs that are not yet the case. Use of spatial memory demonstrates the way successful action often depends on a representation of absent features. Goldfish (*Carassius auratus*) can use landmarks to locate food rewards, even when they approach the landmark from a novel direction (Ingle and Sahagian 1973; Warburton 1990; Rodríguez et al. 1994; Salas et al. 1996). Siamese fighting fish (*Betta splendens*) remember the locations in an eight-arm radial maze that are depleted of food (Roitblat et al. 1982). At high tide gobiid fish (*Bathygobius soporator*) learn the topography of regions surrounding their home pool so that they can jump to an adjacent pool when threatened by a predator during low tide (Aronson 1951, 1971). In all of these cases, appropriate goal-directed behavior indicates that the location of the goal is represented despite its absence from the immediate sensory environment of the fish. Landmarks indicate the absent reward to the goldfish; maze position indicates the presence (and absence) of food to the fighting fish; and the goby springs to safety with no indicators at all save its memory of the position of pools nearby. In each of these cases, the fish discriminates what is now present—the landmark, the maze position, the predator—from what is absent in order to act appropriately to achieve its goals.

These myriad forms of evidence indicate considerable flexibility in fish behavior. A cleaner wrasse will differentially serve or cheat its clients depending on the prospects for reward and punishment. Cichlids react to a novel pair of opponents by inferring their relative fighting strength based on past observations. The six bar wrasse scouts for an ideal rock to break up its food. A goby maps its environment to prepare for an urgent escape. Flexibility depends on an evaluation of the environment as it is now in order to create the environment as it is desired to be. These fish are not simply responding to whatever stimulation appears, they are comparing information about the present situation with information about the past and future in order to respond effectively to the unique demands and prospects of the world as it is now.

3 Neural Structure

To supplement the behavioral evidence for fish consciousness, an examination of neural evidence is the natural next step. But identification of the necessary physical structures for consciousness is tricky. Even in humans, progress on neural correlates of consciousness has been slow and difficult (Metzinger 2000). Reliance on research with mammals assumes that creatures with similar brain structures and functions have similar forms of consciousness. Probably so, but why assume this, and which structures and functions are necessary and sufficient for consciousness? Little theoretical work has been done to justify assumptions that studies on human consciousness apply to nonhuman animals or vice versa. One of the goals of the proposed framework is to help answer these questions.

If the function of consciousness is to represent what is present in order to assess progress toward its goals, a conscious creature's nervous system must do three things. One, it must be capable of acquiring and responding to information about its environment; it must have a sensorimotor system. Two, it must be able to modulate the information it acquires so as to emphasize the goal-related bits and de-emphasize the irrelevant bits; it must have an attentional system. Three, it must be able to selectively act on sensory information; it must have an executive system.

First, a sensorimotor system is the entry-level requirement for mentality of any kind, and certainly for consciousness. The ability to sense and respond appropriately to environmental conditions depends on a representational system that can track beneficial and harmful elements as well as effective and ineffective reactions to them. Recent research has now identified several structures integral to processing pain perception in fish.[11] The same types of nociceptors and specialized fibers that convey information about tissue damage in humans and other mammals have also been described in teleost fish: (1) A-delta fibers are myelinated fibers that are associated with immediate pain; (2) C-fibers are unmyelinated and associated with the longer-lasting form of pain resulting from damage (Sneddon 2002; Sneddon et al. 2003a). The application of noxious stimuli to rainbow trout resulted in diminished appetite, increased evidence of stress (opercula beat rate), and apparent attempts at pain relief by rubbing their snouts on the walls and floor of the tank. Administration of an analgesic has been shown to decrease the negative effect of the noxious stimulus (Sneddon et al. 2003b).

There is also evidence that fish possess the neural structures necessary for rudimentary emotional processes (Braithwaite et al. 2013). Dopaminergic systems essential to reward conditioning, both positively and negatively, have been identified in the fish forebrain (Panula et al. 2010). A section in the fish telencephalon appears to serve the same functional role as the amygdala does in mammals

[11] 'Pain' here refers to the sensory system that registers physical damage and does not imply consciousness. As noted above, at least two other physiological structures are required for conscious pain in addition to pain sensations. For further discussion of animal pain, see Allen 2004; Allen et al. 2005.

(Rodríguez et al. 2005). When this area is lesioned, for example, fish have difficulty learning to avoid a negative, electric shock (Portavella et al. 2002).

Sensory responsiveness is such an important part of life and mind that it is tempting to think that sensation alone is sufficient for consciousness. The worm wriggles across the hot pavement, and we imagine it feels desperation; the bee sucks on the flower, and we can think of it as happy. Watching environmentally effective and often quite complicated behavior generates in human observers an irresistible empathic identification. We put ourselves in the position of the creature and find it difficult to imagine how such behavior could be produced without consciousness.

This anthropomorphic impulse can be reduced somewhat by reflecting on the complex behaviors we humans do unconsciously. The basic processes supporting any habitual activity—walking, driving, speaking—all occur without conscious direction. We are conscious while exercising these habits but not conscious of the body movements, sensorimotor adjustments, or even the word selection and arrangement necessary to execute everyday skills. Given that quite intricate behavior is often unconscious, what accounts for the difference when sensorimotor activity is conscious?

Here again the definition of consciousness in terms of a representation of presence shows the way forward. Sensorimotor activity is conscious when a creature needs to attend to how its actions in the world as it is *now* will best meet its goals. This means there must be a way to select the important information from all the available ongoing sensations and actions in order to make just that information conscious. So the second physical structure necessary for consciousness is some form of attentional system to relay and modulate sensory input. To serve the modulatory function, ramping up relevant stimuli and dampening irrelevant stimuli, some form of feedback loop or recurrent process is needed. In humans and other primates, a thalamocortical circuit figures in most accounts of the neural correlate of consciousness precisely because it is so effectively modulates neural activity across the entire cortex. Developmental studies of the brain of bony fish (teleosts) suggests that fish have preglomerular-pallial pathways that correspond to areas of the thalamocortical system of mammals, however the function of such areas within the fish brain has yet to be determined. Similarly, the insula-anterior cingulated axis has been considered to play a role in modulating consciousness, but as of yet, it is unclear whether a similar structure is found within the fish brain.

The third necessary function that must be performed by the nervous system of a conscious creature is planning and decision making. As argued in Sect. 3, behavioral flexibility indicates consciousness, and where there is flexibility, there is decision making. Fish have no neocortex, which is the brain structure in mammals that subserves executive functions like planning and decision making. This difference has led some to argue that consciousness in fish is impossible, since a brain area essential to mammalian consciousness does not exist in fish (Rose 2002, 2007). More recent evidence suggests that rather than having a layered neocortical structure, fish have more clustered nuclear regions within the telencephalon that have the capacity for some of the functions seen in the mammalian cortex. Indeed, the telencephalon, or forebrain, of fish is seen as a center where information is integrated (Rodríguez et al. 2005). Despite the presence of structures with some similarity in function to those found in mammals, a simple identification of consciousness with a particular physical substrate implausibly limits the possibility of alternative realizations. On our proposed temporal representation account, any physical structure that instantiates the appropriate sort of representations is thereby conscious. Consequently, the question is not whether fish have a neocortex, the question is whether fish have a structure that is capable of the decision making necessary to differential response to novel situations, manipulation of the environment, and representation of absence.[12] We know from other comparative studies between mammals and birds, these two groups of animals process visual information in very different ways, yet we do not deny either group the capacity to visualize the world around them (Braithwaite and Huntingford 2004).

In fish, the telencephalon is the best candidate to fill the role of the executive system, although evidence is still inconclusive. It has been clearly demonstrated that distinct areas in the fish forebrain are responsible for distinct cognitive capacities such as spatial memory and emotional processing (Portavella et al. 2002; Broglio et al. 2003; Broglio 2005; Dúran et al. 2010; Ebbesson and Braithwaite 2012). Also, the relative size of the telencephalon may indicate greater cognitive capacity in the same way it does in mammals and birds. At least in the case of some fish, a relatively large telencephalon, or forebrain, positively correlates with a complex environment (Kotrschal et al. 1998; Gonzalez-Voyer and Kolm 2010). More research is needed to make a convincing case that telencephalon size, or size of certain nuclei within the telencephalon is a decisive indicator of cognitive capacity. There is some evidence that fish living in structurally more complex environments or ranging over a wider territorial area have relatively larger forebrains (Marchetti and Nevitt 2003; Shumway 2010; Costa et al. 2011). In any case, there must be some structure that supports the sophisticated forms of behavior described in Sect. 2. At this point, the telencephalon is the most likely anatomical substrate of these abilities.

4 Conclusion

The foregoing evidence for flexibility in fish and a physical structure to support it is not meant to be the final word on the question of whether and how fish might be conscious, much less on what the ethical consequences of fish consciousness might

[12] Cabanac et al. (2009) have offered a similar account utilizing different criteria. They argue that consciousness depends on the ability to assess pleasure and displeasure. While their comparative and evolutionary account is amenable to the approach we advocate, "pleasure" is too broad a term to effectively isolate all and only conscious states. Pleasurable things may motivate me unconsciously, as when I find myself at the ice cream store again. Neutral things may be consciously represented as well, if there is nothing more interesting going on *now*.

be. It is meant to be the beginning of a cross-disciplinary debate about the sort of framework that will best organize the growing body of data on behavior, development, and anatomy of fish and other nonhuman animals in order to assess the capacity for consciousness. Once we have a means of determining what sorts of animals feel conscious pain, we can more effectively think about ways to minimize or eliminate their suffering.

What we offer here is an operational definition of consciousness in terms of temporal representation, in particular in terms of the phenomenological experience of the world *as present*. Our suggestion is that the representation of presence explains why behavioral flexibility is good evidence for consciousness. In order to respond differentially to the present situation, a creature must be able to identify the salient features of the world *now* as distinct from the way things have been or are desired to be. This critical link between how the world appears to the creature (*as now*) and how we as investigators observe the creature (as behaving in various ways and having certain sorts of anatomical features) forges the connection between phenomenology, cognitive function, and neural structure that promises new insight into the minds of nonhuman animals and new grounds on which to make moral decisions in relation to them.

Acknowledgments Our thinking on the issues raised in this article benefited greatly from discussions with participants at the workshop on Investigating Animal Pain and Consciousness, supported with funding from the Universities Federation for Animal Welfare and the Penn State Institute for the Neurosciences. Thanks to Marian Dawkins, Robert Elwood, Hanno Würbel, Ruud van den Bos, and Bernice Bovenkerk for contributing to the workshop. The cross-disciplinary discussion group Mechanisms of Mind at Pennsylvania State University sparked the conversations that led to this collaboration and has been an important source of inspiration and critique. We are grateful to Colin Allen and Paul Hart for comments on an earlier draft. Financial support was also provided through USDA grant PEN04296 and the Huck Institute of the Life Sciences.

References

Allen C (2004) Animal pain. Noûs 38:617–643. doi:10.1111/j.0029-4624.2004.00486.x

Allen C (2011) Fish cognition and consciousness. J Agric Environ Eth 1–15. doi: 10.1007/s10806-011-9364-9

Allen C, Fuchs PN, Shriver A, Wilson H (2005) Deciphering animal pain. In: Aydede M (ed) Pain: new essays on its nature and the methodology of its study. MIT Press, Cambridge, pp 351–365

Aronson LR (1951) Orientation and jumping behavior in the gobiid fish, *Bathygobius soporator*. Am Mus Novit 1486:1–22. http://hdl.handle.net/2246/3993

Aronson LR (1971) Further studies on orientation and jumping behavior in the Gobiid fish, *Bathygobius soporator*. Ann N Y Acad Sci 188:378–392. doi:10.1111/j.1749-6632.1971.tb13110.x

Baars BJ (1988) A cognitive theory of consciousness. Cambridge University Press, Cambridge

Baars BJ (1997) In the theater of consciousness: the workspace of the mind. Oxford University Press, Oxford

Baars BJ (2003) The global brainweb: an update on global workspace theory. Science and Consciousness Review. http://sciconrev.org/2003/09/the-global-brainweb/. Accessed 22 Sept 2003

Baars BJ, Newnan J (1994) A neurobiological interpretation of global workspace theory. In: Revonsuo A, Kamppinen M (eds) Consciousness in philosophy and cognitive neuroscience. L. Erlbaum Associates, Hillsdale, pp 211–226

Bentham J (1789) Introduction to the principles of morals and legislation. Clarendon Press, Oxford

Blanke O, Metzinger T (2009) Full-body illusions and minimal phenomenal selfhood. Trends Cogn Sci 13:7–13. doi:10.1016/j.tics.2008.10.003

Block N (1995) On a confusion about a function of consciousness. Behav Brain Sci 18:227–287. doi:10.1017/S0140525X00038188

Braithwaite VA (2005) Cognition in fish. Behav Physiol Fish 24:1–37. doi:10.1016/S1546-5098(05)24001-3

Braithwaite VA (2010) Do fish feel pain? Oxford University Press, Oxford

Braithwaite VA, Huntingford FA (2004) Fish and welfare: do fish have the capacity for pain perception and suffering? Anim Welf 13:87–92. http://www.ingentaconnect.com/content/ufaw/aw/2004/00000013/A00101s1/art00012

Braithwaite VA, Huntingford FA, van den Bos R (2013) Variation in emotion and cognition among fishes. J Agric Environ Eth 26:7–23. doi:10.1007/s10806-011-9355-x

Broglio C (2005) Hallmarks of a common forebrain vertebrate plan: Sepcialized pallial areas for spatial, temporal and emotional memory in actinopterygian fish. Brain Res Bull 66:277–281. doi:10.1016/j.brainresbull.2005.03.021

Broglio C, Rodríguez F, Salas C (2003) Spatial cognition and its neural basis in teleost fish. Fish Fish 4:247–255. doi:10.1046/j.1467-2979.2003.00128.x

Brown C (2012) Tool use in fishes. Fish Fish 13:105–115. doi:10.1111/j.1467-2979.2011.00451.x

Bshary R, Grutter AS (2005) Punishment and partner switching cause cooperative behavior in a cleaning mutualism. Biol Lett 1:396–399. doi:10.1111/j.1467-2979.2011.00451.x

Bshary R, Würth M (2001) Cleaner fish Labroides dimidiatus manipulate client reef fish by providing tactile stimulation. In: Proceedings of the Royal Society of London B, vol 268, pp 1495–1501. doi: 10.1098/rspb.2001.1495

Cabanac M, Cabanac AJ, Parent A (2009) The emergence of consciousness in phylogeny. Behav Brain Res 198:267–272. doi:10.1016/j.bbr.2008.11.028

Carruthers P (2000) Phenomenal consciousness: a naturalistic theory. Cambridge University Press, Cambridge

Carruthers P (2004) HOP over FOR, HOT theory. In: Gennaro R (ed) Higher-order theories of consciousness. John Benjamins, Amsterdam, pp 115–136

Chalmers D (1997) Conscious Mind. Oxford University Press, Oxford

Costa S, Andrade R, Carneiro L, Gonçalves E, Kotrschal K, Oliveira R (2011) Sex differences in the dorsolateral telencephalon correlate with home range size in Blenniid fish. Brain Behav Evol 77:55–64. doi:10.1159/000323668

Damasio A (1999) The feeling of what happens: body and emotion in the making of consciousness. Harcourt Brace, New York

Danisman E, Bshary R, Bergmüller R (2010) Do cleaner fish learn to feed against their preference in a reverse reward contingency task? Anim Cogn 13:41–49. doi:10.1007/s10071-009-0243-y

Dawkins MS (2006) A user's guide to animal welfare science. Trends Ecol Evol 21:77–82. doi:10.1016/j.tree.2005.10.017

Dawkins MS (2012) Why animals matter: animal consciousness, animal welfare, and human well-being. Oxford University Press, Oxford

Dehaene S, Changeux J-P (2011) Experimental and theoretical approaches to conscious processing. Neuron 70:200–227. doi:10.1016/j.neuron.2011.03.018

Dehaene S, Naccache L (2001) Towards a cognitive neuroscience of consciousness: basic evidence and a workspace framework. Cognition 79:1–37. doi:10.1016/S0010-0277(00)00123-2

Dennett DC (1995) Darwin's dangerous idea: evolution and the meanings of life. Simon & Schuster, New York

Droege P (2003) Caging the beast: a theory of sensory consciousness. John Benjamins, Amsterdam

Droege P (2009) Now or never: how consciousness represents time. Conscious Cogn 18:78–90. doi:10.1016/j.concog.2008.10.006

Durán E, Ocaña FM, Broglio C, Rodríguez F, Salas C (2010) Lateral but not medial telencephalic pallium ablation impairs the use of goldfish spatial allocentric strategies in a hole-board task. Behav Brain Res 214:480–487. doi:10.1016/j.bbr.2010.06.010

Ebbesson LOE, Braithwaite VA (2012) Environmental impacts on fish neural plasticity and cognition. J Fish Biol 81:2151–2174

Edelman DB, Baars BJ, Seth AK (2005) Identifying hallmarks of consciousness in non-mammalian species. Conscious Cogn 14:169–187. doi:10.1016/j.concog.2004.09.001

Edelman DB, Seth AK (2009) Animal consciousness: a synthetic approach. Trends Neurosci 32:476–484. doi:10.1016/j.tins.2009.05.008

Edelman GM, Tononi G (2000) Reentry and the dynamic core: neural correlates of conscious experience. In: Metzinger T (ed) Neural correlates of consciousness. MIT Press, Cambridge, pp 139–151

Fraser D (2009) Animal Behavior, Animal Welfare and the Scientific Study of Affect. Appl Anim Behav Sci 118:108–117

Gallistel CR (1990) The organization of learning. MIT Press, Cambridge

Gennaro R (2006) Between pure self-referentialism and the (Extrinsic) HOT theory of consciousness. In: Kriegel U, Williford K (eds) Self-representational approaches to consciousness. MIT Press, Cambridge, pp 221–248

Gennaro R (2012) Consciousness paradox: consciousness, concepts, and higher-order thoughts. MIT Press, Cambridge

Goldman A (2006) Simulating minds: the philosophy, psychology and neuroscience of mindreading. Oxford University Press, Oxford

Gonzalez-Voyer A, Kolm N (2010) Sex, ecology and the brain: evolutionary correlates of brain structure volumes in Tanganyikan cichlids. Public Libr Sci One 5:e14355. doi: 10.1186/1471-2148-9-238

Gopnik A (1993) How we know our minds: the illusion of first-person knowledge of intentionality. Behav Brain Sci 16:1–14. doi:10.1017/S0140525X00028636

Griffin DR, Speck GB (2004) New evidence of animal consciousness. Anim Cogn 7:5–18. doi:10.1007/s10071-003-0203-x

Grosenick L, Clement TS, Fernald RD (2007) Fish can infer social rank by observation alone. Nature 445:419–432. doi:10.1038/nature05646

Heider F, Simmel M (1944) An experimental study of apparent behaviour. Am J Psychol 57:243–259

Husserl E (1905) On the phenomenology of the consciousness of internal time (1893–1917) (trans: Brough JB). Kluwer Academic, Dordrecht

Ingle D, Sahagian D (1973) Solution of a spatial constancy problem by goldfish. Physiol Psychol 1:83–84

Kotrschal K, Van Staaden MJ, Huber R (1998) Fish brains: evolution and environmental relationships. Rev Fish Biol Fish 8:373–408. doi:10.1023/A:1008839605380

Kriegel U (2006) The same-order monitoring theory of consciousness. In: Kriegel U, Williford K (eds) Self-representational approaches to consciousness. MIT Press, Cambridge, pp 143–170

Kriegel U (2009) Subjective consciousness: a self-representational theory. Oxford University Press, Oxford

Lycan W (1996) Consciousness and experience. MIT Press, Cambridge

Lycan W (2004) The superiority of HOP to HOT. In: Gennaro R (ed) Higher-order theories of consciousness. John Benjamins, Amsterdam, pp 93–114

Marchetti MP, Nevitt GA (2003) Effects of hatchery rearing on brain structures of rainbow trout, *Oncorhynchus mykiss*. Environ Biol Fishes 66:9–14. http://www.ingentaconnect.com/content/klu/ebfi/2003/00000066/00000001/05088790

Merker B (2005) The liabilities of mobility: a selection pressure for the transition to consciousness in animal evolution. Conscious Cogn 14:89–114. doi:10.1016/S1053-8100(03)00002-3

Merker B (2007) Consciousness without cerebral cortex: a challenge for neuroscience and medicine. Behav Brain Sci 30:63–81. doi:10.1017/S0140525X00028636

Metzinger T (2000) Neural correlates of consciousness. MIT Press, Cambridge

Metzinger T (2009) The ego tunnel: the science of the mind and the myth of the self. Basic Books, Philadelphia

Millikan RG (1984) Language, thought, and other biological categories. MIT Press, Cambridge

Millikan RG (1993) White queen psychology and other essays for Alice. MIT Press, Cambridge

Millikan RG (1998) A common structure for concepts of individuals, stuffs and real kinds: more mama, more milk, and more mouse. Behav Brain Sci 21:55–100. doi:10.1017/S0140525X98000405

Millikan RG (2004) Varieties of meaning. MIT Press, Cambridge

Nagel T (1974) What is it like to be a bat? Philos Rev 83:435–450. doi:10.2307/2183914

Panula P, Chen Y-C, Priyadarshini M, Kudo H, Semenova S, Sundvik M, Sallinen V (2010) The comparative neuroanatomy and neurochemistry of Zebrafish CNS systems of relevance to human neuropsychiatric diseases. Neurobiol Dis 40:46–57. doi:10.1016/j.nbd.2010.05.010

Paśko Ł (2010) Tool-like behavior in the sixbar wrasse, *Thalassoma hardwicke* (Bennett, 1830). Zoo Biol 29:767–773. doi:10.1002/zoo.20307

Perner J (1991) Understanding the representational mind. MIT Press, Cambridge

Portavella M, Vargas JP, Torres B, Salas C (2002) The effects of telencephalic pallial lesions on spatial, temporal, and emotional learning in goldfish. Brain Res Bull 57:397–399. doi:10.1016/S0361-9230(01)00699-2

Rodríguez F, Durán E, Gómez A, Ocaña FM, Álvarez E, Jiménez-Moya F, Broglio C, Salas C (2005) Cognitive and emotional functions of the Teleost fish cerebellum. Brain Res Bull 66:365–370. doi:10.1016/j.brainresbull.2004.11.026

Rodríguez F, Durán E, Vargas JP, Torres B, Salas C (1994) Performance of goldfish trained in allocentric and egocentric maze procedures suggests the presence of a cognitive mapping system in fishes. Anim Learn Behav 22:409–420. doi:10.3758/BF03209160

Roitblat HL, Tham W, Golub L (1982) Performance of *Betta splendens* in a radial maze. Anim Learn Behav 10:108–114. doi:10.3758/BF03212055

Rose JD (2002) The neurobehavioral nature of fishes and the question of awareness and pain. Rev Fish Sci 10:1–38. doi:10.1080/20026491051668

Rose JD (2007) Anthropomorphism and 'mental welfare' of fishes. Dis Aquat Org 75:139–154. doi:10.3354/dao075139

Rosenthal DM (2005) Consciousness and mind. Oxford University Press, Oxford

Salas C, Broglio C, Rodríguez F, López JC, Portavella M, Torres B (1996) Telencephalic ablation in goldfish impairs performance in a spatial constancy problem but not in a cued one. Behav Brain Res 79:193–200. doi:10.1016/0166-4328(96)00014-9

Schuster S, Wöhl S, Griebsch M, Klostermeier I (2006) Animal cognition: how archer fish learn to down rapidly moving targets. Curr Biol 16:378–383. doi:10.1016/j.cub.2005.12.037

Searle JR (1980) Minds, brains and programs. Behav Brain Sci 3:417–424. doi:10.1017/S0140525X00005756

Seth AK, Baars BJ, Edelman DB (2005) Criteria for consciousness in humans and other mammals. Conscious Cogn 14:119–139. doi:10.1016/j.concog.2004.08.006

Shumway CA (2010) The evolution of complex brains and behaviors in African cichlid fishes. Curr Zool 56:144–156. http://www.currentzoology.org/paperdetail.asp?id=11457

Sneddon LU (2002) Anatomical and electrophysiological analysis of the trigeminal nerve in a teleost fish, *Oncorhynchus mykiss*. Neurosci Lett 319:167–171. doi:10.1016/S0304-3940(01)02584-8

Sneddon LU, Braithwaite VA, Gentle MJ (2003a) Do fish have nociceptors? Evidence for the evolution of a vertebrate sensory system. In: Proceedings of the Royal Society London B vol 270, pp 1115–1121. doi: 10.1098/rspb.2003.2349

Sneddon LU, Braithwaite VA, Gentle MJ (2003b) Novel object test: examining nociception and fear in the rainbow trout. J Pain 4:431–440. doi:10.1067/S1526-5900(03)00717-X

Tomasello M (1999) The cultural origins of human cognition. Harvard University Press, Cambridge

Tononi G (2004) An information integration theory of consciousness. BMC Neurosci 5:42. doi:10.1186/1471-2202-5-42

Tononi G (2008) Consciousness as integrated information: a provisional manifesto. Biol Bull 215:216–242. doi:10.2307/25470707

Tononi G, Koch C (2008) Neural correlates of consciousness: an update. Ann N Y Acad Sci 1124:239–261. doi:10.1196/annals.1440.004

Van Gulick R (2004) Higher-order global states (Hogs): an alternative higher-order model of consciousness. In: Gennaro R (ed) Higher-order theories of consciousness. John Benjamins, Amsterdam, pp 67–92

Warburton K (1990) The use of local landmarks by foraging goldfish. Anim Behav 40:500–505. doi:10.1016/S0003-3472(05)80530-5

Telos, Conservation of Welfare, and Ethical Issues in Genetic Engineering of Animals

Bernard E. Rollin

Abstract The most long-lived metaphysics or view of reality in the history of Western thought is Aristotle's teleology, which reigned for almost 2,000 years. Biology was expressed in terms of function or *telos*, and accorded perfectly with common sense. The rise of mechanistic, Newtonian science vanquished teleological explanations. Understanding and accommodating animal *telos* was essential to success in animal husbandry, which involved respect for *telos,* and was presuppositional to our "ancient contract" with domestic animals. *Telos* was further abandoned with the rise of industrial agriculture, which utilized "technological fixes" to force animal into environments they were unsuited for, while continuing to be productive. Loss of husbandry and respect for *telos* created major issues for farm animal welfare, and forced the creation of a new ethic demanding respect for *telos.* As genetic engineering developed, the notion arose of modifying animals to fit their environment in order to avoid animal suffering, rather than fitting them into congenial environments. Most people do not favor changing the animals, rather than changing the conditions under which they are reared. Aesthetic appreciation of husbandry and virtue ethics militate in favor of restoring husbandry, rather than radically changing animal *teloi.* One, however, does not morally wrong *teloi* by changing them—one can only wrong individuals. In biomedical research, we do indeed inflict major pain, suffering and disease on animals. And genetic engineering seems to augment our ability to create animals to model diseases, particularly more than 3,000 known human genetic diseases. The disease, known as Lesch–Nyhan's syndrome or HPRT deficiency, which causes self-mutilation and mental retardation, provides us with a real possibility for genetically creating "animal models" of this disease, animals doomed to a life of great and unalleviable suffering. This of course creates a major moral dilemma.

B. E. Rollin (✉)
Department of Philosophy, Colorado State University, Fort Collins, CO 80523-1781, USA
e-mail: Bernard.Rollin@colostate.edu

Curr Topics Behav Neurosci (2015) 19: 99–116
DOI: 10.1007/7854_2014_279

Published Online: 5 February 2014

Perhaps one can use the very genetic engineering which creates this dilemma to ablate consciousness in such animal models, thereby escaping a moral impasse.

Keywords Teleology · *Telos* · Mechanistic explanation · Animal husbandry · Genetic engineering · Transgenic animal models

Contents

1 Telos

Expressed in very simple terms, a metaphysics is a set of concepts in terms of which we understand the world; a frame through which we organize what we mean by reality. Far and away, the most long-lived metaphysics that ever held sway in the Western world is Aristotle's teleology, which saw the world and what took place in the world in terms of ends, functions, purposes and what Aristotle called final causes. This worldview reigned supreme in the Western world from the time of Aristotle until the scientific revolution of the seventeenth century, almost 2,000 years. In Aristotelian terms, biological organisms represented the model according to which all other organisms and processes in the physical world were to be understood. Just as was the case with living things, all natural and artifactual entities possessed a *telos*, or final cause or end or nature or purpose, which determined its function, and thereby its nature. Rather than biology being subsumed under mechanistic causation, efficient causes, even putatively "dead matter" had a nature or unique function by which it was to be explained. The function of a rock, for example, was, unless impeded, was to fall toward the center of the Earth, which was also the center of the universe. Hold a rock in your hand, and you feel it's tendency to move downward if all obstacles to such movements are removed.

The science of biology, for Aristotle, was very simply understanding how living things fulfilled the functions of any living thing—sensation, nutrition, locomotion, growth, and reproduction. The sum total of how an organism does so is constitutive of its *telos* or nature. Every living thing was to be explained in terms of how it fulfilled its *telos*. *Telos*, in modern terminology, is roughly what is encoded in an animal's genetics, as expressed in its normal environment—the pigness of the pig, the dogness of the dog, what common sense recognizes as "fish gotta swim; birds gotta fly."

The fact that nature was to be explained teleologically or functionally did not presuppose that the functions in question were consciously adhered to by an organism, or consciously designed, even though Aristotelian teleology was adopted by the Catholic Church to fit a theological purpose, namely that all of nature had been designed by God. We can explain the sharp edge of a knife by reference to what a knife does, namely cut, without assuming consciousness on the part of the knife. In a similar manner, we can explain the building of dams by beavers in terms of such dams increasing the likelihood of catching fish without assuming either that beavers have a conscious purpose in mind when they build, or that they were consciously designed to do so; evolution by natural selection is perfectly adequate as an explanation, especially of the latter.

Seeing the world in terms of functions and purposes, particularly seeing living things, is totally compatible with ordinary experience and a common sense view of the world. (For this reason, Aristotle is often viewed as the greatest philosopher of common sense.) But, in the seventeenth century, when both common sense and Aristotelian philosophy were challenged by the solidly mechanistic scientific revolution of Galileo, Descartes, and Newton, teleological explanations were dealt with a serious blow, at least as far as scientists and philosophers were concerned. As is well-known, Descartes strongly challenged the world shown to us by our senses, and assumed to be real by common sense. Less well known, but even more destructive to commonsense and teleological explanations was Spinoza's vicious (and ultimately unfair) attack on thinking in teleological terms. Since teleology was completely rejected by adherents of the scientific revolution, it significantly waned in importance in modern thought, except for the Catholic Church and other theologians.

Teleology, at the hands of these theologians became equated with what is today known as "intelligent design"—evidence of a superior power who planned the world in an intentional, carefully designed manner. While teleology certainly fits for explaining artifactual creations, that is not necessarily its role as an explanatory model. Consider the adrenal gland: from an explanatory perspective, the adrenal gland exists to alert the body, and prepare for "fight or flight." Suppose a human infant dies before the adrenal gland has a chance to function. It is still reasonable to explain its presence in the body in terms of that role, even if it is never, in fact, actualized. Similarly, with the reproductive system, it is there to effect reproduction, even if the person lives a celibate life, and never in fact even attempts to reproduce. The same holds of animal *telos*—a pig, for example, embodies a certain set of functions constitutive of its life as a pig, even if it dies at birth; a certain set of potentials, which are real and explanatory regardless of whether that set is ever actualized, to use Aristotle's clear principles. There is nothing mystical, or even theological, in invoking *telos* as the blueprint or template for a certain form of life, actualized or not. The fact that the Scientific Revolution restricted itself to mechanistic explanations is totally irrelevant to the question of the utility or coherence of teleological explanations. As we shall show, *telos* has recently emerged as an explanatory concept for animal ethics and for genetic engineering.

One may in fact, look at the situation in the following way: If one is scientifically oriented in the mechanistic, post-Renaissance sense, and thereby believes with Descartes that biology is and should be reducible to the physics of particles, one may eschew functional or teleological explanations strictly in favor of mechanistic, efficient causes. On the other hand, those who view the world in terms of common sense, not only do have any problem with teleological causation, but actually are compelled to see the world that way. Imagine trying to have a pet dog living with you and not being able to say "He wants to go out." Historically, however, understanding our companion animals was of little significance compared to understanding agricultural animals. Domestication of animals is more than 10,000 years old, and understanding these animals' *telos* represents both a cause and an effect of domestication. We could never have domesticated them if we failed to understand at least the basics of their *telos*, and as we domesticated them, we changed their *teloi* to suit domestication, making them more docile and tractable, and more dependent on us.

2 Violation of *Telos*

For virtually all of the history of domestication (99 % plus), we successfully managed the animals we employed for food, fiber, work, and transport by understanding their natures and respecting their *teloi*, in what has been called "the ancient contract" between humans and the animals that made civilization possible. It is arguable that the development of human civilization was directly dependent on the creation of a secure and predictable food supply. Such a food supply freed people from the uncertainties and vagaries of depending on hunting and gathering, and enabled the establishment of communities. Predictability regarding food was assured by the development of both plant and animal agriculture, which operated synergistically. Cultivation of crops and plants secured human ability to depend on (barring catastrophes of weather) foods of plant origin, and on a steady and local source of animal feed. Animal agriculture, in turn, provided a source of labor for crop production, as well as a predictable reservoir of animal protein for human consumption. The secure food supply ramified in the ability to develop manufacturing, trade, commerce, and in Hobbes's felicitous phrase, the "leisure that is the mother of philosophy," construed in the broadest sense as speculative thought, science, technological innovation, art, and culture.

Presuppositional to the development of both agricultures was the concept of sustainability, i.e., assurance that the conditions and resources necessary to them were indefinitely renewable. As children, many of us learned about balanced aquariums. If we wished to keep a fish tank where the fish lived and we didn't want to keep tinkering with it, we needed to assure that the system in question was as close to a "perpetual motion" machine as possible, a system that required little maintenance because all parts worked together. That meant including plants that produced oxygen and consumed carbon dioxide, enough light to nourish the plants, or rather plants that thrived in the available light source, water that was properly constituted chemically, scavengers to remove wastes, and soon. When such a system worked, it required minimal maintenance. If something were out of balance, plants and animals would die, and require constant replacement. The fish tank aims at being a balanced ecosystem, and thus represents a model of traditional approaches to cultivation of land, wherein one sought to grow plants that could be grown indefinitely with available resources, which conserved and maximized these resources, and which would not die out or require constant enrichment. Hence, the beauty of pastoral agriculture, where pasture nourished herbivores, and herbivores provided us with milk, meat, and leather, and their manure enriched the pasture land in a renewable cycle.

Cultivation of land evolved locally with humans. If one did not attend to the constraints imposed by nature on what and how much could be grown in a given region, the region would soon cease to yield its bounty, by virtue of salinization, or depletion of nutrients or overgrazing, or insect infestation. Thus, over time, humans evolved to, as one book put it, "farm with nature," which became, like animal husbandry, both a rational necessity and an ethical imperative. Local knowledge, accumulated over a long period of trial and error, told us how much irrigation was too much; what would not grow in given soils; what weeds left standing protected against insects; where shade and windbreaks were needed, and so on. Thus, accumulated wisdom was passed on—and augmented—from generation to generation, and was sustainable, i.e., required minimal tweaking or addition of resources. The genius of agriculture was to utilize what was there in a way that would endure. If the land did not thrive, you did not thrive. Traditional agriculture, then, was inherently sustainable; by trial and error over long period of time it evolved into as close to a "balanced aquarium" as possible.

Not surprisingly, precisely isomorphic logic applied to sustainability in animal production. The maxim underlying continued success in rearing animals was good husbandry, which represented a unified synthesis of prudence and ethics. Husbandry meant, first of all, placing the animals into the optimal environment for which they had been bred, and where they could maximally fulfill their *telos*—their physical and psychological needs and natures. Having done so, the husbandman then augmented animals' ability to survive and thrive by watching over them—by providing protection from predators, food during periods of famine, water during times of drought, shelter during extremes of climate, assistance in birthing, medical attention, and generally ministering to whatever needs the animals had. So powerfully ingrained was this imperative in the human psyche, that when the Psalmist searches for a metaphor for God's ideal relationship to human beings, he can do no better than seizing upon the conceit of the Good Shepherd. The Shepherd serves as far more than merely a herdsman, but more as a guardian and protector of the sheep under his aegis:

> The Lord is my shepherd; I shall not want
> He maketh me to lie down in green pastures.
> He leadeth me beside the still waters.
> He restoreth my soul. (Psalm 23)

We want no more from God than what the Good Shepherd provides for his flock. As we know from other passages in the Old Testament, a lamb on its own would live a miserable, nasty, and short life by virtue of the proliferation of predators—hyenas, raptors, wolves, bears, lions, foxes, jackals, and numerous others. With the care and ministrations of the Shepherd, the animal lives well until such time as humans take its life, in the meantime supplying us with milk, wool, and in the case of some domestic animals, the labor that became indispensable to the working of land for crops.

The power of this symbiotic image cannot be overestimated in the history of Western civilization. In Christian iconography, for thousands of years, Jesus is depicted both as Shepherd and as lamb, a duality built into the very foundations of human culture. The pastor, a word harking back to pastoral, tends to his flock; the members of his congregation are his sheep. And when Plato discusses the ideal political ruler in the Republic, he deploys the shepherd-sheep metaphor: The ruler is to his people as the shepherd is to his flock. Qua shepherd, the shepherd exists to protect, preserve, and improve the sheep; any payment tendered to him is in his capacity as wage-earner. So too the ruler, again illustrating the power of the concept of husbandry on our psyches.

Animal agriculture was indispensable to the subsequent development of society and culture. Husbandry agriculture is the ancient contract that was presuppositional to that entire evolutionary process. In one of the most momentous ironies in the history of civilization, this ancient contract with the animals, as well with the Earth, in terms of sustainability, contained within it the seeds of its own undoing. It was in virtue of a secure and predictable food supply that humans could proceed with trade, manufacturing, invention, and the general flourishing of culture.

By the late nineteenth century, industrial proliferation and innovation had reached a point where sustainability and good husbandry seemed to be no longer essential presuppositions of civilization. The ancient contract, which we may characterize as husbandry with regard to animals, and stewardship with regard to the land, was the presuppositional bedrock upon which economics, art, and culture rests. Yet, with the profound hybrids of an Icarus who challenged inherent human limitations, with blind and abiding faith in the humanly crafted tools which repeatedly show themselves as impotent in the face of natural disaster, we thumbed our noses at both morality and prudence. As the ancients crafted the tower of Babel, so we began to overreach the constraints imposed on us by the natural world. In both crop and animal agriculture, the ancient values of sustainability, stewardship, and husbandry inexorably gave way to modernist values of industrialization, productivity, and efficiency. The symbiotic partnership between humans and the Earth, and between humans and animals, was rapidly transmuted into patent exploitation with no respect or attention to what priceless elements were lost.

The coming of the Industrial Revolution gave us the tools to break our husbandry contract with domestic animals. No longer were we helping the animals fit into the natural environment for which they were adapted. The values of good husbandry, and coexistence with animals for mutual benefit were, as soon as possible, replaced by an overwhelming emphasis on productivity and efficiency. No longer did the animals' *telos* need to fit in the environment. Technology gave us the ability to force animals into deleterious environments that did not fit their *teloi*, and thus greatly damaged their welfare, but at the same time, did not affect their productivity. The happy unity of animal welfare and productivity was rendered asunder, with animals forced to remain productive while losing any chance of a good life. Consider any aspect of industrialized agriculture. Traditionally, if one crowded thousands of animals into high confinement, the animals would have sickened and died, and the producer would likewise have suffered. But, with the advent of technological manipulations, we could force animals into environments where they did not fit, without loss of productivity. The need for agriculturalists to understand animal natures in order to ensure both welfare and productivity disappeared with the advent of antibiotics, vaccines, air handling systems, hormones, all of which severed welfare from productivity. Understanding *telos*, historically presuppositional to agricultural success, ceased to matter.

As long as respect for *telos* made animal agriculture possible, there was little need for an ethic of animal welfare. For, if one failed to respect animal nature, the animals failed to produce. The only ethic extant was a prohibition of deliberate, intentional, sadistic, purposeless, and deviant cruelty. If society wished to preserve fair treatment of animals in the face of relentless industrialization, an articulated ethic for animal treatment to guide behavior, a new ethic was required, one that would preserve and restore the proper treatment of animals presuppositional to good husbandry.

3 A New Ethic for Animals Based in Telos

Various philosophers proposed different approaches to creating a new ethic. Most noteworthy, perhaps, was Peter Singer's attempt to deduce a new ethic for animals from Utilitarianism (Singer 1975), the theory based on maximizing pleasure and minimizing pain, and creating "the greatest happiness for the greatest number." While ingenious and compelling, there were problems in Utilitarianism, which rendered it unacceptable to many people. In particular, it seemed artificial to express all the harms we do to animals, from removing their babies too early; to keeping social animals isolated from conspecifics; to depriving them of the ability to move or forage; to performing surgery on them with no anesthesia or analgesia; to stressing and frightening them in transport; to beating them; to preventing them from utilizing the natural powers they have evolved to survive, along a simple quantitative spectrum of pleasure and pain. The second problem arises from what is famously known as "the tyranny of the majority," the fact that Utilitarianism seems to allow the good of the majority to outweigh basic interests of the minority. We will return to this point.

The resurrection of the concept of *telos* was largely accomplished by this author in his attempt to create an animal ethic not subject to the criticisms directed at

Utilitarianism (Rollin 2006a, b, 2011a, 2013). In my view, any successful ethic for animals had to compel acquiescence from virtually everyone in society, who would see it as a consequence of beliefs they already held. I based this position on a point insisted upon by Plato, namely in order to convince people of ethical ideas, one needed to *remind*, not to *teach* (Rollin 2011b). This strategy was in turn developed by Martin Luther King, who realized that all Americans would accept two ethical premises fundamental to US government: (1) All humans were entitled to be treated as equals and (2) Black people were humans. Segregationists had just not bothered to draw the conclusion. If the conclusion was "written large" as Plato said, people would "recollect" and acquiesce to it.

In my reasoning, I argued that what we did to animals mattered to them, and that such mattering was best expressed in terms of violating their nature or *telos*. Furthermore, the ethical theory underlying American democracy was a blend of Utilitarianism, and protection of individuals who might be in the minority on certain issues. Certainly, in the USA and other democratic societies, we make our social decisions by reference to the benefit of the majority. But, we protect individuals in the minority by building protective fences around their human nature, or *telos,* as characterized in the Bill of Rights. Humans are beings who wish not to be tortured, to express themselves freely, to worship as they wished, to assemble with like-minded others, to hold onto their possessions. These basic human desires are protected by rights, which serves as a check on unbridled Utilitarianism.

It is part of normal common sense to view animals as having natures, the "pigness of the pig," the "dogness of the dog." And common sense recognizes the need to respect animal *telos*; "fish gotta swim, birds gotta fly." Given the basic ethic built into society, we ought to protect the fundamental interests of animals from encroachment. As I was told once in 1980, in the course of addressing all Canadian government ministers whose brief included animals, "we need a Bill of Rights for animals." Furthermore, common sense is quite capable of identifying fundamental aspects of animal nature. Ordinary people know full well that cows belong on pasture not concrete; that pregnant sows do not belong in cages. (Smithfield Farms, the world's largest swine producer, acknowledged this and announced their abandonment of gestation crates when they surveyed their customers at my suggestion and found that 78 % unequivocally rejected such high confinement.) And the public in Europe and the USA has further shown their willingness to encode such respect for animal nature in law if it no longer follows naturally from good husbandry.

My account of the emerging societal ethic for animals, based in our universally accepted ethic for humans, seems to capture some fundamental features of social thought. And the concept of *telos* is a foundational concept of that ethic. Granting that respect for animal nature or *telos* is a critical part of animal ethics, what does that ethic tell us about changing *telos* with the advent of genetic engineering? Unfortunately, a number of writers in this area have concluded that basing animal ethics on *telos* means that genetic engineering is inherently wrong with regard to animals. This may be true on some interpretations of *telos*, but emphatically not on my common sense version. In the Catholic tradition, and in the biblical account, animal nature is permanent, fixed, immutable, and set by God. On the other hand, the notion of *telos* we have explained is perfectly compatible with what biology tells us. An animal's nature is a snapshot of a constantly dynamic, developing process of evolution. There is nothing prima facie wrong in itself with humans participating in that process, as we have done with domestication. It is estimated that 70 % of grasses and 40 % of flowering plants represent new species created by humans through hybridization, cultivation, preferential propagations, and other means of artificial selection. So, as I have extensively argued, there is nothing *intrinsically* wrong with genetically engineering animals.

I have asserted that, given an animal's *telos*, and the interests that are constitutive thereof, one should not violate those interests. I never argued that a given *telos* itself could not be changed. If the animals could be made happier or less miserable by changing their natures, I see no moral problem in doing so (unless, of course, the changes harm or endanger other animals, humans, or the environment). *Telos* is not sacred; what is sacred are the interests that follow from it.

4 Changing *Telos* to Avoid Suffering

Can one then use genetic engineering of *telos* as a remedy for the serious animal welfare issues that emerge from modern industrialization of agriculture or other animal uses? (Rollin 1995). Consider a case where one might indeed be tempted to change the *telos* of an animal chickens kept in battery cages for efficient, high-yield egg production. It is now recognized that such a production system frustrates numerous significant aspects of chicken behavior under natural conditions, including nesting behavior, and that frustration of this basic need or drive results in a mode of suffering for the animals. Let us suppose that we have identified the gene or genes that code for the drive to nest. In addition, suppose we can ablate that gene or substitute a gene (probably *per impossibile)* that creates a new kind of chicken, one that achieves satisfaction by laying an egg in a cage. Would that be wrong in terms of the new ethic I have described?

If we identify an animal's *telos* as being genetically based and environmentally expressed, we have now changed the chicken's *telos* so that the animal that is forced by us to live in a battery cage is satisfying more of its nature than is the animal that still has the gene coding for nesting. Have we done something morally wrong? I would argue that we have not. Recall that a key feature, perhaps the key feature of the new ethic for animals I have described, is a concern for preventing animal suffering and augmenting animal happiness, which I have argued involves satisfaction of *telos*. One can also argue that the primary, pressing concern is the former, the mitigating of suffering at human hands, given the proliferation of suffering that has occurred in the twentieth century. I have also argued that suffering can be occasioned in many ways, from infliction of physical pain to prevention of satisfying basic drives. So, when we engineer the new kind of chicken that prefers laying in a cage and we eliminate the nesting urge, we have removed a source of suffering. Given the animal's changed *telos*, the new chicken is now suffering less than its predecessor and is thus closer to being happy, that is, satisfying the dictates of its nature.

Mill asks "is it better to be a satisfied pig or a dissatisfied Socrates?" His response, famously inconsistent with his emphasis on pleasure and pain as the only morally relevant dimensions of human life, is that it is better to be a dissatisfied Socrates. In other words, we intuitively consider the solution to human suffering offered, for example, in *Brave New World,* where people do not suffer under bad conditions because they are high on drugs, to be morally reprehensible, even though people feel happy and do not experience suffering. Why then, would we consider genetic manipulation of animals to eliminate the need that is being violated by the conditions under which we keep them to be morally acceptable?

This is an interesting and important objection, amenable to a number of different responses. Let us begin with the *Brave New World* case. Our immediate response to that situation is that the repressive society should be changed to fit humans, rather than our doctoring humans (chemically or genetically) to fit the repressive society. It is, after all, more sensible to alter clothes that do not fit than to perform surgery on the body to make it fit the clothes. And it is certainly possible and plausible to do this. So we blame the *Brave New World* situation for not attacking the problem.

This is similarly the case with the chickens. We know that laying chickens lived happily and produced eggs under conditions where they could nest for millennia. It is our greed that has forced them into an unnatural situation and made them suffer—why should we change them, in order to succumb to greed? This seems to be a simple point of fairness.

A disanalogy between the two cases arises at this point. We do not accept any claim that asserts that human society must be structured so that people are totally miserable unless they are radically altered or their consciousness distorted. Given our historical moral emphasis on reason and autonomy as nonnegotiable ultimate goods for humans, we believe in holding on to them, come what may. Efficiency, productivity, wealth—none of these trump reason and autonomy, and thus the *Brave New World* scenario is deemed unacceptable. On the other hand, were Mill not a product of the same historical values but was rather truly consistent in his concern only for pleasure and pain, the *Brave New World* approach or otherwise changing people to make them feel good would be a perfectly reasonable solution.

In the case of animals, however, there are no ur-values like freedom and reason lurking in the background. We, furthermore, have a historical tradition as old as domestication for changing (primarily agricultural) animal *telos* (through artificial selection) to fit animals into human society to serve human needs. We selected for nonaggressive animals, animals that depend on us not only on themselves, animals disinclined or unable to leave our protection, and so on. Our operative concern has always been to fit animals to us with as little friction as possible—as discussed, this assured both success for farmers and good lives for the animals. If we now consider it essential to raise animals under conditions like battery cages, it is not morally jarring to consider changing their *telos* to fit those conditions in the way that it jars us to consider changing humans.

Why then does it appear to some people to be prima facie somewhat morally problematic to suggest tampering with the animal's *telos* to remove suffering? In large parts, I believe, because people are not convinced that we can't change the conditions rather than the animal. (Most people are not even aware how far confinement agriculture has moved from traditional agriculture. A large East Coast chicken producer for many years ran television ads showing chickens in a barnyard and alleging that he raised "happy chickens.") If people in general do become aware of how animals are raised, as occurred in Sweden and later all over Europe, and as animal activists are working to accomplish here, they will be in doubtless demand, just as the Swedes did, first of all a change in raising conditions, not a change in the animals.

On the other hand, suppose the industry manages to convince the public that we cannot possibly change the conditions under which the animals are raised or that such changes would be outrageously costly to the consumer. And let us further suppose that people still want animal products, rather than choosing a vegan lifestyle. There is no reason to believe that people will ignore the suffering of the animals. If changing the animals by genetic engineering is the only way to assure that they do not suffer (the chief concern of the new ethic), people will surely accept that strategy, though doubtless with some reluctance.

From whence would stem such reluctance, and would it be a morally justified reluctance? Some of the reluctance would probably stem from slippery slope concerns—what next? Is the world changing too quickly, slipping out of our grasp? This is a normal human reflexive response to change—people reacted that way to the automobile. The relevant moral dimension is consequentialist; might not such change have results that will cause problems later? Might this not signal other major changes we are not expecting?

Closely related to that is a queasiness that is at root aesthetic. The chicken sitting in a nest is a powerful aesthetic image, analogous to cows grazing in green fields. A chicken without that urge jars us. But when people realize that the choice is between a new variety of chicken, one without the urge to nest and denied the opportunity to build a nest by how it is raised, and a traditional chicken with the urge to nest that is denied the opportunity to build a nest, and the latter is suffering while the former is not, they might well accept the removal of the urge, though they are likelier to be reinforced in their demand for changing the system of rearing and, perhaps, in their willingness to pay for reform of battery cages.

The most significant justified moral reluctance would probably come from a virtue ethic component of morality. Genetically engineering chickens which no longer want to nest could well evoke the following sort of musings: "Is this the sort of solution we are nurturing in society in our emphasis on economic growth, productivity, and efficiency? Are we so unwilling to pay more for things that we do not hesitate to change animals that we have successfully been in a contractual relationship with, since the dawn of civilization? Do we really want to encourage a mind-set willing to change venerable and tested aspects of nature at the drop of a hat for the sake of a few pennies? Is tradition of no value?" In the face of this sort of component to moral thought, I suspect that society might well resist the changing of *telos*. But at the same time, people will be forced to take welfare concerns more seriously and to decide whether they are willing to pay for tradition and amelioration of animal suffering, or whether they will accept the "quick fix" of *telos* alteration. Again, I suspect that such musings will lead to changes in husbandry, rather than changes in chickens—"raise the bridge, don't lower the river."

Some people have argued that my suggestion for changing chickens "degrades and diminishes the *teloi* found in the world." In my discussions of what sorts of genetic engineering of animals are morally acceptable, I have argued for what I call "the principle of conservation of welfare," (Rollin 1995) a moral principle that asserts that genetically engineered animals should be no worse off in terms of welfare than the parent stock from whence they have been derived. It might appear to my critics that my idea for ablating from the chicken genome many of the interests we fail to meet in modern agriculture is in stark violation of this principle, since the engineered chickens have an impoverished *telos*. This would constitute a logical error. The welfare I insist be conserved is that of the *individual* chickens, animals which suffer considerably if they retain the traditional chicken *telos*. The *telos* itself does not have welfare, or non-welfare. Thus, "simplifying" the chickens' nature to avoid suffering does not create negative welfare or suffering; in fact it *ablates* suffering by removing the conflict between the animals' lives and their natures. One can certainly adopt a metaphysical perspective, affirming that the more complex the *teloi*, the better the world, but that would be difficult indeed to argue in the face of considerable suffering accompanying greater complexity, and difficult to argue without invoking theology. At best, the objection raised against me is an aesthetic one, claiming that a simplified universe is an uglier universe. I would respond by acknowledging this point, but also arguing that suffering of sentient beings is far more morally reprehensible than ugliness. (In any case, there will always as a matter of fact be those consumers who demand the traditional chicken, even as a small number of people buy "heritage turkeys" at hugely inflated prices, thereby preserving the *telos* in question, albeit with far fewer instances.)

In any case, attributing qualities that apply to individual *teloi* to the generic concept of *telos* represents a classic logical fallacy, *the fallacy of division*. Just because mammals have hair and give milk does not mean that the concept of "mammal" has hair and lactates. In fact, of course, it does not.

Using genetic engineering to fix the welfare problems emerging from the industrialization of agriculture, as discussed with regard to the poultry industry, seems to be, as we saw, an unnecessary self-indulgence, as we raised poultry for eons with great success by respecting their *telos*, rather than modifying it for what essentially amounts to reasons of expediency resulting from greed and selfishness rather than necessity. When we attend to biomedical uses of animals in experimentation, we confront a far more difficult problem. In the research area, the purpose of the enterprise is to ameliorate and eliminate human (and animal) suffering resulting from disease, be it environmental, microbial, or genetic. It is for this reason that the research community has long sought animal models for disease conditions. But this project generates a fundamental ethical dilemma that emerges from biomedical science, namely what entitles us to create disease, with attendant severe pain, suffering, and distress in primordially innocent animals? Even if the research is aimed at benefiting other animals, it is difficult rationalizing causing misery in Peter in order to cure Paul.

Historically, the response to this ethical problem is to ignore it, or minimize its significance by stressing the absolute and unequivocal value and meaning of human life over that of animals, generally by invoking theological justifications. But recent cultural history and moral thinking has cast a shadow of doubt upon this facile justification. Fifty years of the development of animal ethics has provided good reason to doubt human superiority as a trump card. In addition, this ethical reasoning has also undercut the convenient ideological maxims employed as a justification by scientists, namely that "science is value-free in general, and ethics-free in particular," and that science must be agnostic regarding the presence of consciousness in animals, including pain and suffering (Rollin 2006b). In the face of recent progress by both ethicists and scientists in explicating animal mind, thoughts, and feelings, as well as exponential growth in societal ethical concerns about animal treatment, the denial of subjective experience in animals can no longer be utilized as a justification for hurting them.

The response to these recent realizations has taken various forms. For one thing, the search for what Russell and Burch postulated as *alternatives to live animals,* Replacement, Reduction, and Refinement, has developed considerably but as yet remains very limited as regards eliminating animal use by substituting non-animal use. Most plausible and practicable have been *Refinements,* aimed at minimizing pain and suffering attendant upon animal use in research. These refinements have been driven by regulatory change, such as the mandates to control animal pain, embodied in the USA, the UK, EU regulations that have developed since the 1980s. In particular, one can cite proliferation of analgesic use in science. In 1982, a literature search I undertook for the U.S. Congress revealed only two papers covering analgesia in the scientific literature. Two years ago, the same search yielded almost 12,000 papers. But let us recall a number of caveats about analgesia. Most notably, analgesia does not *eliminate* pain. What it does do is raise the *pain tolerance threshold,* which is significantly different from eliminating pain altogether. While this is certainly a step toward mitigating the dilemma of invasive animal use, it is far from a complete answer.

At first blush, genetic engineering compounds the problem of invasive animal use entailed by animal research. This is because transgenic technology provides us with the potential for creating "animal models" of vast number of diseases and conditions hitherto not researchable on animals, in particular, genetic diseases. I am referring to the creation and maintenance of seriously genetically defective animals developed and propagated to model some human genetic disease. This was traditionally accomplished through identification of adventitious mutations and selective breeding. Transgenic technology allows for accomplishing the same goal far more quickly, and in a far wider range of areas. Thus, one can, in principle, essentially replicate any human genetic disease in animals. And therein lies the major ethical concern growing out of transgenic technology in the research area. It is a true dilemma, because there are strong moral pulls on both sides of the issue.

A chapter in a book devoted to transgenic animals helps to focus the concern:

> There are over 3,000 known genetic diseases. The medical costs as well as the social and emotional costs of genetic disease are enormous. Monogenic diseases account for 10 % of all admissions to pediatric hospitals in North America....and 8.5 % of all pediatric deaths...They affect 1 % of all live born infants...and they cause 7 % of stillbirths and neonatal deaths...Those survivors with genetic diseases frequently have significant physical, developmental or social impairment....At present, medical intervention provides complete relief in only about 12 % of Mendelian single-gene diseases; in nearly half of all cases, attempts at therapy provide no help at all (Karson 1991).

This is the context in which one needs to think about the animal welfare issues growing out of a dilemma associated with transgenic animals used in biomedical research. On the one hand, it is clear that researchers will embrace the creation of animal models of human genetic disease as soon as it is technically feasible to do so. Such models, which introduce the defective human genetic machinery into the animal genome, appear to researchers to provide convenient, inexpensive, and most important, high fidelity models for the study of the gruesome panoply of human genetic diseases outlined in the over 3,000 pages of text comprising the sixth edition of the standard work on genetic disease, *The Metabolic Basis of Inherited Disease*. Such "high fidelity models" may occasionally reduce the numbers of animals used in research, a major consideration for animal welfare, but are more likely to increase the numbers as more researchers engage in hitherto impossible animal research. On the other hand, the creation of such animals can generate inestimable amounts of pain and suffering for these animals, since genetic diseases, as mentioned above, often involve symptoms of great severity. The obvious question then becomes the following: Given that such animals will surely be developed wherever possible for the full range of human genetic disease, how can one assure that vast numbers of these animals do not live lives of constant pain, suffering, and distress? Further, given the emerging ethic we outlined above, control of pain and suffering is a *sine qua non* for continued social acceptance of animal research.

In today's moral ethos, it is simply not the case that any possible human benefits will outweigh any amount of animal suffering. If a genetic disease is rare, affects only small number of people, and can be prevented by genetic screening and what Kelley and Wyngaarden call in reference to Lesch–Nyhan's Syndrome "therapeutic abortion," (Kelley and Wyngaarden) it is not clear that society will accept the long term suffering of vast numbers of animals as a price for research on the disease. More and more, a cost-benefit mind-set is emerging vis à vis animal use in science just as it is legally mandated for research on humans—though it is by no means clear how one rationally weighs animal cost against human benefit!

In order to flesh out our discussion with a real example, let us examine the very first attempt to produce an animal "model" for human genetic disease by transgenic means, i.e., the development by embryonic stem cell technology of a mouse which was to replicate Lesch–Nyhan's disease, or hypoxanthine-guanine phosphororibosyltransferase (HRPT) deficiency. Lesch–Nyhan's disease is a particularly horrible genetic disease leading to a "devastating and untreatable neurologic and behavioral disorder." Patients rarely live beyond their third decade, and suffer from spasticity, mental retardation, and choreoathetosis. The most unforgettable and striking aspect of the disease, however, is an irresistible compulsion to self-mutilate, usually manifesting itself as biting fingers and lips. The following clinical description conveys the terrible nature of the disease:

> The most striking neurological feature of the Lesch–Nyhan syndrome is compulsive self-destructive behavior. Between 2 and 16 years of age, affected children begin to bite their fingers, lips and buccal mucosa. This compulsion for self-mutilation becomes so extreme that it may be necessary to keep the elbows in extension with splints, or to wrap the hand with gauze or restrain them in some other manner. In several patients mutilation of lips could only be controlled by extraction of teeth.
>
> The compulsive urge to inflict painful wounds appears to grip the patient irresistibly. Often he will be content until one begins to remove an arm splint. At this point, a communicative patient will plead that the restraints be left alone. If one continues in freeing the arm, the patient will become extremely agitated and upset. Finally, when completely unrestrained, he will begin to put the fingers into his mouth. An older patient will plead for help, and if one then takes hold of the arm that has previously been freed, the patient will show obvious relief. The apparent urge to bite fingers is often not symmetrical. In many patients, it is possible to leave one arm unrestrained without concern, even though freeing the other would result in an immediate attempt at self-mutilation.
>
> These patients also attempt to injure themselves in other ways, by hitting their heads against inanimate objects or by placing their extremities in dangerous places, such as between spokes of a wheelchair. If the hands are unrestrained, their mutilation becomes the patient's main concern, and effort to inflict injury in some other manner seems to be sublimated (Kelley and Wyngaarden 1983).

At present, "there is no effective therapy for the neurologic complications for the Lesch–Nyhan's syndrome". Thus, Kelley and Wyngaarden, in their chapter on HPRT deficiency diseases, boldly suggest as alluded to earlier, "the preferred form of therapy for complete HPRT deficiency [Lesch–Nyhan's syndrome] at the present time is prevention," i.e., "therapeutic abortion." This disease is so dramatic that I predicted in 1976 that it would probably be the first disease for which genetic researchers would attempt to create a model by genetic engineering. Researchers have, furthermore, sought animal models for this syndrome for decades and have in fact created rats and monkeys which will self-mutilate by administration of caffeine drugs. It is thus not surprising that it was the first disease genetically engineered by embryonic stem cell technology. But to the surprise of the researchers, these animals, although they lacked the HPRT enzyme, were phenotypically normal and displayed none of the metabolic or neurologic symptoms characteristic of the disease in humans. The reason for the failure of this transgenic "model" has been suggested to be the presence of a backup gene for xanthine metabolism in mice, though other research has cast doubt on this notion. Though an asymptomatic mouse is still a useful research animal, for example to begin to test gene therapy, clearly a symptomatic animal would, as a matter of logic, represent a higher fidelity model of human disease, assuming the relevant metabolic pathways have been replicated. Presumably too, it is simply a matter of time before researchers succeed in producing symptomatic animals—I have been told in confidence of one lab that seems to be close to doing so, albeit in a different species of animal. One may perhaps need to move up to monkeys to achieve replication of the behavioral aberrations.

The practical moral question which arises then is clear: Given that researchers will certainly generate such animals as quickly as they are able to do so, how can one assure that the animals live lives which are not characterized by the same pain and distress that they are created to model, especially since such animals will surely be used for long-term studies of the development of genetic diseases. Or should such animal creation be forbidden by legislation, the way we forbid multiple use of animals in unrelated surgical protocols in the U.S. or the British forbid learned helplessness studies?

There is, admittedly, no absolute or direct proof that U.S. society at least will reject the creation of such animals. The proof is indirect, based on George Gaskell's survey in Europe which morally rejected genetic engineering of animal models of disease (Gaskell 1997) and was also based on the incompatibility of creating such animals with the direction in which worldwide attitudes and laws regarding animal research are moving. At the very least, however, it would be prudentially unwise for the research community to forge ahead cavalierly with the creation of long term use of such animals. For, if U.S. attitudes are analogous to European ones, such proliferation of suffering animals could well evoke significant legislative restriction or even banning of any transgenic animal work, including the sort of work where lifelong suffering can be avoided by early endpoints, anesthesia, etc.

In a dialectical reversal worthy of Hegel or Marx, the very enterprise of genetic engineering that creates this difficult animal welfare problem, may contain within itself the seeds of the solution. Perhaps, one could, through the use of genetic engineering, create a nature for the genetically engineered animal model in which, similar to the case of the chicken discussed earlier, mentation in the Lesch–Nyhan animal could be eliminated to the extent that whatever symptoms are created in the defective animal, these symptoms no longer *matter* to the animal. In other words, however horrible the symptoms may be, they do not enter into the consciousness of the animals so designed. What would occur in the animal *telos* is rendering the animal functionally nonconscious. Since we are creating an animal where suffering is inevitable, removing consciousness, and thus removing the ability to suffer from its *telos*, does not violate our *principle of conservation of welfare*, since the Lesch–Nyhan animal has in essence been created to suffer, and removing that capability creates an animal with a *telos* meeting the human purpose for that animal devoid of the level of consciousness that makes its creation morally problematic. Even if, for all intents and purposes, this animal's *telos* renders it incapable of awareness, it is still constructed so as to enjoy a better life than it would if its experiences mattered to it in a significantly negative way.

Everything we have argued, of course, depends upon the assumption that the scientific community *will* create such defective and suffering animals in order to study human disease. But this seems to be a safe assumption, given the history of biomedical research. Here, of course, the motivation for creating such animals in the first place is far more laudable than the patent greed underlying the chicken example, in that what is primarily at stake is the elimination of human suffering, not additional profit. Clearly, if one is going to hurt animals for the benefit of humans, it is better that the ability to suffer on the part of those animals be ablated. Monstrous though this may appear at first blush, it still results in a better universe than if the animals *can suffer*.

The only question that remains is whether one could create such animals while they continue to serve as high fidelity models for the disease in question, since Lesch–Nyhans may indeed involve some element of consciousness in its very nature. On the other hand, there are a multiplicity of genetic diseases that do not involve consciousness whose study could benefit from the living but nonconscious animals we are postulating. Once again, the principle of conservation of welfare is not violated, as the *telos* of the animal rendered nonconscious should be compared to the conscious Lesch–Nyhans animal, whose life includes the defects generating suffering, not to a normal animal.

The modification of *telos*, by way of combining genetic engineering with behavioral neuroscience as a remedy for practices that cause pain or suffering by violation of *telos*, represents a whole new approach to intractable problems of animal welfare that emerge from contemporary animal use. Once the behavioral genetics of pain, fear, loneliness, distress, and other forms of suffering are understood, it is reasonable to expect that, in those cases where we insist on causing such suffering in pursuit of human benefit, one could transgenically remove either consciousness as a whole, or the ability to feel pain (such people are in fact born), or the particular mode of consciousness resulting from the animal need that is being violated, leaving the animal vegetatively alive but incapable of experiencing suffering as a result of the violation. It is an open question whether society will accept such radical changing of nature at human hands, as we discussed earlier, but it is also plain that society is likely to choose such a modality if the alternative is creating animals experiencing a lifetime of misery.

References

Gaskell G et al (1997) Europe ambivalent on biotechnology. Nature 387:845ff

Karson EM (1991) Principles of gene transfer and the treatment of disease. In: First N, Haseltine FP (eds) Transgenic animals. Butterworth-Heinemann, Boston

Kelley WN, Wyngaarden JB (1983) Clinical syndromes associated with hypoxanthine-guanine phosphororibosyltransferase deficiency. In: Stanbury JB et al (eds) The metabolic basis of inherited disease, 5th edn. McGraw-Hill, New York. Chapter 51

Rollin BE (1995) The Frankenstein syndrome. Cambridge University Press, New York

Rollin BE (2006a) Animal rights and human morality, 3rd edn. Prometheus Books, Buffalo

Rollin BE (2006b) Science and ethics. Cambridge University Press, New York
Rollin, B.E.(2011a) Animal rights as a mainstream phenomenon. Animals 1:102–115
Rollin BE (2011b) Putting the horse before descartes. Temple University Press, Philadelphia
Rollin BE (2013) Telos. In: Wathes C et al (eds) Veterinary and animal ethics. Wiley-Blackwell, Oxford
Singer P (1975) Animal liberation. New York Review of Books, New York

Would the Elimination of the Capacity to Suffer Solve Ethical Dilemmas in Experimental Animal Research?

Adam Shriver

Abstract The use of genetic engineering to enhance the welfare of laboratory animals can reduce the amount of suffering in current neuroscience research paradigms. In particular, for some forms of basic research, we can use welfare-enhanced animals to reduce harms to animals without sacrificing any of the scientific validity. In another group of experiments, we can use welfare-enhanced animals to dramatically reduce the number of unprotected animals enduring aversive procedures. Many of the objections to using welfare-enhanced animals for food production do not apply to their use in research, since genetic knockout techniques are already used routinely in research for human ends and since there is no risk for human health. Furthermore, examples of recent knockout experiments suggest that we already have, or are very close to having, the capacity to reduce suffering in laboratories via genetic engineering. If we are truly committed to balancing the advancement of science with the welfare of animals, this option should be further explored.

Keywords Genetic engineering · Suffering · Animal research · Animal welfare · Research ethics

Contents

A. Shriver (✉)
Department of Medical Ethics and Health Policy, University of Pennsylvania,
Philadelphia, PA, USA
e-mail: ashriver@wustl.edu

Curr Topics Behav Neurosci (2015) 19: 117–132

DOI: 10.1007/7854_2014_318

Published Online: 3 June 2014

1 Introduction

Invasive research on nonhuman animals has played an important role in the recent tremendous growth in our understanding of the nervous system. From Hodgkin and Huxley's early investigation of action potentials in squid giant axons to our modern understanding of neurotransmitters, neuroanatomy, neurology, and just about anything else with the prefix "neuro," most of our knowledge about how brains work can be traced, at least in part, to investigations involving other species.

Understanding of the brain is both one of the great frontiers of science and immensely valuable in practical terms. Phenomenal consciousness is currently one of the greatest puzzles, arguably *the* greatest puzzle, in the biological sciences, and it is unlikely we will be reach any firm conclusions about the nature of consciousness without a thorough understanding of how the brain works (see Braithewaite and Droege, this volume). On the more practical side, brain diseases like Alzheimer's and Parkinson's affect millions of people worldwide and cures would have the potential to greatly enhance the lives of many people. Moreover, the greatest feelings of elation as well as the most intense forms of suffering, as well as any experience in between, are all mediated by the operations of the brain. Thus it is no surprise that modern science has invested a huge amount of resources on investigation of the mind.

On the other hand, few who have looked into the issue would disagree that neuroscience research has been responsible for at least some suffering in the unwilling animals that have served as the subjects in tests. Animals are poked, prodded, shaken, spun, "sacrificed," dissected, and held captive as a means to human ends. Thus, the research has led to great benefits and, in some cases, great costs, with the costs born by nonhuman animals and the benefits accrued almost exclusively by humans. (But see Ohl et al., this volume).

The desire to advance modern medicine and our understanding of the mind comes into frequent conflict with the desire to not cause suffering, and we are left both with a constantly evolving set of regulations and with, at times, advocates at two sides of a debate that are deeply distrustful of one another. The current status quo is not satisfactory to many animal advocates, yet any changes that further restricted research would likely be vigorously opposed by the scientific community.

What if, however, we could eliminate many of the harms caused by the current system while preserving the momentum of the recent cognitive revolution? Wouldn't it be clear that we should take this opportunity?

I will argue that we already have, or are extremely close to having, the capacity to dramatically reduce the amount of suffering caused in biomedical research via genetic modification of the animals used in research. The significance of this possibility will vary according to different views; for those who think that the advancement of knowledge and human well-being trumps all concerns about animal welfare, there will be an opportunity to eliminate large amounts of animal suffering without any impairment in the forward march of science. For those who think that progress needs to be carefully balanced against any harms inflicted on animals, the implications for current practices will be even more dramatic. But, most importantly from my perspective, on almost any view that takes the elimination of unnecessary suffering to be a valuable end of moral conduct, there are options available that can reduce suffering with minimal or even, I would argue, nonexistent cost.

My plan in setting forth this case begins with providing an overview of different ways that animals can suffer in neuroscience research, in particular classifying the results based on the relationship between the capacity to suffer and the potential knowledge gained from the research. I will then provide a brief review of experiments that have *already* arguably reduced the capacity to suffer in rats and mice, the species that make up the overwhelming majority of research animals. Next I argue that, consistent with any view that considers nonhuman suffering morally significant, we should change from the current status quo to a system where welfare-enhanced (more on this terminology below) animals are used whenever doing so does not impede the likelihood of gaining knowledge. I will also argue that we can further reduce suffering by using welfare-enhanced animals in initial tests in order to reduce the number of nonenhanced animals who suffer. I finally examine how my view fares against various objections, starting first with criticisms that have been raised against using welfare enhanced animals in agriculture, and then moving on to consider other objections.

2 Suffering in Neuroscience Research on Animals

Before proceeding, I need to flag a couple of assumptions that are central to my arguments. First, I assume that all vertebrates are sentient. Though there are still some skeptics who argue that only humans have phenomenal consciousness, I think the best evidence to date suggests that vertebrates, and particularly mammals, are likely to be conscious (Report of the Committee on Recognition and Alleviation of Pain in Laboratory Animals 2009, see also Kaldewaij and Bovenkerk, this volume). I further assume that though it is conceptually possible for an organism to be sentient but not capable of suffering (as will be important later), vertebrates and especially mammals are also capable of suffering.

Suffering is an evocative term. Many researchers prefer to use the terms "distress" or even "pain" when referring to nonhuman animals to avoid the connotations of suffering and to leave open the possibility that there might be differences between human and nonhuman experience. However, I use the term to refer to unpleasant overall experiences, which should be unobjectionable to anyone who agrees that some nonhuman animals are sentient. We can remain neutral on the question of whether human suffering is more intense or more profound than that of other species while still agreeing that it is bad from the perspective of the sentient organism to be in a state of suffering as it is defined here.

I link suffering to "overall experience" because there may be individual aspects of experience that are unpleasant but which are outweighed by other features. For example, runners might feel some aches and pains while jogging which are trivial compared to the overall rush of endorphins. Or, as Rollin (2012) has pointed out, dogs may choose to undergo a mildly painful procedure because they are looking forward to the reward at the end. So when I use the term suffering, I don't intend to refer to mild or short-lived experiences that are part of overall positive experiences; rather, I mean experiences sufficiently long and intense to cause the overall experience to be unpleasant.

On this definition, many animals suffer in a wide variety of behavioral neuroscience experiments. Mammals have been used as animal models of acute pain, depression, anxiety, fear, nausea, disgust, startle, as well as a number of pathological conditions and models of neurological disease such as Alzheimer's disease, Parkinson's, chronic pain, and so on. In many experiments, the animals' condition is managed with anesthetics that render the animal unconscious or with analgesics that diminish the pain (at least during invasive procedures; Carbone 2011 has pointed out that there is often very little information in studies about how pain is managed post-surgery). In other cases, it is deemed necessary for the experiment that the animals do not receive pain management. Of course, it is worth noting that many neuroscience experiments do not involve any procedures that would necessarily cause any suffering.

Much of the harm to animals via the institution of animal research might come not from the experimental procedures themselves, but rather from aversive experiences caused by conditions the animals live in, their handling, their transportation, the technology used to measure brain activity, and any imperfect execution of the welfare policies put in place by various laboratories. Rollin (see this volume) has emphasized that the confined environments of laboratory animals does not allow them to exhibit a wide range of species typical behavior, or *telos* as he puts it. A related point can be made purely by focusing on experience; first, pleasure often exerts inhibitory influence on unpleasant feelings (Leknes and Tracey 2010), so an environment that provides very little opportunity for pleasurable experiences might amplify any potential negative feelings. And second, the inability to act on biological drives might itself be a negative experience. Thus, the mere existence of animals being used for research, even if the experimental design does not itself cause any discomfort, may count as harming the animals, if the unpleasantness of their conditions outweighs any positive aspects of well-being included in their life.

Putting aside the harms that may occur outside of the experimental context, it will be useful for my purposes to sort various neuroscience experiments into three categories. The first category of experiment is one where any suffering experienced by the animal is completely irrelevant for the scientific outcome that is being tested. For example, one set of experiments involves measuring the neural activity in rhesus monkeys as they are restrained and presented with visual information. Surgery is performed to implant a skullcap on the monkeys, which facilitates the direct measurement of individual neurons or groups of neurons. Animal activists and researchers strongly disagree about whether this set up is inhumane and causes aversive experiences for the animals. I have no intention of weighing in on this debate here; however, I think it is clear in this situation that suffering is in no way necessary for the results of the experiment. Assume that the experiments were performed on rhesus monkeys that lacked the ability to feel pain, anxiety, and fear. There is no (or rather very, very little) reason to think that an inability to suffer would in any way interfere with early visual neurons' response to their stimulation. Thus, the validity of these results would not be compromised by using animals incapable of experiencing those feelings but otherwise identical to the animals that were used. I will refer to these sorts of experiments as Aversion Independent.

A second category is experiments where a form of suffering or a behavior correlated with a form of suffering is a direct object of study. In these cases, the scientific validity of the experimental design depends on the capacity of the animals to suffer. Experiments investigating the neural underpinning of pain, depression, and learned helplessness are examples of this. In these cases, if the animal's capacity to suffer is eliminated, then the experiment will no longer be capable of obtaining the information it is intended to measure. Clearly, for example, it would not make much sense to study the efficacy of a proposed new analgesic on an animal incapable of feeling pain. I will refer to these types of experiments is Aversion Dependent.

And between these two categories exists another category where states associated with suffering are not the object of study, but nevertheless might be plausibly thought to indirectly contribute to the mechanisms or capacities under investigation. Consider research that investigates the effects of a certain drug on a neurodegenerative disease. While the direct object of study is not directly related to suffering, one might think that interfering with certain negative emotions could potentially cause downstream effects that influenced the strength of the model. For example, say that mild stress influences the body's reaction to the drug, or the immune system's response, or the rate of degeneration; in these cases, the elimination of suffering could impact the validity of the results, although the extent is not entirely clear. I will refer to this type of study as Aversion Related.

Thus we are left with three types of experiments: those where the direct object of study is a form of suffering or a mechanism associated with suffering, those where the object of study might be indirectly influenced by the capacity to suffer, and those where the object of study is entirely unrelated to the capacity to suffer.

3 The Elimination of Negative Experiences

Neuroscientists now have the ability to "knock out" certain capacities via the insertion of altered DNA sequences with inactivated genes into embryos that develop into adult animals. This technique is routinely used to study the development of certain capacities and the cellular building blocks that compose various mechanisms governing behavior. Though few researchers make claims as dramatic as saying they have eliminated an aspect of consciousness, I think it will be clear from the range of capacities that have already been eliminated under various circumstances that we are already close to having a tremendous opportunity to diminish the capacity for suffering in rats and mice, the animals that make up the vast majority of nonhuman research subjects.

As I have detailed elsewhere (Shriver 2006), there are at least two dissociable components of pain experience. The sensory dimension of pain includes representation of the location of the experience, the intensity of the experience, and the modality of the pain (whether it is a burning pain, a cutting pain, a pinching pain, etc.). The affective dimension of pain in humans involves an assessment of how unpleasant the pain is. Humans with various lesions to areas of the brain associated with the affective dimension of pain will report still feeling the pain but no longer finding it unpleasant. Likewise, many analgesics used to ease the suffering of people will not eliminate the pain entirely, but rather will reduce the unpleasantness of the pain.

Researchers using knockout technology have produced rats that still have the sensory features of pain but appear to lack the affective component, at least using the measure that several authors have claimed is the best model for affective pain (Shriver 2009). Thus, assuming the measures are accurately capturing the affective dimension of pain, the rats would be lacking the "unpleasantness" of pain experience while still showing normal withdrawal from the immediate stimulus. This is especially helpful since the animals would still presumably be able to protect themselves from acutely noxious stimuli via immediate reactions, but unable to experience the unpleasantness associated with pain.

Nor is pain unique as an experience that has been diminished in experimental animals. Haenisch and Bonisch (2011 GE2) reported that mice with noradrenaline knockouts showed behavior in the tail-suspension test and the forced swim test similar to mice given antidepressants. Wild-type mice (no knockout) often respond to stressors such as restraint and social defeats by staying immobile longer, showing less willingness to continue swimming when placed in a pool of water, and consuming less sucrose. However, the knockout mice performance on these tasks were not affected by these stressors.

In the case of anxiety, researchers used Fmr1 knockouts in mice to reduce anxiety behavior (Eadie et al. 2009). Mice with Fm1 knockouts travelled a significantly longer distance in an open field and defecated less than wild-type mice. The mice were also more willing to enter the open arms of an elevated plus maze than wild-type mice, which is also taken to be a sign of reduced anxiety. Finally,

Eadie et al. also found that the mice had lower levels of cortisol compared to controls after being restrained. Importantly, however, the authors suggest that Fmr1 knockout mice may have increased *social* anxiety compared to controls. I will return to this complication below.

This is not intended as a comprehensive review of all of the ways in which presumably negative states have been diminished in mice. These are just a tiny sample of many, many experiments searching for animal models of negative states in humans that have used knockout technology to alter the experience of animals. The interpretations of all of the results I mentioned are complicated and should not be read as straightforward evidence that a particular experience has been completely eliminated. However, I do think this sample shows that across a broad swath of behavioral indicators of negative experience, knockouts already exist in mice that can eliminate the behaviors.

Clearly, additional steps would need to be taken before carelessly assuming that these changes have eliminated aspects of conscious experience. Most researchers do not weigh in on whether conscious aversive states such as "depression" or "pain" are truly missing in knockout animals, and they often assume they do not need to do so because the behaviors can serve as models for human conditions even if the animals lack higher-order capacities. However, I think this perspective is a mistake, and one guided by a fear of taking a stance of the issue of animal consciousness. Pain researchers have recently been lamenting how poorly pain treatments that appeared to work in animal models have translate to humans. But, as many have pointed out, this is because much of the research has used spinally mediated withdrawal reflexes as a measure of pain without considering whether this behavior is actually dependent on conscious awareness (it isn't).

Thus, though I think additional research would need to be performed before we could have confidence that any particular knockout was truly eliminating a form of suffering, this research needs to be performed anyway to confirm the possible relevance for humans. Knowing whether a particular behavior is truly indicative of a form of suffering, or at least having a scientifically informed best guess as to whether this is the case, would not be a diversion from current research but rather is a necessary addition to current research if we want to truly understand how it relates to humans. So any additional testing needed to confirm that some particular form of suffering is knocked out by a procedure is research crucial for our understanding of the brain, and as such should not count as an additional "cost" of using welfare-enhanced animals.

I have been focusing on knocking out suffering as an example of welfare enhancement. But of course an additional way to improve the welfare of an animal over the course of its life would be to use genetic modifications to cause the animal to experience more and greater positive experiences. Since whether or not we have harmed an animal by bringing it into existence depends on the course of experience over the animal's life, enhancing positive emotions could also presumably go a long way toward lowering the "costs" of animal research.

In what follows, I will refer to both suffering diminishment and enjoyment enhancement as welfare enhancement, since I think both would improve the overall quality of laboratory animals' lives. It may sound strange to use the term "welfare-enhancement" to describe animals who lack the capacity to suffer. After all, we are subtracting, rather than adding, to the animals' capacities. Moreover, animals that lack a capacity to suffer would presumably fare worse than other members of their species were they to try to survive in a species-typical environment. Perhaps, one might argue, we should refer to animals with suffering knockouts as "diminished" rather than "enhanced."

However, I think the terms "enhancement" or "diminishment" can be used in this debate only in relation to certain contexts. An anxiety-free mouse might be poorly suited for avoiding predators in the wild. However, they might fare better than mice with anxiety on health measures in laboratory settings where they have no risk of encountering predators but are handled frequently. As such, I think the term "enhancement" only is useful if it means "enhanced relative to a particular context," and when I use the term "welfare-enhanced" I don't mean to imply the animals are enhanced full stop or somehow superior to their wild-type relatives, but rather that they are enhanced specifically in their well-being.

4 The Argument(s) for Diminishing Animals in Experimental Models

With that background, I now turn to the arguments for using welfare enhancement. The least assailable arguments for using welfare-enhanced animals in research would show that there are positive benefits that result from it and no serious costs. It will not apply to all research designs, or maybe even most, since as I have previously stated the validity of certain experiments would be undermined if the animals in the experiments lacked the capacity for negative experiences. This is not to imply that we should always choose the advancement of science when it is in conflict with animal well-being. However, I think there is a significant portion of research where the validity of the tests would not be diminished at all by using welfare-enhanced animals.

I hope all readers will agree that we should avoid causing unnecessary suffering, especially when we can do so with minimal effort. Replacing fully sentient beings with welfare-enhanced animals in the Aversion Independent experiments would not interfere with the validity of results for these experiments nor cause any other problems that impair the benefits of the research. Furthermore, as I suggested above, the change can be facilitated without the use resources that could be put to better use. Therefore, continuing to use fully sentient animals in aversion-independent experiments causes unnecessary suffering and is therefore wrong. So for one subset of experiments, we clearly ought to use welfare-enhanced animals.

This argument, I believe, should have force for anyone who places any weight on animal suffering. Even if one was committed to the view that the benefits of scientific advancement are so important that they always trump the disvalue of animal suffering, he or she could accept the use of welfare-enhanced animals in these experiments.

On the other hand, what about the Aversion Related experiments where there is some chance that the diminishment would influence the results, although not in a direct fashion? Those who believe (as I do) that the benefits of research should be balanced against animal welfare might think that in certain cases it is acceptable to take a slightly increased risk of the model failing in exchange for clear improvement in the well-being of animals. Anyone truly committed to a balancing act, rather than simply thinking all research is justified if it advances human knowledge, should at least be open to this possibility.

But there is another option that again should appeal even to those who think human values trump all. In toxicity testing, often initial tests are performed on tissue rather than in live animals. Researchers generally still perform tests on live animals to ensure safety, but the number of animals needed is drastically reduced because of the initial tests. Similarly, in cases where there is some slight possibility that welfare-enhanced animals will not serve as effectively as models as nonenhanced animals, initial tests could be performed on welfare-enhanced animals, with follow up experiments on a much smaller number of nonenhanced animals used simply to ensure that there are no major differences. Thus, the use of welfare-enhanced animals would not eliminate suffering in these experiments, but potentially could greatly reduce it.

I will make one further point in support of these arguments. I admit that most people find the suggestion of welfare enhancement through genetic engineering to be a shocking and radical proposal. But consider that many of the effects of knockouts are similar, in a sense, to the effects of giving an animal an analgesic drug like morphine throughout the course of it's life (minus some of the side effects of the drug, of course). Given that we do not object to blocking pain or suffering in animals via drugs, we should not object to providing a more thorough form of relief via genetic modification.

5 Comparison with Using Diminished Articles in Agriculture

There has been some debate regarding the proposal that modern factory farms should rely on welfare-enhanced animals in order to reduce suffering (see Rollin, this volume, Shriver 2009). Given that there are more animals grown and killed for human food than in animal research by several orders of magnitude, and also considering that the protections for animals in laboratories is generally far stronger than those in agricultural settings, the potential reduction of suffering from diminishment in agriculture is far greater than that in research animals. Nevertheless, it is worth noting that there are many objections to welfare-enhanced animals in agricultural settings that do not seem to apply to their use in research.

The most obvious difference, of course, is that the animals in laboratories are not consumed by humans, so any safety or health concerns about eating genetically engineered animals will clearly not apply. Likewise, it seems unlikely that there's much risk of laboratory animals mingling with wild animals and causing environmental disruption. And using knockouts on livestock would require developing techniques for new species, whereas the examples I cited have already been shown to work on mice. But there are other important differences to note as well.

The central difference to a lot of practical objections is that in animal research, unlike in agriculture, genetic engineering is already used routinely. The research community and a public that has tacitly (though tentatively) accepted the research has shown that it believes that this is a tool that can be effectively used to study the brain. And if, as is often said, we are committed to a balance between welfare and advancement, then it follows that we should also be willing to use the same technology to improve the well-being of animals in the experiments. Arguments about whether genetic modification should take place at all are essentially irrelevant in the modern context; the only question is whether it will be used to eliminate suffering and enhance welfare in addition to studying animal models in hopes of learning about human cognition.

So consider arguments that genetic manipulation of animals is "playing God" or "unnatural." Fiester (2009) has shown that there are equally plausible interpretations of religious texts and "naturalness" that do not prohibit genetic modification. But even putting those concerns aside, clearly people advocating these arguments could consistently claim only that we ought to prohibit *all* genetic engineering on animals in research settings. These ideas do not provide any reason for thinking that genetic engineering for the advancement of human knowledge is permissible but genetic engineering for animal welfare is impermissible. Thus, these arguments cannot be used as a defense of the status quo.

There is a familiar slippery slope argument applied to the use of genetic engineering in animal husbandry which also does not seem to apply to research. If, for example, we start engineering animals to improve their welfare on the farm, what's to stop us from engineering absurdly large animals, or animals with environmentally friendly excrement, or animals that have exotic flavors? This is a real concern for animal agriculture, as genetic engineering is not yet prevalent in the field. However, if there is a slippery slope in the animal laboratory, we are already on it, and there is no reason to think that using engineering for welfare purposes would do anything to make future, more problematic changes, occur. If anything, taking steps specifically to improve welfare would set an important precedent of taking animal welfare especially seriously.

Thus the strongest objections to genetically engineering farm animals do not seem to apply to neuroscience research, where the practice is already widespread. There are, however, additional objections that need to be considered.

6 Additional Objections to Welfare-Enhancement

One common thought is that by removing the capacity to suffer, we might also render animals unable to experience positive emotions. This idea is common in folk psychology, expressed in ideas like "you need to experience the bad in order to appreciate the good." It also is implicit in a number of philosophical views that take enjoyment and suffering to be different ends of a single spectrum of experience. On this view, we might actually be harming animals by eliminating their capacity to suffer, because we would also be taking away their capacity for enjoyment.

However, though this is a popular form of new-age folk wisdom, I don't think there are good reasons to believe it. I have argued elsewhere the pleasure and pain are two separate neural systems that contribute to well-being in very different ways (Shriver 2014). There is no reason to expect that removing the capacity to feel pain would also impair the ability to feel pleasure, or vice versa. In fact, medical literature is full of examples of people with select deficits in one but not the other capacity. People with anhedonia can still suffer; those with congenital indifference to pain can still feel enjoyment. Thus, there is no reason to expect animals whose capacity to suffer has been diminished to be unable to have positive experiences. So I do not think we could plausibly be claimed to be harming them with the procedures I am suggesting.

Up until now, I have considered objections that would favor the current status quo over the change to use suffering-diminished animals in research. However, I will also consider objections to the project of genetic engineering in general. As already mentioned, I think Fiester (2008) has usefully dismissed the claims that genetic engineering is "unnatural" or "playing God." In that article, she also notes that the public's initial intuitive discomfort with genetic engineering, often referred to as the "yuck factor," should not count as an argument against engineering unless they are anchored by rational justification. However, Fiester does believe that the public's negative opinion of genetic modulation should be taken as a warning flag that requires further examination, and she criticizes consequentialist views for failing to capture the views expressed in public sentiment, in particular those views holding genetic manipulation to be wrong even if it does not lead to increased suffering.

Fiester develops her own account that she thinks captures public sentiment while also being grounded in legitimate concerns. She calls for a "Presumption of Restraint" that must be met in order to justify genetic alteration and requires that any proposed genetic alteration demonstrates:

(1) a pressing reason to take the dramatic step of genetically altering life;
(2) careful consideration of the potential consequences of the project, including the "big picture" concerns of how we are radically, possibly irreversibly, altering our world;

(3) a recognition that unbridled animal biotechnology could create a world we no longer recognize or want to live in, which means that our animal biotechnology projects must be carefully, reflectively chosen;
(4) a clear regard for the basic tenet of animal research, i.e., that the benefit must far outweigh the cost; and
(5) a strong resistance to debasement of sentient life (2008).

I will consider (5) below as I discuss considerations raised by Thompson, but regarding (1), insofar as we are committed to animal research but also recognize the need to prevent unnecessary suffering, I think there is clearly a pressing need to take a step to reduce the amount of suffering in the world. Though I believe most researchers care about the well-being of animals and are committed to minimizing suffering in the current system, modern regulations are not perfect and even if they were there would still be mistakes made in the transportation, housing, and care of animals as well as some suffering caused by the experimental procedures themselves. As such, we owe it to the animals involuntarily subjected to the experiments to minimize their suffering to the extent that we can. Thus, the procedure meets the first criteria. Likewise regarding (4), the emphasis on cost-benefit analysis, the procedure is nothing if not an attempt to reduce the moral costs of the animal research program.

(2) and (3) can be considered together as they express the need to be aware of possible "down the road" consequences of engaging in the alteration. Certainly, I agree that a world in which humans have no hesitation whatsoever to manipulate the genetic code of other organisms for entertainment or even on a whim would be very seriously flawed. However, the proposal for welfare enhancement is one that does not disregard the feelings of animals; rather, it emphasizes how important it is to protect them from harm. So while genetic engineering as a whole might be thought to lead to a seriously flawed world, the move to protect animals via welfare enhancement is a positive step that leads to greater, rather than less, consideration for sentient life.

Regarding Fiester's criticisms of consequentialist reasoning, I recognize that my proposal does not fit with the intuitions of many and that a strong "yuck" reaction is a likely result. However, many public reactions to relatively new ideas, or even some old ones, are not trustworthy. Thirty years ago, a majority of the public in the United States was opposed to gay marriage and presumably (sometimes explicitly) were guided by aversive emotions. Today, legalizing gay marriage is popular and growing in popularity as it should be. Thus, I agree with Fiester that intuitions can serve as warning flags, but in the case of genetic engineering I see no reason to think the intuitions being expressed are reliable.

Furthermore, as Fiester notes, people's resistance to genetic engineering decreases when clear benefits for humans from the procedure can be shown (for example, growing human organs in pigs that can be used for transplants). If it turned out that the public was comfortable with genetic engineering to grow kidneys for humans, but not with genetic engineering to grow kidneys for dogs, or to enhance animal welfare in other ways, could this be regarded as anything other than an expression of bare speciesism that should cause us to question the underlying intuitions?

One of the underlying challenges to the idea of genetic engineering, related to Fiester's (5) above, is the idea that we are harming the animals by changing their species-typical essence. On this view, it is a mistake to assume that subjective well-being is the only morally valuable trait in nonhuman animals. We can harm them by changing their constitution in a way that makes them less of what members of their species would typically be. Paul Thompson, citing Blandford and Fulponi 1999 and Musschenga 2002, describes the view as follows:

> In this view, animals should live a "natural life," so that the life typical of a species becomes a model having normative force. Animals whose lives deviate from that model to a significant degree can be said to have been deprived to a morally significant degree. The standard can be applied to the living conditions in which the animal lives, but it would also apply to animals that do not have species-typical traits. Blindness or deafness, for example, would be characterized as a deficit" (2010, p. 818).

Thompson (2010) does not claim this view for himself, and neither of the articles he cites explicitly claim that blindness or deafness is a deficit; nevertheless, I think this way of framing the discussion helps to reveal the problems with this account. Using blindness as an example is a reference to a debate that ensued after a poultry breeding operation produced blind hens. The interesting twist of the story, at least in theory (Sandoe et al. 2014 have questioned the empirical claims behind the idea), is that blind hens tend to peck each other much less and, as such, presumably live a life of "improved welfare" compared to species-typical hens. So "blinding" hens could surprisingly lead to improved well-being in hens as measured by their experiences,[1] yet most people seem to think there is something deeply troubling about it.

Thus, the claim that blindness is a deficit expressed by Thomson seems to be implicit in this argument from Webster (2011):

> Recognize the telos, i.e., the fundamental biological and psychological essence of any animal; in simple terms "the pigness of a pig." A pregnancy stall for sows that denies them the freedom to express normal behaviour is an insult to telos, even if we cannot produce evidence of physical or emotional stress. If you disagree with this concept (and many do), consider two more extreme possible manipulations of farm animals in the interests of more efficient production: breeding blind hens for battery cages, or genetically engineering pigs to knock out genes concerned with perception and cognitive awareness (in essence, to destroy sentience).

With this explicit example of blindness, it seems to me that this debate can learn much from the problematic history of bioethicist assumptions about disability. It was often assumed in the field that various disabilities would dramatically lower

[1] If the increased aggression is caused by stress, it's true that making the hens blind does not remove the stress, and hence shouldn't be thought to remove all of the welfare problems. However, insofar as increased aggression and pecking behavior causes *more* stress and suffering in these conditions, then eliminating it still amounts to an improvement in welfare.

the well-being of the disabled individuals. This assumption could be quantified by asking able-bodied individuals to decide whether they would prefer to, for example, live five more years in their current state, or live 20 years with a disability (be it blindness, deafness, being confined to a wheelchair, etc.).

The problem (or rather, one of the problems) with these assumptions is that able-bodied laypeople, philosophers and bioethicists turn out to be rather terrible at imagining what it would be like to be in many of the conditions. As it happens, blind people report being happy, quadrapalegics report being happy, and in general most people with disabilities are in fact fairly happy (Diener and Diener 1996). A contemporary example of this comes from locked-in syndrome, a condition where patients are unable to move except, in some cases for slight eye movements. Many people have assumed that being in such a state would be "worse than death." However, according to a recent study, 47 people with locked-in syndrome reported being happy, compared to only 18 who reported being unhappy (Bruno et al. 2011).

What are we to say about these cases? Do we really want to claim that, in spite of what the people themselves feel and report, their lives are actually morally *worse* than the lives of able-bodied humans? That though they report high levels of well-being, they are actually importantly less well-off than they believe? I think such conclusions would be ridiculous, and we have just as much reason to accept the self-assessments of those labeled disabled as we do the self-reports of able-bodied individuals.

But if we don't have good evidence that blindness, deafness, etc. make human lives worse, why should we believe it fundamentally makes the lives of hens worse? The answer, is seems to me, is that we shouldn't. The hen's lives are better from their point of view, which is the only point of view that matters for assessing their welfare. Stubbornly declaring that there needs to be more suffering in the world in order to better fit our aesthetic preference for a certain "species typical" traits is a mistake of the same type, though not the same magnitude, as making assumptions that disabilities necessarily decrease well-being.

7 Conclusion

Undoubtedly, great caution is needed to avoid the possibility of causing more suffering with these techniques. Some of our current animal models for negative states and psychiatric conditions seem primitive, at best. If we mistakenly assume that a behavioral test is a reliable indicator of a form of suffering and focus on knockouts that eliminate the behavior, we run the risk of mistakenly treating certain knockout animals as suffering-free. And the example above of a knockout of Fmr1 knockouts decreasing general anxiety but increasing social anxiety highlights the risk that genetic manipulations might have unexpected consequences for animals' welfare. Thus, the utmost care is needed to ensure that we are using the best tests possible to measure the states in question, and to ensure that we truly are eliminating suffering and producing a net benefit in welfare.

But the current reality is that researchers are already using knockout techniques to investigate the neural mechanisms of negative affective states on nonhuman animals in order to better understand these states in humans. The publications that result from this research generally remain officially agnostic as to whether suffering has truly been diminished in the animals themselves. Yet knowing whether suffering is truly eliminated is relevant both for our potential to apply the research to humans and for the opportunity to decrease the total "costs" of animal research borne by the animals. If we truly are committed to balancing the advancement of science with the concern for animals, then we have a duty to ensure that research investigating the mechanisms of negative states in humans is also put to use eliminating the suffering of animals in laboratories.

We have, or are extremely close to having, the capacity to dramatically reduce the amount of suffering in animal research. Many knockouts that could likely benefit welfare have already been performed in search of models of human conditions. Moreover, the ability to enhance animals' welfare can be done without creating additional risks and without harming the animals themselves. As such, I believe we have a moral obligation to use genetic modification to improve the welfare of nonhuman animals in research settings.

References

Blandford D, Fulponi L (1999) Emerging public concerns in agriculture: domestic policies and international trade com- mitments. Eur Rev Agric Econ 26:409–424. doi:10.1093/erae/26.3.409

Bruno MA, Bernheim JL, Ledoux D et al (2011) A survey on self-assessed well-being in a cohort of chronic locked-in syndrome patients: happy majority, miserable minority. BMJ Open 1:e000039

Carbone L (2011) Pain in laboratory animals: the ethical and regulatory imperatives. PLOS One 6(9). doi:10.1371/journal.pone.0021578

Eadie BD, Zhang WN, Boehme F, Gil-Mohapel J, Kainer L, Simpson JM, Christie BR (2009) Neurobiol Dis 36:361–73

Fiester A (2008) Justifying the principle of restraint in animal biotechnology. Am J Bioeth 8(6):36–44

Haenisch B, Bonisch H (2011) Depression and antidepressants: insights from knockout of dopamine, serotonin or noradrenaline re-uptake transporters. Pharmacol Ther 129:352–68

Leknes S, Tracey I (2010) Pain and pleasure: masters of mankind. In: Berridge K, Kringlebach M (eds) Pleasures of the brain. Oxford University Press, Oxford

Musschenga AW (2002) Naturalness: beyond animal welfare. J Agric Environ Ethics 15:171–186

Report of the Committee on Recognition and Alleviation of Pain in Laboratory Animals (2009) From the Institute for Laboratory Animal Research of the National Academy of Sciences

Rollin B (2012) The moral status of invasive animal research. Hastings Center special report on animal research ethics

Sandoe P, Hocking PM, Forkman B, Haldane K, Kristensen HH, Palmer C (2014) The Blind Hens' challenge. Environmental values. http://www.whpress.co.uk/EV/papers/Sandoe.pdf

Shriver A (2006) Minding mammals. Philos Psychol 19(4):433–42

Shriver A (2009) Knocking out pain in livestock: can technology succeed where morality has stalled? Neuroethics. doi:10.1007/s12152-009-9048-6

Shriver A (2014) The asymmetrical contributions of pleasure and pain to subjective well-being. Rev Philos Psychol. http://link.springer.com.proxy1.lib.uwo.ca/article/10.1007/s13164-013-0171-2

Thompson PB (2010) Why using genetics to address welfare may not be a good idea. Poult Sci 89:814–21

Webster J (2011) Husbandry and animal welfare. In: Webster J (ed) Management and welfare of farm animals: the UFAW farm handbook. Wiley-Blackwell, Chichester

Part II
Clinical Research

Ethical Issues in Behavioral Neuroscience

Ethics of Human Research in Behavioral Neuroscience: Overview of Section II

Grace Lee

Contents

This volume, Ethics in Behavioral Neuroscience, gathers fresh new perspectives on how the ethical and rational pursuit of knowledge informs the neurobiological approach to the study of behavior. The first section of the volume focuses on ethical challenges for experimental approaches in behavioral neuroscience research using nonhuman subjects. It represents the ethical challenges of experimental animal research on how the brain drives external behaviors as well as the internal processes underlying these behaviors, such as responses to stimuli from the environment, learning, memory, emotion, and perception. Despite the difficulties of directly translating results from experiments with animal models to the human condition, the knowledge gained from basic research provides deep insights into the processes underlying behavior. The chapters in the first section provide authoritative reviews of commonly used experimental approaches to study behavior, including the creation of behavioral deficits via genetic manipulation, selective breeding, pharmacologic interventions, or invasive surgical procedures. The chapters each further provide scholarly discussion of the ethical problems that arise from considerations associated with these experimental approaches.

As a segue to the first section of the volume, the second section of the volume brings together nine chapters from seven different countries and covers a wide range of neuroscience research in the area of human behavior. Cassaday starts this section with a discussion on important ethical issues related to inducing illness in experimental subjects to model neuronal disorders, and emphasizes the differences between neuroscience and other biomedical research. Christen and Müller present a framework for understanding the structure of moral agency, discuss how brain

G. Lee (✉)
National Core for Neuroethics, Department of Medicine, Division of Neurology, University of British Columbia, 2211 Wesbrook Mall, Koerner Pavilion S-124, Vancouver, BC V6T 2B5, Canada
e-mail: mail@drgracelee.ca

Curr Topics Behav Neurosci (2015) 19: 135–136
DOI: 10.1007/7854_2014_343

Published Online: 3 September 2014

lesions produce changes in moral behavior, and identify ethical challenges for investigating these shifting phenomena.

Two chapters focus on neuroimaging interventions that are currently being developed for use in health care. Volume editors Lee and Illes report findings from a qualitative study of the ethics of brain imaging and genetic testing for predicting and diagnosing mental illness in youth. We report that imaging and genetic testing may potentially provide clarity about mental illness and more accurate diagnoses. These benefits are balanced against the complexities of interpreting test results in the mental health context and the potential negative impact on a young person's self-esteem. Farisco, Laureys, and Evers review recent advancements in neuroimaging research to assess residual consciousness in patients with disorders of consciousness and reflect upon the ethical impact of these advances on informed consent and self-determination. Their chapter expands from prior work on the neuroscience of disorders of consciousness by offering neurophilosophical and neuroclinical perspectives of the possibilities and limits of neuroimaging in this domain.

Cabrera discusses how the ability for cognitive enhancement affects human values and uses the interplay between enhancing and valuing to argue for social responsibility around enhancement practices. Racine, Bell, and Zizzo discuss the ethical and clinical challenges of deep brain stimulation as an evolving technology for neurological and neuropsychiatric conditions. Together, these two chapters cover both ends of the spectrum in the conversation about the ethical use of brain technology in health and disease.

Altis, Elwood, and Olatunji review the empirically supported treatments for anxiety disorders under the category of exposure therapy, discuss related ethical concerns, and suggest strategies for how to minimize risk during exposure. Their suggestion that risk management improves patient outcomes during the course of exposure therapy is particularly salient in terms of ethical considerations such as anxiety symptom exacerbation, inadequate training of therapists, and the risk of physical harm.

Maney discusses current examples of publicly misrepresented findings from studies of sex differences, argues how such misrepresentation may lead to a crisis in public health, and offers recommendations to the research community for addressing this important problem. The arguments presented in this chapter remind researchers about how responsible science communication can have a positive impact on attitudes and actions in healthcare, education, and other aspects of society.

Eaton, Kwon, and Scott focus on the ethics of clinical trials, and they specifically examine the ethical and social effects that arise when biopharmaceutical companies prematurely end their clinical trials for financial reasons. They offer patient-centered recommendations that rest on corporate social responsibility and a collective research ethic.

Taken together, these original contributions highlight the need to deepen the ethical discourse as research in behavioral neuroscience continues. Pragmatic anticipation and examination of ethical issues are critical to assure the most beneficial translation of findings in behavioral neuroscience research for the promotion of public health.

What's Special about the Ethical Challenges of Studying Disorders with Altered Brain Activity?

Helen J. Cassaday

Abstract Where there is no viable alternative, studies of neuronal activity are conducted on animals. The use of animals, particularly for invasive studies of the brain, raises a number of ethical issues. Practical or normative ethics are enforced by legislation, in relation to the dominant welfare guidelines developed in the United Kingdom and elsewhere. Guidelines have typically been devised to cover all areas of biomedical research using animals in general, and thus lack any specific focus on neuroscience studies at the level of the ethics, although details of the specific welfare recommendations are different for invasive studies of the brain. Ethically, there is no necessary distinction between neuroscience and other biomedical research in that the brain is a final common path for suffering, irrespective of whether this involves any direct experience of pain. One exception arises in the case of in vitro studies, which are normally considered as an acceptable replacement for in vivo studies. However, to the extent sentience is possible, maintaining central nervous system tissue outside the body naturally raises ethical questions. Perhaps the most intractable challenge to the ethical use of animals in order to model neuronal disorder is presented by the logical impasse in the argument that the animal is similar enough to justify the validity of the experimental model, but sufficiently different in sentience and capacity for suffering, for the necessary experimental procedures to be permissible.

Keywords Reduction · Refinement · Replacement · Neuroscience · Cost–benefit analysis · Speciesism

H.J. Cassaday (✉)
School of Psychology, University of Nottingham, University Park, Nottingham NG7 2RD, UK
e-mail: helen.cassaday@nottingham.ac.uk

Curr Topics Behav Neurosci (2015) 19: 137–157
DOI: 10.1007/7854_2014_333

Published Online: 10 September 2014

Contents

Pre-clinical studies of the brain may be conducted on both animal subjects and human participants. Thus, neuroethics cover human neuroimaging and psychopharmacology, for example, as well as the direct study of human disorders with altered neuronal activity. Here, the focus will be on pre-clinical work of the kind that is argued to necessitate the use of animals.

The ethical challenges of experimentally inducing illness in a subject or experimental species for the benefit or potential benefit of the agent or experimenter species are many. For present purposes, I will focus on practical or normative ethics, as enforced by legislation, in relation to the guiding principles of reduction, refinement and replacement (the 3Rs; Russell and Burch 1959). These are applied to animal work in the United Kingdom, embedded as Article 4 in the new European Directive 210/63/EU (European Commission 2010) and promoted as a key concept in the US Guide for the Care and Use of Laboratory Animals (National Research Council 2011). The importance of evidence-based welfare follows from due consideration of species typical behaviour. Finally, returning to ethics in its broader sense, I will consider the perception that there is an ethical demand to ease human (and animal) suffering through scientific advance, which may only be possible through the use of animals. However, scientific advances may also be used to improve functions that are already in the normal psychological range, or to alleviate arguably self-inflicted conditions such as drug addiction. Contemporary views of the ethics of animal use in the neurosciences may take into account, for example perceptions of need for the treatment, as well as human culpability in relation to the development of mental illness.

1 Ethics and Legislation

The use of cannabis, even for medical reasons, is still illegal in many countries or states. In contrast, the general use of excess alcohol, at doses that result in a range of social and health costs, is legal in most countries. Specific actions with potentially fatal consequences such as driving when drunk are generally illegal, particularly where others may be harmed. In contrast, driving after a sleepless night might involve an equivalent risk of accident but drivers (and their employers in the case of shift workers) are much less likely to be prosecuted. In other words, appropriate ethical codes are not necessarily enforced by legislation and are subject to contextual factors. A full discussion of the general issue of the rights and wrongs of using animals—as companion animals, in food production, as well as in biomedical research—is beyond the scope of this current topic. Briefly, influential positions include the view that the use of animals amounts to 'speciesism', reflecting a discrimination similar to racism and nepotism (Ryder 1975), and that if animals are considered to have rights (Regan 1984), then actions such as killing animals for any purpose are intrinsically wrong. Alternatively, if science is to progress through the study of living organisms, then perhaps experiments on both humans and animals should be considered on an equivalent basis. The fact that sequences of the human genome have been found in other animals has been argued to lend support to the argument that to sacrifice the 'non-human' for the sake of the 'human' animal cannot be legitimate (Hoeyer and Koch 2006). The utilitarian position takes the consequences of progressing science through the use of animals (or not conducting these experiments) into account (Singer 1975).

With respect to utility, the distinction between pure and applied research will not be addressed. In any case, with increasing emphasis on translation to practical benefit through the consideration of impact, as required by many research funding bodies, much fundamental 'curiosity-driven' research in the life sciences may be viewed as pre-clinical in so far as its implications for future clinical benefits are in sight. Similarly, increased ethical regulation and legislation has an impact on the study of animal behaviour for its own sake, yet in the longer term, further developments will be essential both for animal welfare science and to further inform public debate as to the legitimacy of animal use in general (Dawkins 2006; Barnard 2007; Patterson-Kane et al. 2008).

The ethical codes applied to animal use are practical or normative in that all are enforced by legislation, with current European Union guidelines considered gold standard. The general area of biomedical ethics is of still broader scope, covering also non-neuroscience animal work to which the same considerations apply. Conversely, many of the ethical issues raised by work in the neurosciences are of course generic, applying to any in vivo research, rather than specific to in vivo studies of the effects of altered neural activity. Moreover, as the brain provides a final common path for the perception of suffering, distinctions based on how that suffering has been induced may not be pertinent to the outcome from the animal's point of view. In other words, the perception of suffering will be the same irrespective of how the underlying neural substrates have been activated, though the likely benefits of the research may well vary depending on the field of study. The challenges presented by the legislation applied to enforce appropriate ethical standards are in part technical, for example whether the anaesthetic regime is optimal for the species and procedure in use (Fornari et al. 2012; Ideland 2009). There are also practical challenges given that resources will be limited. For example, continuous out-of-hours monitoring on an individual animal basis might be desirable after some kinds of procedure, but even the best research facilities are unlikely to have the resources to provide a level of care beyond that routinely provided for sick humans. The ethical guidance provided by the 3Rs (Russell and Burch 1959) and their application to neuroscience research (Blakemore et al. 2012) will be considered in relation to the feasibility of using non-invasive techniques developed for use in human, either by way of replacement of animal work or as a refinement. As it is the ultimate goal of those ethically opposed to animal experimentation, the replacement of such use will be considered first.

1.1 Replacement

Replacement is the most challenging of the 3Rs as applied to neuroscience. Altered neuronal activity can be studied directly in human participants using the non-invasive techniques of the cognitive neurosciences, such as electroencephalography (EEG), which reveals patterns of association between the electrical activity of the brain and behavioural changes, and functional magnetic resonance imaging (fMRI), to measure brain activity in so far as this is reflected in blood flow. These approaches are for the most part correlational in that possible brain substrates, which are identified without any neural intervention, and the data recorded provide only indirect measures of neural activity and with limited spatial and temporal resolution (Logothetis 2008). Invasive experimental studies of the human brain are conducted using techniques that apply stimulation to the scalp rather than surgical intervention. Although the spatial resolution is limited, areas of the brain can be temporarily inactivated in normal participants by means of transcranial magnetic stimulation (TMS) or transcranial direct current stimulation (tDCS). Thus, TMS and tDCS can be used to model altered neuronal activity.

Over the last three decades, an explosion of work conducted in human participants claims to relate recorded neuronal activity to a bewildering variety of psychological processes. This work has even gone so far as to include ethical reasoning: the 'neuroscience of ethics' as distinct from the ethics of neuroscience (Funk and Gazzaniga 2009; Kahane et al. 2011). Beyond the localisation of specific or more likely non-specific psychological processes to specific brain regions or networks, it is not clear what such studies necessarily add to our theoretical understanding of psychology (Sarter et al. 1996; Coltheart 2006). However, the contribution of such methods to the field of neuroscience is more widely accepted. Moreover, in principle, disorders characterised by altered neuronal activity can be studied directly in clinical populations. However, such observations may be confounded by the use of medication and, whatever precautions are in place, in cases of psychological and psychiatric disorder, the ability to give informed consent may be compromised.

In the short term, the continued use of animal models has been argued to be essential to our understanding of the relationships between neuronal activity and behaviour, for example the mechanisms of learning and memory and their disorder (Blakemore et al. 2012). Only in animals and in vivo can we conduct direct manipulations of a brain system to test its role in psychological processes (in vitro tests cannot substitute for behavioural tests of psychological responses to drugs and lesions). This approach is complementary to those approaches that involve measuring neural changes in human subjects, but the animal work is necessary because the human evidence is largely correlational and therefore inconclusive on its own, for example if we study human subjects who take drugs, we cannot know whether the effects we observe are a consequence of the drug or of psychiatric illness. TMS and tDCS techniques are promising but unsuitable for deep brain structures. Compared to controlled intervention studies in animals—using techniques such as microdialysis and electrophysiology—fMRI has limited temporal and spatial resolution. Computer simulations cannot substitute for experiments until we have sufficient data to successfully model the real nervous system. Thus, for some purposes, it has been argued that the use of animals cannot be replaced.

Related to the principle of replacement, further justification of precisely which animal species has been selected for a programme of work is required. Neuroscientific studies in which the nervous system is directly manipulated typically use rats rather than mice or some other small mammal to make use of the huge body of evidence already collected on the rat (both behavioural and neuroanatomical). There are excellent stereotaxic atlases for rats and a wealth of behavioural studies provides a sound basis for the selection of experimental parameters. Rats are also a hardy species, well able to tolerate the mild food or water deprivation necessary to motivate responding in order to test the behavioural consequences of altered neuronal activity. Some behavioural tests of activity or exploration are unconditioned and require no motivation for their expression but learning can only be demonstrated by testing the effects of a conditioned cue on a motivated response.

Arguably, the mouse has yet to demonstrate the same level of behavioural sophistication as the rat, in part because many mouse strains are hyperactive and aggressive and therefore difficult to work with. For example, being much smaller than the rat, the mouse is less well able to tolerate the deprivation schedules that can be essential to motivate reliable response rates. However, excellent progress is nonetheless being made in adapting benchmark tests of learning for use in the mouse (Schmitt et al. 2003, 2004; Deacon 2006; Bonardi et al. 2010). Mice remain the species of choice for studies of the effects of genetic modifications and cognitive effects have been clearly demonstrated in relation to genotype (Schmitt et al. 2003, 2004). However, for studies that manipulate neural activity directly, the smaller brain of the mouse can make some brain lesions and injections harder to restrict to their intended locations than is the case in the rat. Overall rodent species give quite a good trade-off between complexity of brain (necessary to meet the scientific objectives) and the need to consider phylogenetic position. Although invertebrates may suffer more than is commonly believed (Sherwin 2001; Crook and Walters 2011), animals in 'higher' phylogenetic positions are generally considered to have an increased capacity for suffering. Such judgements in relation to level of species are reflected in the introduction of legal protection (UK Animals [Scientific Procedures] Act 1986; European Directive 2010/63/EU) at the level of more neurologically complex invertebrates such as the octopus, as well as in the special considerations that apply to mammals of the primate genus. Thus, the use of rodents can be viewed as a replacement for the use of primates.

In addition to the scientific limitations of in vitro studies of nervous function raised above, the demarcation between in vivo and in vitro is dubious in the case of brain tissue. Indeed, one early study reported the use of an isolated whole brain preparation in the rat, which on some criteria was still alive up to 5 h after removal from the rest of the animal: in addition to metabolic activity showing glucose utilisation, there was both spontaneous EEG activity and an EEG response to drug administration as well as to a loud sound (Andjus et al. 1967). More recently, an isolated guinea pig whole brain has been reported viable as a preparation for the study of the auditory system (Babalian et al. 1999) and to provide a useful in vitro model of cerebral ischaemia (Breschi et al. 2010). Again to the extent such an in vitro whole brain preparation shows viable physiological activity, conscious perception cannot be assumed to have been removed by decerebration. Logically, the use of smaller samples of brain tissue may present similar challenges. The olfactory-hippocampal circuit of the guinea pig has similarly been reported to be viable in vitro and over an even longer time frame, at least with respect to its electrophysiological properties (de Curtis et al. 1991). This preparation can be seen as a significant scientific advance on the use of traditional slice preparations to study smaller samples of brain tissue and has clearly had translational impact for our understanding of temporal lobe epilepsy (Paré et al. 1992). However, maintaining parts of a brain, such as emotional or pain centres, or even a collection of nerve cells from such a region in vitro clearly poses ethical challenges that are different from working with, for example, an isolated heart. Thus, in the case of nervous tissue, it should be emphasised that replacement by way of in vitro tests raises particular issues.

The use of immature forms of vertebrates can also be presented as replacement. However, particularly for studies of the nervous system, there is compelling evidence that age matters. Even adolescent organisms respond quite differently from those of adults, and this constrains interpretation of both in vitro tissue studies as well as in vivo studies of juvenile systems (McCutcheon and Marinelli 2009).

Finally, replacement is not a logical objective in areas of animal science, where the animals are the object of study rather than acting as a model for a human condition (Barnard 2007). In this sense, studies of animal behaviour, which may include investigation of its underlying neural substrates, should have special status.

1.2 Reduction

Rigorous peer review of applications for funding, as well as of articles submitted for publication, should ensure that animal studies are well designed and appropriately analysed statistically. However, reduction is not simply a matter of using fewer animals. Rather the objective is to use a sample appropriate to detect the effect size of interest, otherwise statistically small effects that are nonetheless of potential scientific importance will remain undetected. Potential clinical significance is also a consideration: a small improvement to a serious illness such as Alzheimer's disease, or a delay in the onset of symptoms could represent an important advance. With appropriate statistical advice, reduction within any particular experimental protocol is achievable and generally considered best practice. However, to achieve an overall reduction in the number of animals entering regulated procedures is more challenging because of rapid progress in the development of genetically modified mouse models. These are providing vital information with respect to both normal function such as learning and memory and disorders such as neurodegenerative diseases. A consequence of this success has been an increase in the number of laboratory animals used in neuroscience as well as other forms of biomedical research (Blakemore et al. 2012).

1.3 Refinement

General improvements to laboratory animals' conditions are discussed in Sect. 2 below. The most obvious refinement specific to studies of altered neuronal activity would be to adopt the cognitive neuroscience techniques used in human studies to make all studies of altered neuronal activity, including those conducted in animals, non-invasive. However, as discussed in Sect. 1.1 above, these techniques are insufficiently advanced to allow the replacement of animal experimental subjects with willing human participants. In common with all neuroscientific techniques, the presently available non-invasive methods to study brain function in animals also have technical limitations which restrict their usefulness, in animal studies in particular. One particularly important limiting factor is the level of spatial resolution, which can be achieved. Functional imaging techniques are insufficiently advanced to allow us to address the anatomical subdivisions of interest, for example the distinction between shell and core sub-regions of nucleus accumbens. This is because the resolution is too poor for deep structures, and resolution <1 mm would be required. Anatomically, it is possible to achieve resolution of the order of 1 mm with a standard scanner. However, for functional imaging, which is necessary to address functional questions, it is very difficult to get images with voxels this small. Moreover, the temporal resolution of fMRI is at best around 1 s, which is insufficiently precise to capture neuronal activity in relation to behavioural reaction times, which are of the order of milliseconds. Relatedly, the question as to what the activity measured in functional imaging studies reflects remains controversial because blood flow is an indirect measure of neural activity (Logothetis 2008).

Therefore, although the same non-invasive (EEG and fMRI) or less invasive (TMS and tDCS) techniques can in principle be applied in animals, there would be no particular advantage to this line of work for its own sake and some additional disadvantages. For example, animals typically have smaller brains and do not keep still without the use of anaesthetic or restraint. However, structural imaging in animals will allow for refinement in so far as it can be used to verify experimental lesion placements prior to assessment of the brain post-mortem. Additionally, pharmacological MRI can be combined with the administration of experimental drugs to animals to delineate their effects without the need for any stressful procedure beyond the administration of the drug itself and the anaesthetic or restraint required for the MRI.

Animal work to study altered brain activity typically involves the use of invasive surgical procedures, which cannot be used experimentally in humans, to allow examination of the effects of experimental manipulation of neuronal activity on behaviour. The adverse effects resulting from these procedures can be broadly categorised into unintended or incidental effects, as distinct from the intended experimental effects intrinsic to the changes in neuronal activity induced. The routine management of these adverse effects is described below.

1.3.1 Incidental Effects

Without proper precautions, rats could experience pain during or after the surgical procedures necessary to access the brain. This is avoided by authorising only trained and competent staff to administer the most suitable anaesthetic for the species in use, under veterinary guidance for current best practice. Analgesics are routinely administered to minimise post-operative discomfort. Long-lasting systemic analgesics administered pre-operatively are ideal, in that pain relief will be in place immediately after the anaesthetic wears off. As an additional precaution to ensure long-term pain relief, local anaesthetic may be applied peri-operatively to the region of the wound. Animals showing subsequent signs of pain or discomfort are given a follow-up treatment systemically and treated topically if the operation wound is scratched.

Post-operative experimental procedures commence only once animals have made a full recovery from surgery. Animals are typically checked at least daily by the experimenters and the technicians and at more frequent intervals when an animal is sick. Malaise is recognised as, for example lethargy, loss of appetite, or poor coat condition. As a last resort, animals showing recognised signs of illness or discomfort that do not respond to treatment may be humanely killed. In particular, any animals showing gross locomotor deficits or serious impairment of the special senses, or that show other symptoms that exceed the severity limit of the agreed programme of work, are put down immediately.

The majority of the invasive techniques used in the neurosciences are classed as moderate under the UK legislation as they require surgery with recovery. However, animals, usually rodents, generally recover rapidly from these surgeries and the established techniques used have no long-term impact on the health and welfare of the animals. The combination of surgery techniques with systemic or localised pharmacological manipulations is unlikely to impose any additional health risks, and in all cases, animals are fully recovered from surgery at the time of any drug administration. Even after an animal has made a full recovery from surgery, it might in consequence of that surgery show altered sensitivity to some other treatment.

For example, it might show a shifted-dose response to a drug treatment and the objective might be to determine whether lesion-induced deficits can be reversed with drug treatments. Interactive effects that result in suffering or malaise for the animal typically occur relatively rarely. Predicting when such interactive effects will occur remains challenging. However, in general, the successful management of unwanted side effect of experimental treatments, together with ongoing improvements to husbandry, is a matter of routine in institutions authorised to conduct experimental work with animals. Refinement is perhaps the most readily achievable principle of the 3Rs and at the same time improves the quality of the science.

1.3.2 Intended Effects

Some aspects of the adverse effects seen post-operatively are an inevitable consequence of the scientific objective, in the case of the current topic, to study altered neuronal activity. Behavioural changes seen post-operatively after brain surgeries can include hyperactivity and increased aggression. These changes are usually relatively innocuous (e.g. hyperactivity) and can be within the species typical range (e.g. slightly increased aggressive behaviours). Such non-specific changes typically subside as the animal recovers, and if not veterinary treatment may be indicated. Additionally, it may be necessary to cage separately any rats which show increased aggression post-operatively.

Hyperactivity or other alterations in typical behaviour can also be seen as a lasting effect of some experimental brain treatments. Some of these effects are functionally related to the psychological changes under experimental investigation, and in this case, the incidence should be high (approaching 100 %) because the changes induced specifically relate to the scientific objectives. These adverse effects present an ethical challenge: to the extent they are integral to the scientific programme (the defined purpose for which the legal authority to conduct the work has been granted), they are of necessity left untreated. Such an experimental programme must be legal, but nonetheless represents a significant challenge ethically. The successful simulation of distressing psychological, psychiatric or neurological disorders, such as anxiety, schizophrenia or Huntington's disease, requires sufficient comparability in the level of suffering induced, in order for the science to be valid.

1.4 Rules and Recommendations: The Need for Flexibility

There is a clear difference between a rule and a recommendation and applying the 3Rs as a routine prescription may not work as intended when a number of considerations need to be taken into account. Viable strategies for replacement are insufficient for reduction to meet this target, and the ethical gap may effectively set reduction against refinement (Olsson et al. 2011). In other words, reuse or continued use in order to achieve reduction results in more harm on fewer animals, rather than the alternative of less harm on more animals to achieve the same experimental objectives in a more refined way.

More specific challenges arise when one proposed refinement can be seen to work against another. For example, with respect to the outcome to be learned about, there may be grounds to motivate conditioning procedures using aversive (e.g. mild foot shock) rather than appetitive (e.g. food reward) stimuli. At first sight, the selection of an aversively motivated procedure might seem to represent an unnecessary increase in the overall severity of the procedure. However, such aversively motivated procedures typically use mild foot shocks, just sufficient to produce reliable associative learning and within just two conditioning trials (Nelson et al. 2011a, b). This rate of learning is much faster than the equivalent appetitively motivated procedures in which the outcome is food reward (Cassaday et al. 2008; Horsley et al. 2008). Thus, aversive procedures allow the refinement of studies that require the use of microinjection procedures (in order to examine the effect of localised drug administrations) because the number of injections that can be administered without causing local damage at the point of infusion is limited (Nelson et al. 2011a, b).

Similar considerations arise in that proposed refinements can work against reduction if important experimental baselines are shifted. For example, studies investigating the neural substrates of associative learning require that a behavioural response first be established (in order that changes in associative strength can be detected). Food-motivated responses such as lever pressing can provide suitable baseline responses but have the disadvantage that they take some time to establish. Associative learning has also been investigated using licking for water as the motivated response, and these variants have the advantage that the licking response is readily established. In principle, these procedures could be refined to exclude the requirement for water deprivation, by the use of sweetened milk or sucrose solution as a food reward. However, there can be barriers for making such a switch: most importantly, to introduce the use of high incentive rewards would increase the behavioural baseline response. The incentive value of rewards as demonstrated behaviourally is known to be significantly affected by quite minor changes to experimental procedure such as a change in the reinforcer in use (Randall et al. 2012). Behavioural analyses of reinforcement-value measure responding on schedules requiring animals to make progressively more and more responses (such as pressing a lever within a Skinner box) to secure the same level of food reward. This provides a measure of their level of motivation for different reinforcers, in other words, their reinforcing strength relative to other 'less rewarding' reinforcers. Systematic comparisons of responding for different reinforcers on progressive ratio schedules, controlling for calorific content, suggest that the level of sucrose determines the reinforcing properties of novel foods that contain a mix of nutrients and flavours (Naleid et al. 2008). Moreover, the neural activity underlying the processing of reinforcers can show differences in relation to the reinforcer in use. For example, antagonists at both dopamine D_1-like and D_2-like receptors reduce the incentive value of sucrose, whereas the incentive value of corn oil is more sensitive to blockade of D_2-like than D_1-like receptors (Olarte-Sánchez et al. 2013).

Thus, there is a particular issue with respect to shifts in the baseline behavioural response in studies, which directly or indirectly manipulate dopaminergic neuronal activity in a manner likely to result in changes in hedonic tone (Wise 2008). When tasks are adapted to run with different reinforcers, direct comparability between task variants is compromised and there may be a substantial body of work completed with the reinforcer originally adopted. Moreover, where the neuronal activity under study modulates incentive salience and this is not the objective of the study, any shift in the behavioural baseline response would be predicted to compromise identification of the associative learning effects of interest. Whilst the above examples were selected from behavioural neuroscience studies, of course similar considerations arise in other areas of biomedical research.

Particularly where recommendations may have an unforeseen impact on the quality of the scientific outcomes, a two-way dialogue is essential. For example, refinements such as 'environmental enrichment' might seem unlikely to affect experimental outcomes. However, depending on the nature of the study, statistical power may be affected (Baumans and Van Loo 2013). Statistical power could be improved to the extent variability is reduced in animals better accustomed to novelty and change but results might be more variable between laboratories if standardisation of more varied environments is harder to achieve. For example, depending on strain and previous housing conditions, increased cage size and other forms of enrichment can significantly increase aggression in some male mice, most likely because of increased territoriality (Barnard 2007). Increased aggression can be a particular problem in studies involving some neural manipulation but could equally adversely affect the outcome of other kinds of biomedical research.

Importantly, institutional ethical review procedures debate such issues. However, it must be acknowledged that the effectiveness of such committee ethics has been questioned on a number of grounds. The general barriers to the debate and implementation of best practice include lack of resources and administrative burden (Illes et al. 2010). Additionally, researchers actively engaged in animal research, and others who may be seen to have a vested interest in animal research, have been suggested to be over-represented on such committees in the USA (Hansen 2013). The proportion of lay members on the equivalent committees in the United Kingdom is comparable, but in Sweden, for example, animal ethics committees have a much higher proportion of laypersons, including animal rights activists (Ideland 2009). However, even with such wider representation, interview methods confirm that such committees remain focused on refinement and optimisation of experimental protocols rather than questioning whether the research should be done in the first place. Thus, the context of the committee meeting may be sufficient to constrain the scope of its effectiveness (Ideland 2009). Moreover, non-specialists are unlikely to have sufficient knowledge to predict the effects of proposed refinements, either on other aspects of refinement or on the experimental outcomes that relate to the objectives of the study. Thus, lack of representation by other neuroscientists with relevant expertise extending to the behavioural techniques in use, could be a particular issue with respect to the evaluation of experimental programmes to study altered neuronal activity.

2 Species Typical Behaviour and Evidence-Based Welfare

Species differences mean that welfare guidelines should be evidence-based rather than rely on anthropomorphism. Moreover, consideration of species typical behaviour is fundamental to the assessment of potential suffering or lasting harm, which may be inflicted in the course of neuroscientific studies of any particular species of laboratory animal.

Laboratory housing conditions are the most important non-specific factor, affecting the well-being of laboratory animals. In the past, caging for laboratory animals was primarily designed on the basis of practical requirements such as construction and maintenance costs, space limitations and convenience of use for the experimenter. These practical considerations are still important and budgets for upgrading facilities are a precious resource. Since animal welfare is a major driver for upgrading laboratory housing, it is vital to be clear about the costs and benefits of proposed innovations from the animals' point of view. For example, modern split-level cages allow greater opportunity for exploration and separate areas provide the opportunity for the animal to retreat to hiding places. Moreover, they are suitable for animals with brain implants such as indwelling cannulae.

Within these improved caged environments, further opportunities can be provided. Standard laboratory feeding regimes deny the animal the opportunity to forage which in a natural habitat would take a high proportion of their time. Additionally, the provision of ad libitum food results in shortened life span due to overfeeding and inactivity. Environmental refinement refers to modifications to the housing of laboratory animals intended to enhance welfare, for example by simulating natural foraging conditions as far as possible or through the provision of other stimuli appropriate to the animals' species-specific needs (Baumans and Van Loo 2013). Other species typical behaviours include nest building and a variety of opportunities for social contact. Nesting and chewing materials can be provided as part of the environmental refinement. The five freedoms, first established by the Brambell Committee as a set of guiding principles to promote the welfare of farm animals, are specifically framed in terms of the 'freedom adequately to react to' a variety of aversive situations including injury and stress, in addition to the freedom to display normal species-specific behavioural patterns. However, breeding is not desirable in standard experimental colonies. Similarly, aggressive encounters may be part of the animal's repertoire but cause problems in the laboratory environment because they inflate the severity banding. Yet adaptive cost is not necessarily tantamount to suffering in that defending a territory is a normal behaviour for many species and one that would ordinarily confer reproductive advantage (Barnard and Hurst 1996; Dawkins 2006; Ohl and Staay 2012).

Knowledge of an animal's natural habitat and behaviour provides an excellent starting point for laboratory animal husbandry. For example, species such as the African mole rat, which lives in dark burrows, should be provided with burrowing and foraging opportunities in the laboratory. Moreover, there is evidence to suggest that such environmental refinement may be an important determinant of their cognitive performance in experimental studies (du Toit et al. 2012). Conversely, exposure to novel stimulation of the wrong kind, particularly under brightly lit conditions, would most likely result in stress rather than 'enrichment' for such a subterranean species. However, in general, anthropomorphism provides an unreliable basis from which to gauge animal welfare and we lack insight into how the animal in question would normally wish to spend its time. Animals' choices may result in short-term discomfort yet make excellent functional sense in terms of 'adaptive self-expenditure' (Barnard 2007). Since the same refinements will not be appropriate for all species, it is essential that the effectiveness of environmental refinements be evaluated, for example through the use of preference tests and other behavioural and physiological parameters (Chmiel and Noonan 1996; Dawkins 2006; Fitchett et al. 2006; Patterson-Kane et al. 2008; Baumans and Van Loo 2013).

Neuroscience studies do not raise special challenges with respect to general refinements to standard animal husbandry practices within the laboratory environment. However, additional considerations do arise with respect to the deprivation schedules used to motivate some behavioural neuroscience studies of learning and memory. Such studies may, for example, rely on stable baseline response rates in order to assess the degree of learning to a conditioned stimulus. For example, conditioned suppression of drinking provides a reliable measure of conditioned fear: to the extent animals (typically rats or mice) are fearful of the conditioned stimulus, they should be hesitant to drink. The experimental induction of fear and thirst, compounded by the trade-off between emotion and motivation inherent to the use of conditioned suppression of drinking to measure learning and memory, can be seen to raise concerns from an anthropomorphic perspective.

The justification for refinement, however, depends on the evidence that the water deprivation schedule in use results in adverse effects. The weights of rats on water deprivation are closely monitored daily since restricted water access tends to reduce food intake and routine welfare checks include the examination of skin elasticity, to check for any signs of dehydration. Additionally, the evidence base includes a systematic study of the health effects of restricted access to water: schedules of deprivation typical of those used in conditioned suppression studies have been reported to have no adverse physiological effects on rats and, moreover, to be appropriate to the experimental objectives (Rowland 2007; Hughes et al. 1994). In the wild, rat species inhabit a wide range of environments including desert, and the deprivation schedules adopted in laboratories may represent little in the way of deviation from the species typical range of intake patterns. Similarly, there is no evidence that the foot shocks used in such conditioned suppression studies result in lasting trauma in that when tested, the animals do not show total suppression, either to the experimental context or the conditioning cue (Nelson et al. 2011a, b).

3 Ethical Demand to Ease Human and Animal Suffering

The legitimacy of essential medical research is widely accepted amongst the general public and also a dominant theme at ethical review committees (Ideland 2009) and amongst researchers who use animals (Hobson-West 2012). Indeed, the ethical guidelines arising from the 1947 Nuremberg Code require that experiments should be based on the results of animal experiments, to minimise unnecessary human suffering. There was a historic context to this directive and contemporary views on the ethics of animal experimentation take into account (for example) perceptions of need for the treatment, as well as human culpability. For normal individuals, cognitive enhancers may be seen as inessential psychological cosmetics. Individuals who suffer addiction to drugs or who become obese could be argued to be less worthy of research effort necessitating the use of animals (see Sect. 4). Thus, the interpretation and implementation of the objective of the code—to minimise unnecessary human suffering—varies between counties, and for many disorders, there is no universally accepted animal model (Nature Neuroscience Editorial 2010).

Advances in veterinary science that alleviate animal suffering are also dependent on experimental studies of other (laboratory) animals. The animals that principally benefit are companion, farm and laboratory animals; thus, such advances can still be argued to be of benefit to the human owners, compounded by potential commercial gain in the case of farm and laboratory animals. However, curiosity-driven work in animal science is essential to an understanding of the normal behavioural repertoires, which should as far as possible be made available to any captive animal. This provides the evidence base for evolutionarily salient welfare (Barnard 2007; Ohl and Staay 2012).

Many scientists and lay persons would share the view that the capacity for feelings, both positive and negative, is of central concern (Balcombe 2009). That animals should have a comparable level of sentience is essential to the validity of models of psychological and psychiatric disorder. However, it is precisely this comparability, especially in respect of the capacity to suffer pain, which raises the issue as to whether animal experiments should be conducted in the first place. At the same time, points of difference in cognitive and other capacities can be argued to justify the demarcation of ethical responsibility in relation to species. For example, neuronal correlates of almost every imaginable facet of higher order processing are now being extensively studied in human participants, including ethical decision-making itself (Funk and Gazzagina 2009; Kahana et al. 2011). Cognitive processes unique to ethical decision-making are beyond the scope of animal models. However, non-human primates in particular show compelling behavioural evidence of a variety of cognitive capacities that provide rational justification for their continued protection (Mameli and Bortolotti 2006). At the same time, the use of pigs in neuroscience research has increased (Lind et al. 2007). In turn, the scientific advantage of the resemblance of the pig to the human brain raises ethical concerns. The use of pigs may be seen as ethically preferable to the use of primates but their use in neuroscientific studies is likely to remain less acceptable than the use of rodents. This use of 'sentientism' has been argued to be formally analogous to speciesism (Würbel 2009). Furthermore, the majority of judgements of sentience are clouded by prejudice based on species, for example pigs are widely perceived as intelligent emotional animals. Whilst a high proportion of individuals may empathise with pigs, for many empathy breaks down with 'pest animals' such as rodents (Würbel 2009).

Some of the same considerations apply to other areas of biomedical research, but the issue is particularly sensitive where sentience is the direct object of study as is the case in studies of altered neuronal activity. Moreover, particularly in the case of disorders that might have been avoided, cost–benefit analyses take human culpability into account.

4 Getting a Grip: Human Culpability for Behavioural Disorders

Animal work to test cosmetics for recreational use, as distinct from dermatological products for what might be seen as medical use, receives relatively little public support. Similarly, research to identify cognitive enhancers suitable for general use in normal individuals could be viewed as less ethically defensible than that directed towards identifying treatment for age-related cognitive decline. In extreme form, the former could amount to intellectual vanity. In contrast, the latter can manifest as severe dementia, resulting in significant human suffering and economic cost. However, such a distinction is blurred in that many of the new treatments for neurological diseases are also likely to have uses for people without disease, to the extent they can also improve normal brain function via their effects on cognition or affect (Chatterjee 2004). In practice, controlling the use of drugs (with or without prescription) is difficult. Prozac, whether obtained under prescription or purchased online, is already widely used in cases of mild depression and to some extent in individuals unlikely to meet contemporary diagnostic criteria.

Animal work intended to alleviate the consequences of 'self-inflicted' problems such as those related to alcohol consumption and cigarette smoking is already falling into a similar category: this despite the increasing recognition of addiction as a disease process. Obesity is similarly a disorder with a recognised neuronal component that could to some extent be argued to be self-inflicted, thus raising additional questions as to the acceptability of animal models in obesity research. This widening concern with the use of animals for laboratory research, which aims to alleviate human suffering which could have been avoided through behavioural change, could be further extended to raise questions with respect to a range of stress-related psychological and psychiatric disorders (Lund et al. 2013). Arguably, human individuals should take some responsibility for their exposure and reactions to stressors. Similarly, in addition, to the direct risks associated with drug taking, from overdose to accidents in consequence of impaired judgement, drugs too can increase the risk of psychological and psychiatric disorders. For example, there is good evidence that cannabis use increases the risk of psychosis (Verdoux et al. 2003; Moore et al. 2007), there is some evidence that the use of MDMA ('Ecstasy') is a risk factor for depression (Parrott 2001) or at least acute mood swings (Baylen and Rosenberg 2006). In short, psychological and psychiatric disorders are commonly seen in relation to substance use and direction of causality can be extremely difficult to establish (Verdoux et al. 2003; Soar et al. 2006; Moore et al. 2007). Head injuries are preventable to the extent that they result from engaging in sport, riding a bicycle without a helmet, driving a car without due care and attention. Thus, a wide range of disorders based on altered brain activity have some lifestyle aspect. Accidents aside, given what we now know about the importance of the epigenetic processes that determine gene expression in relation to environmental exposures, it would be surprising if they did not. However, to dismiss sufferers of conditions to which their own behaviour could be seen to be a contributing factor would raise further questions about individual responsibility in relation to social factors such as economic deprivation and level of education, as well as early environmental effects (such as the pre-pregnancy body weight of the mother), which obviously could not be controlled at the level of the affected individual (Lund et al. 2013). Obesity in companion animals is also relatively commonplace. The same arguments can be seen to apply to the owners of obese companion animals: arguably, they should know better, but their capacity effectively to take responsibility for their animal's diet may again be affected by economic deprivation and level of education.

5 Conclusions

Pre-clinical studies involving animal use face many of the wider challenges of neuroethics: not all neuronally mediated treatments or improvements are necessarily ethical in the wider sense, particularly in cases when there is no underlying disease in need of treatment. Thus, one commonly raised issue is whether we necessarily want to advocate the use of drugs by way of 'cosmetic' cognitive enhancements that might—like any performance-enhancing drug—permit unfair advantage advantages in assessment situations (Farah 2012). Such challenges are compounded to the extent advances can be seen to derive from invasive animal work. Surgical interventions to the brains of animals allow the precise experimental manipulation of neuronal activity in order to establish its effects under controlled experimental conditions. This kind of work presents additional ethical considerations in that it involves direct manipulation of animals' emotional and cognitive systems. Direct experimental manipulation of the brain might seem more ethically dubious than invasive studies of other essential organs such as the heart. Certainly, human patients needing invasive medical procedures may be justified in having a greater fear of brain compared with open-heart surgery: the brain is more identifiable with the human sense of self than is the heart; assuming they survive, the side effects of brain surgery are more difficult to predict with any certainty. However, peripheral procedures can impact on the brain, for example if altered sensory experience or suffering result from the procedure. Pain and suffering are mediated by a network of brain areas, which thus provide a final common path for suffering arising in consequence of all aspects of animal usage, including neuroscientific studies, invasive biomedical research on other organ systems, as well as non-invasive work which may nonetheless result in suffering or distress. Yet pain is not a direct consequence of tissue damage in the brain in that there are no pain receptors in the brain itself. Therefore, the ethical guidelines to be followed are general rather than specific to the organ system or behaviour, which is the subject of study. The legislation surrounding all such work ensures that animals' experience of pain and suffering is the minimum necessary to achieve the scientific objectives and moreover limited in relation to the likely benefits of the programme of work. One important exception to the applicability of the 3Rs arises in the case of in vitro studies that are normally considered as an acceptable replacement to in vivo studies. However, to the extent sentience is possible, maintaining central nervous system tissue outside the body raises ethical questions.

The debate around the moral justification for the ethical norms in place is another matter. Indeed, recognising the difficulty inherent in identifying moral absolutes applicable under every conceivable circumstance, Aristotle's 'virtue ethics' focused on the character of the moral agent rather than the fundamental ethical principles underlying the available guidance. In particular, virtue ethics point to the extent to which the agent—in this case, the experimenter using animal subjects—can be seen to reflect morally on his or her actions.

Many of the key questions surrounding the ethics of research involving animals were raised in the comprehensive 2005 report published by the Nuffield Council on Bioethics. This document remains an excellent summary. From the researchers' perspective, the fundamental challenge is presented by the logical impasse in the argument that the animal is similar enough to justify the validity of the experimental model, but sufficiently different in sentience and capacity for suffering, for the necessary experimental procedures to be in principle permissible (their implementation being highly regulated). The evidence of continuity provided by functional genomics has been used to support the argument that research has undermined its own legitimising principle (Hoyer and Koch 2006).

Distinctions drawn on the basis of species have of course been central to some of the ethical arguments made *against* animal use, principally that such use amounts to speciesism, similar in connotation to racism (Ryder 1975). However, although the term speciesism was intended to highlight discrimination against animals in a negative way, some researchers do now nonetheless describe themselves as speciesist in Ryder's sense (Hobson-West 2012). Moreover, distinctions drawn on the basis of species can also be an inevitable part of the justification for such animal use, based on criteria that indicate level of sentience. Essentially, cost–benefit analyses seek to quantify the suffering experimentally inflicted on 'lower' animals and offset this against potential benefit for the human species. Thus, the legislation concerning animal experimentation could be described as inherently speciesist in that special protection is afforded to primates and all but one of the invertebrates are excluded. More generally, the law could be said to be speciesist in that euthanasia is enforced for sick animals likely to be suffering in excess of what is considered acceptable. The regulatory frameworks require the use of a humane endpoint, whereas the very option of euthanasia of terminally ill humans is highly controversial. Indeed, speciesism could be said to be widespread in that, for example, the vast majority of individuals of both our own and other species only attempt to mate with members of their own species. As a species, we do not love other animals in the same way that we love other people. Any matings with a member of another species that do occur are by definition unsuccessful in a biological sense in which any viable offspring will not be fertile. Similarly, the conservation of endangered animal species attracts far more public attention than does the conservation of rare plant species. This wider consideration of what it might mean to be speciesist is not intended to trivialise the discussion: the acknowledgement of the role of speciesism seems essential to the logic of arguments for as well as against the use of animals in neuroscience. By definition, humanism is 'species-centric' to the extent its philosophies and morality are centred on human interests and needs. As an ethical stance, biocentrism that recognises the value of all non-human life in nature may very well be more ethically defensible. However, rightly or wrongly, the vast majority of human activity promotes human interests and needs. This is the context in which the ethics of animal use, for experimental neuroscience as well as for other human purposes, are situated.

Sentience is not a uniquely human attribute and sentientism or using the ability to feel and perceive as a criterion for the level of protection an animal should receive can also amount speciesism. With the exception of those presented by in vitro studies of altered neuronal activity, ethical challenges are not unique to the use of animals in neuroscience studies. Naturally, the ethical challenges of animal work are particularly emotive when sentience is the direct object of study, as is the case in studies of altered brain activity.

Acknowledgments Thanks to Pru Hobson-West, Tobias Bast, Gareth Hathway and Denis Schluppeck for comments and advice. The experimental work conducted in the author's laboratory was supported by the Wellcome Trust (ref. 082940) and the BBSRC (ref. BB/K004980/1).

References

Andjus RK, Suhara K, Sloviter HA (1967) An isolated, perfused rat brain preparation, its spontaneous and stimulated activity. J Appl Physiol 22:1033–1039

Babalian AL, Ryugo DK, Vischer MW et al (1999) Inhibitory synaptic interactions between coclear nuclei: evidence from an in vitro whole brain study. Neuroreport 10:1913–1917

Balcombe J (2009) Animal pleasure and its moral significance. Appl Anim Behav Sci 118:208–216

Barnard C (2007) Ethical regulation and animal science: why animal behaviour is special. Anim Behav 74:5–13

Barnard CJ, Hurst JL (1996) Welfare by design: the natural selection of welfare criteria. Anim Welf 5:405–433
Baumans V, Van Loo PLP (2013) How to improve housing conditions of laboratory animals: the possibilities of environmental refinement. Vet J 195:24–32
Baylen CA, Rosenberg H (2006) A review of the acute subjective effects of MDMA/ecstasy. Addiction 101:933–947
Blakemore C, MacArthur Clark J, Nevalainen T et al (2012) Implementing the 3Rs in neuroscience research: a reasoned approach. Neuron 75:948–950
Bonardi C, Bartle C, Bowles K et al (2010) Some appetitive procedures for examining associative learning in the mouse: Implications for psychopathology. Behav Brain Res 211:240–247
Breschi GL, Librizzi L, Pastori C et al (2010) Functional and structural correlates of magnetic resonance patterns in a new in vitro model of cerebral ischemia by transient occlusion of the medial cerebral artery. Neurobiol Dis 39:181–191
Cassaday HJ, Finger BC, Horsley RR (2008) Methylphenidate and nicotine focus responding to an informative discrete CS over successive sessions of appetitive conditioning. J Psychopharmacol 22:849–859
Chatterjee A (2004) Cosmetic neurology—the controversy over enhancing movement, mentation, and mood. Neurology 63:968–974
Chmiel DJ, Noonan M (1996) Preference of laboratory rats for potentially enriching stimulus objects. Lab Anim 30:97–101
Coltheart M (2006) Perhaps functional neuroimaging has not told us anything about the mind (so far). Cortex 42:422–427
de Curtis M, Paré D, Linás RR (1991) The electrophysiology of the olfactory-hippocampal circuit in the isolated and perfused adult mammalian brain in vitro. Hippocampus 1:341–354
Crook RJ, Walters ET (2011) Nociceptive behaviour and physiology of molluscs: animal welfare implications. ILAR J 52:185–195
Dawkins MS (2006) A user's guide to animal welfare science. Trends Ecol Evol 21:77–82
Deacon RMJ (2006) Appetitive position discrimination in the T-maze. Nat Protoc 1:13–15
du Toit L, Bennett NC, Nickless A et al (2012) Influence of spatial environment on maze learning in an African mole-rat. Anim Cogn 15:797–806
Editorial (2010) Ethical neuroscience. Nat Neurosci 13:141
Commission European (2010) Official J European Union L 276:33–79
Farah MJ (2012) Neuroethics: the ethical, legal, and societal impact of neuroscience. Annu Rev Psychol 63:571–591
Fitchett AE, Barnard CJ, Cassaday HJ (2006) There's no place like home: cage odours and place preference in subordinate CD-1 male mice. Physiol Behav 87:955–962
Fornari RV, Wichmann R, Altsak P et al (2012) Rodent stereotaxic surgery and animal welfare outcome improvements for behavioural neuroscience. J Vis Exp 59:e3528
Funk CM, Gazzaniga MS (2009) The functional brain architecture of human morality. Curr Opin Neurobiol 19:678–681
Hansen LA (2013) Institution animal care and use committees need greater ethical diversity. J Med Ethics 39:188–190
Hobson-West P (2012) Ethical boundary-work in the animal research laboratory. Sociology 46:649–663
Hoeyer K, Koch L (2006) The ethics of functional genomics: same, same, but different? Trends Biotechnol 24:387–389
Horsley RR, Moran PM, Cassaday HJ (2008) Appetitive overshadowing is disrupted by systemic amphetamine but not by electrolytic lesions to the nucleus accumbens shell. J Psychopharmacol 22:172–181
Hughes JE, Amyx H, Howard JL et al (1994) Health effects of water restriction to motivate lever-pressing in rats. Lab Anim Sci 44:135–140
Ideland M (2009) Different views on ethics: how animal ethics is situated in a committee culture. J Med Ethics 35:258–261

Illes J, Tairyan K, Federico CA et al (2010) Reducing barriers to ethics in neuroscience. Front Hum Neurosci 4:Article 167
Kahane G, Wiech K, Shackel N et al (2011) The neral basis of intuitive and counterintuitive moral judgment. Soc Cogn Affect Neurosci 7:393–402
Lind NM, Moustgaard A, Jelsing J et al (2007) The use of pigs in neuroscience: modelling brain disorders. Neurosci Biobehav Rev 31:728–751
Logothetis NK (2008) What we can do and what we cannot do with fMRI. Nature 453:869–878
Lund TB, Sorensen TI, Olsson AS et al (2013). Is it acceptable to use animals to model obese humans? A critical discussion of two arguments against the use of animals in obesity research. J Med Ethics. doi:10.1136/medethics-2011-100368
Mameli M, Bortolotti L (2006) Animal rights, animal minds, and human mindreading. J Med Ethics 32:84–89
McCutcheon JE, Marinelli M (2009) Technical spotlight: age matters. Eur J Neurosci 29:997–1014
Moore THM, Zammit S, Lingford-Hughes A et al (2007) Cannabis use and risk of psychotic or affective mental health outcomes: a systematic review. Lancet 370:319–328
Naleid AM, Grimm JW, Kessler DA et al (2008) Deconstructing the vanilla milkshake: the dominant effect of sucrose on self-administration of nutrient-flavor mixtures. Appetite 50:128–138
National Research Council (2011) Guide for the care and use of laboratory animals, 8th edn. National Academies Press, Washington DC
Nuffield Council on Bioethics (2005) The ethics of research involving animals. Nuffield Council on Bioethics, London
The Nuremberg Code (1947) In: Mitscherlich A, Mielke F Doctors of infamy: The story of the Nazi medical crimes. 2008 Kessinger Publishing, Whitefish
Nelson AJD, Thur KE, Marsden CA et al (2011a) Dopamine in nucleus accumbens: salience modulation in latent inhibition and overshadowing. J Psychopharmacol 25:1649–1660
Nelson AJD, Thur KE, Horsley RR et al (2011b) Reduced dopamine function within the medial shell of the nucleus accumbens enhances latent inhibition. Pharmacol Biochem Behav 98:1–7
Ohl F, van der Staay (2012) Animal welfare: at the interface between science and Society. Vet J 192:13–19
Olarte-Sánchez CM, Valencia-Torres L, Cassaday HJ et al (2013) Effects of SKF-83566 and haloperidol on performance on progressive-ratio schedules maintained by sucrose and corn oil reinforcement: quantitative analysis using a new model derived from the Mathematical Principles of Reinforcement (MPR). Psychopharmacology (Berl) 230:617–630
Olsson IAS, Franco NH, Weary DM et al (2011) The 3Rs principle—mind the ethical gap! ALTEX Proceedings, 1/12, Proceedings of WC8:333–336
Paré D, deCurtis M, Linás RR (1992) Role of the hippocampal-entorhinal loop in temporal lobe epilepsy: extra- and intracellular study in the isolated and guinea pig brain in vitro. J Neurosci 12:1867–1881
Parrott AC (2001) Human psychopharmacology of Ecstasy (MDMA): a review of 15 years of empirical research. Hum Psychopharmacol 16:557–577
Patterson-Kane EG, Pittman M, Pajor EA (2008) Operant animal welfare: productive approaches and persistent difficulties. Anim Welf 17:139–148
Randall PA, Pardo M, Nunes EJ et al (2012) Dopaminergic modulation of effort-related choice behavior as assessed by a progressive ratio chow feeding choice task: pharmacological studies and the role of individual differences. PLoS ONE 7:e47934
Regan T (1984) The case for animal rights. Routledge, London
Rowland NE (2007) Food or fluid restriction in common laboratory animals: balancing welfare considerations with scientific inquiry. Comp Med 57:149–160
Russell WMS, Burch RL (1959) The principles of humane experimental technique, 2nd edn, 1992 UFAW. Methuen, London
Ryder R (1975) Victims of science: the use of animals in research. Open Gate Press, London
Sarter M, Berntson GG, Cacioppo JT (1996) Brain imaging and cognitive neuroscience—toward strong inference in attributing function to structure. Am Psychol 51:13–21

Schmitt WB, Deacon RMJ, Seeburg PH et al (2003) A within-subjects, within-task demonstration of intact spatial reference memory with impaired spatial working memory in glutamate receptor-A-deficient mice. J Neurosci 23:3953–3958
Schmitt WB, Deacon RMJ, Reisel D et al (2004) Spatial reference memory in GluR-A-deficient mice using a novel hippocampal-dependent paddling pool escape task. Hippocampus 14:216–223
Sherwin CM (2001) Can invertebrates suffer? or, how robust is argument-by-analogy? Anim Welf 10:S103–S118
Singer P (1975) Animal liberation: a new ethics for our treatment of animals. Avon Books, New York
Soar K, Turner JJD, Parrott AC (2006) Problematic versus non-problematic ecstasy/MDMA use: the influence of drug usage patterns and pre-existing psychiatric factors. J Psychopharmacol 20:24–417
Verdoux H, Gindre C, Sorbara F et al (2003) Effects of cannabis and psychosis vulnerability in daily life: an experience sampling test study. Psychol Med 33:23–32
Wise RA (2008) Dopamine and reward: the anhedonia hypothesis 30 years on. Neurotox Res 14:169–183
Würbel H (2009) Ethology applied to animal ethics. Appl Anim Behav Sci 118:118–127

Effects of Brain Lesions on Moral Agency: Ethical Dilemmas in Investigating Moral Behavior

Markus Christen and Sabine Müller

Abstract Understanding how the "brain produces behavior" is a guiding idea in neuroscience. It is thus of no surprise that establishing an interrelation between brain pathology and antisocial behavior has a long history in brain research. However, interrelating the brain with moral agency—the ability to act in reference to right and wrong—is tricky with respect to therapy and rehabilitation of patients affected by brain lesions. In this contribution, we outline the complexity of the relationship between the brain and moral behavior, and we discuss ethical issues of the neuroscience of ethics and of its clinical consequences. First, we introduce a theory of moral agency and apply it to the issue of behavioral changes caused by brain lesions. Second, we present a typology of brain lesions both with respect to their cause, their temporal development, and the potential for neural plasticity allowing for rehabilitation. We exemplify this scheme with case studies and outline major knowledge gaps that are relevant for clinical practice. Third, we analyze ethical pitfalls when trying to understand the brain–morality relation. In this way, our contribution addresses both researchers in neuroscience of ethics and clinicians who treat patients affected by brain lesions to better understand the complex ethical questions, which are raised by research and therapy of brain lesion patients.

Keywords Brain injury · Brain lesion · Neurodegenerative diseases · Moral agency · Neuroscience of ethics · Neuroethics

Abbreviations

DBS	Deep brain stimulation
DLPFC	Dorsolateral prefrontal cortex

M. Christen (✉)
University of Zurich, University Research Priority Program Ethics, Zollikerstrasse 117, 8008 Zürich, Switzerland
e-mail: christen@ethik.uzh.ch

S. Müller
Charité—Universitätsmedizin Berlin, Forschungsbereich Mind and Brain, Charitéplatz 1, 10117 Berlin, Germany
e-mail: mueller.sabine@charite.de

Curr Topics Behav Neurosci (2015) 19: 159–188
DOI: 10.1007/7854_2014_342

Published Online: 15 August 2014

FTD	Frontotemporal dementia
NMDA	N-Methyl-D-aspartate
PET	Positron emission tomography
PFC	Prefrontal cortex
SPECT	Single-photon emission computed tomography
ToM	Theory of mind
VMPFC	Ventromedial prefrontal cortex

Contents

1 Introduction

Case 1 The accident changed everything. One moment of inattention, and (Madison) fell down the scaffold, resulting in a severe head trauma. A complicated surgery and weeks of rehabilitation followed, until Madison could be discharged from the hospital. But Madison was not the same person any more. Soon, her marriage dissolved and she was unable to continue her work. Madison underwent several neuropsychological assessments to settle her health insurance claims. One time, she arrived in tears, because her grandfather just died. The neuropsychologist was very sorry and offered to cancel the meeting—but then Madison giggled, saying that this was just a joke. She sat down in her shorts, although it was winter. Her mood changed every minute—from exorbitant joy to deep sadness. She confabulated and was sometimes verbally aggressive. She insulted the neuropsychologists as one of the many incompetent physicians she had met so far, unable to help her; and the next minute, argued that everything was fine with her and that she needed no help. Somehow she managed her life—she initiated relationships, but the relationships never lasted long. Sometime later, the neuropsychologist tried to contact her again, but the trail grew cold. None of the social workers who tried to help her after her accident knew where she was. No officials had any clue whether she was still living in town.

Case 2 Doctor Tolliver was a popular pediatrician—until police caught him in the act of abusing a 7-year-old girl during a medical examination. The police knew about other reports on sexual abuse of girls aged 3–12 years by Tolliver, some of them were even filmed by Tolliver. The investigators also found child pornography on the desktop computer of Tolliver. During interrogation, Tolliver claimed that since a year ago, he sometimes had an unstoppable drive to touch girls during medical examinations. While in investigative custody, neurological problems emerged and a brain tumor was diagnosed. Tolliver was successfully operated on before his trial. During trial, the defense counsel argued that the tumor caused pedophilia and additionally deficits in impulse control and emotion recognition, which were responsible for Tolliver's abusive actions toward children. He outlined a temporal correlation between tumor growth and the documented incidences of child abuse. However, the judge did not agree with this argument, since the scientific literature did not demonstrate a sufficiently deterministic relation between this type and location of brain tumor and delinquent behavior. Furthermore, Tolliver demonstrated an excellent ability to plan and organize the abuse and performed well as a pediatrician. Tolliver was found guilty and sentenced to 8 years in prison.

Case 3 It was almost 10 years ago when Sten was diagnosed with Parkinson's disease. In the early phase, the symptoms were well controlled by medication—but the disease progressed and it became increasingly difficult to avoid dyskinesia phases. In line with disease progression, Sten became depressed and apathetic; and his wife took care of him. She reduced her employment substantially and finally became his nurse. One day, Sten's physician explained that he might be a candidate for deep brain stimulation (DBS) and that this therapy could help diminish the side effects of medication. An assessment—demonstrating that Sten did not have a history of psychiatric disorders besides his Parkinsonism-related depression—demonstrated that DBS indeed was a suitable therapy for him. Sten decided on this option, and the surgery went well. The result was amazing, in particular for Sten. He felt that he gained a new life—but his wife could barely recognize him. Now, Sten often wanted to go out without her and he came back late. When checking his credit card bill, Sten's wife realized that her husband frequently visited strip clubs. Confronted with this fact, Sten admitted that he regularly visited prostitutes; he felt he had to catch up with all the life he had missed in the preceding years. He also admitted that somehow the DBS device might influence his new desires—but he rejects any change to his stimulation settings.

These three cases—all of them inspired by real patients—outline the complex relationship between changes in the brain and aberrations in morally relevant behaviors. This complexity is present on both sides of the brain–behavior relationship. Behavioral changes may result from sudden injuries of the brain, slowly progressing brain diseases, or therapies intended to counteract brain disorders. While some behavioral changes are reversible by neurosurgery, medication, rehabilitation, natural healing processes, or adequate social surroundings, others are irreversible.

For some aberrant behaviors of brain lesion patients, it is difficult to evaluate the moral component of behavioral changes objectively, since no consensus exists about the morality of certain behaviors within a given society, and least of all between different cultures. Rather, moral evaluations of different behaviors depend on a given cultural context, differ between subcultures, and undergo transformation processes. By way of example, slavery is nearly undisputedly considered immoral today, whereas visiting prostitutes is discussed more controversially.

How to treat individuals living with damaged brains that influence their behavior poses ethical questions. Behavioral changes in individuals with a frontal lesion may be stressful for families and caregivers who live with them—and the social services and financial benefits in most modern societies may be inadequate for these patients.

This reminder of the complexity of the relationship between the brain and moral behavior is an important caveat against overly straightforward causal explanations of immoral behavior. Such a simplification is exemplified by the iconic figure of Phineas Gage—the railroad construction foreman who suffered in 1848 from one of the most prominent traumatic brain injuries in history. While using an iron-tamping rod to pack explosive powder into a hole, the powder detonated and the rod penetrated Gage's left cheek, tore through his brain, and exited his skull. Gage survived this accident, but became according to the popular narration, a different person. In the words of Dr. Edward H. Williams, the physician who treated Gage's injuries: "He is fitful, irreverent, indulging at times in the grossest profanity (which was not previously his custom), manifesting but little deference for his fellows, impatient of restraint or advice when it conflicts with his desires (…). His mind was radically changed, so decidedly that his friends and acquaintances said he was 'no longer Gage'" (Harlow 1868). Although most accounts of Gage's life after 1848 are strange mixtures of slight fact, considerable fancy and downright fabrication (Macmillan 2000), his case became a widely used example of how brain and moral behavior are related—that the dysfunction of some parts of the brain, namely the right orbitofrontal or ventromedial prefrontal cortex, inevitably leads to major aberrations in moral behavior. The case of Phineas Gage is frequently mentioned in the introduction of papers that discuss the relationship between brain and moral behavior.

Seen from a historical perspective, this relationship between brain lesions and (anti)social behavior is a recurrent topic in brain research. In 1888, Leonore Welt, the first woman in Switzerland who was allowed to study medicine at the University of Geneva, published what today would be called a review paper on character change after frontal lesions (Welt 1888). She discussed 11 cases—among them a case of her own clinical practice and the crowbar case referring to Phineas Gage—where frontal lesions were associated with negative changes in moral behavior. She also discussed 47 other cases, where such lesions did not have such effects. Certainly, degree and localization of these injuries were much harder to describe when neuroimaging was not yet available. Nevertheless, Welt urged for caution when deterministically associating brain lesions with character changes.

To what extent this diagnostic caution is present today can be questioned. Experimental and clinical studies demonstrating that focal lesions in the right frontal cortex lead to specific changes in moral and social behavior generate almost four times more citations when compared to studies that describe the complexity of behavioral changes and social adaptations after frontal lobe injuries (Christen and Regard 2012). This citation bias may indicate an ethical dilemma associated with the relationship between brain and morality, namely that research may promote a neurodeterministic view of moral agency that is not sufficiently supported by the current state of knowledge.

In outlining this ethical dilemma, we have to resolve difficult measurement problems when analyzing the possible effects of brain injuries on moral agency, which denote the ability of individuals to act in reference to right and wrong. On the side of behavior, standardized questionnaires and tests (e.g., Iowa Gambling task, moral dilemmas) are available, but they may not reflect sufficiently the behavior changes and their effects in real life. And on the side of the brain, although sophisticated imaging techniques are used today, it is still difficult to directly assess residual functionalities in the affected brain tissue, particularly in cases involving neurodegenerative diseases, and the potential of neuroplasticity.

We structure the investigation as follows:

1. We need a detailed understanding of moral agency. This involves both an empirical part—namely outlining mental competencies and the related physiological conditions—and a normative part. The latter is needed to evaluate the legitimacy of moral claims toward the behavior of others.
2. We need a typology of brain injuries and their known behavioral sequelae. A major issue is the variability of behavioral sequelae of brain lesions. This variability may be partly explained by difficulties of assessing the exact location of lesions and their effects on neural networks. Furthermore, the variability could be based on individual differences in neuronal plasticity and differences in rehabilitation measures as well as on differences in the pre-lesion personality and social relationships.
3. We have to keep in mind that the endeavor of relating brain and behavior itself has a history and may be driven by different agendas. Ethics research is not a purely objective or rational science. Ethical justifications appeal to intuitions that have both natural and cultural histories. Thus, the ethical framework used to investigate moral behavior itself needs to be reflected upon as well.

The structure of our contribution to this volume follows this basic outline: In Sect. 2, we introduce the notion of moral agency both with respect to its normative and empirical dimension. In Sect. 3, we provide a typology of effects of brain lesions on moral behavior. In Sect. 4, we discuss ethical pitfalls of relating brain lesions with moral behavior changes. Section 5 concludes our contribution with some preliminary thoughts on using knowledge on the relation between brain and morality to restore moral behavior that is compromised by a brain lesion.

2 Moral Agency and the Brain

Human beings possess the ability to act with reference to right and wrong, which is framed as moral agency. The structure of moral agency as we construe it here is threefold (Christen and Alfano 2013). First, moral agency requires a specified set of competencies that the agent must have. Second, it involves a normative reference frame to which the agent has at least partial access. Third, moral agency is always situated in a context that consists of other agents and physical boundary conditions that constrain behavior. Competencies, normative frame, and context thus form the structural components of moral agency. A particular empirical investigation of moral agency may refer to just one or two of these structural components or to the interaction of two or all three components.

2.1 Conceptual Issues of Moral Agency

A relevant problem refers to the prescription of agency. In theory, most would agree that behaviors caused by defined brain lesions or abnormalities are not under the control of the individual and thus cannot be called acts or considered as an expression of moral agency. But in practice, brain lesions or other abnormalities often do not prove a lack of understanding the wrongness of a certain act at the time of commission (Batts 2009). Neither does it prove an absent ability to act according to one's insight, which would justify a diminished or missing criminal responsibility in most European countries. Therefore, in most cases, it is not possible to draw a clear causal line from a brain lesion or other abnormality to a missing moral agency and thus to a missing moral and legal responsibility. Another controversial issue is to what extent patients with ventromedial frontal damage possess moral concepts or relevant moral beliefs (Cholbi 2006; Roskies 2006), a discussion that relates to the philosophical debate on internalism and externalism (for an overview see Björklund et al. 2012). These important issues refer to the broad discussion on free will and the determination of behavior by the brain. This problem can have practical consequences, as our second case with Tolliver outlines, where the counsel was unable to convince the judge that the defendant's brain tumor caused him to abuse children sexually. We will come back to this point in Sect. 4.

For the following explanations, we propose to relate moral agency to the fact that patterns of moral behavior are displayed by persons whose behavior is somehow regulated by a normative framework that includes an idea of good and bad. We use the term "patterns of moral behavior" rather than "moral actions" in this context because we do not want to restrict it to punctate actions. This is consistent with the clinical practice of therapy and rehabilitation that does not address specific acts but behavioral patterns and dispositions.

The first thing to specify is what the term "moral" should denote. A simple fact about morality is that people are disposed to react to issues according to what they consider right or wrong, good or bad. This implies the existence of some normative frame and its connection with the real world in the sense that it guides thought, feeling, deliberation, and behavior of people. Another basic fact is that morality is situated in a social world of actions, judgments, negotiations, and other kinds of expressions made by social beings. This social world is embedded in a history, and its evolution is driven by many different factors. This means that acts, norms, and virtues that we may call moral are subject to fuzziness in two respects: First, within a society, there are actions that are undisputedly either moral or immoral, whereas other actions are less clear in that respect. Second, across societies and during history, the moral condemnation of some behaviors seems to be stable, whereas others undergo remarkable changes. Thus, moral evaluations of given actions differ both with respect to inner-societal agreement and evaluation stability over time.

Figure 1 illustrates these two dimensions with exemplary cases, although the precise location of these acts in this scheme can be debated. Morally, condemning the murder of innocent people is relatively stable both within a society as well as during the course of time. Slavery was for a long time morally accepted within societies but lost acceptance in a relatively short time span and is now regarded as absolutely inacceptable (Appiah 2010). Abortion has a long history of moral disagreement, whereas each position is relatively stable in time. Finally, the degree of moral acceptance of homosexuality shifted several times in history and to date the inner-societal disagreement is still high in many countries.

When evaluating the changes in moral behavior of patients, both the evaluation instability of moral behaviors and the inner-societal disagreement about them have to be taken into account. For the following general discussion, we define morality very broadly as a set of norms, principles, values, and virtues that are governed by an orientation toward the good. As such, morality reflects respect and concern for oneself and for other entities (persons, animals, or environment) and is embedded in a justification structure. We are aware that understanding one's moral decision-making

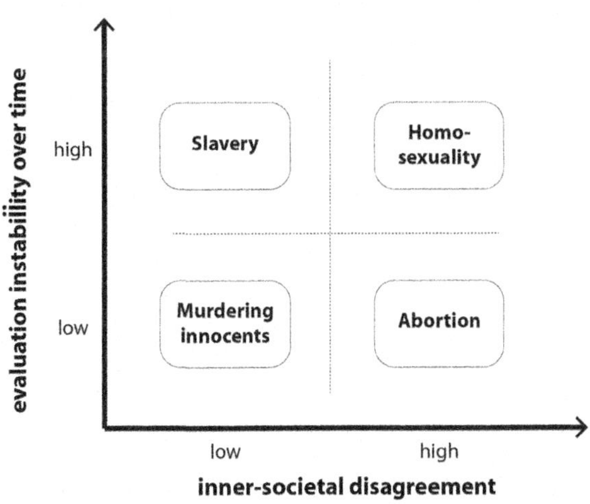

Fig. 1 Exemplary cases of moral valuation structured along the dimensions inner-societal disagreement and evaluation instability over time. The figure only identifies ideal cases in the four quadrants of the scheme, separated by a *dotted line*

and behavior requires an analysis of the agent's understanding of morality and on what he or she considers right or wrong. In addition, one would have to assess the actual justifications and their adequateness for an analysis of arguments.

2.2 *Moral Intelligence as a Psychological Working Model*

In our topic, a natural focus would be on the competencies and their foundation in the brain. This requires a theoretical framework that summarizes our knowledge on how agents reason, decide, and act morally. The major source of this knowledge is still moral psychology, which underwent a remarkable development in the last few years. In the following, we propose to use the concept of moral intelligence (Tanner and Christen 2013) as a theoretical framework.

Moral intelligence is defined as the capability to process moral information and to manage self-regulation in any way that desirable moral ends can be attained. It refers to the set of skills the moral agent needs in order to align her behavior with the moral ends she has set for herself, using the broad understanding of morality defined above. It is thus a skill-based conception of moral behavior, analogous to the concept of emotional intelligence that describes the ability to deal with emotions. The framework describes the sequential logic of moral behavior along with the associated underlying psychological processes, and the way in which implicit and explicit knowledge of morality and its justifications are included. These elements underlie the five competencies of moral intelligence (see also Fig. 2):

- *Moral compass*: This metaphor encompasses the set of moral schemata whose content is responsible for orienting the subject's behavior (Narvaez 2005). As such, it is concerned with mental representations of both declarative and procedural knowledge, each of which is accessible to the subject in varying degrees.
- *Moral commitment*: The ability to activate or sustain a motivation for the inclusion of moral considerations in the process of perception, decision-making, and action. In contrast to the typical process logic of moral behavior (perception → decision → motivation → action; Rest 1986), moral commitment is a

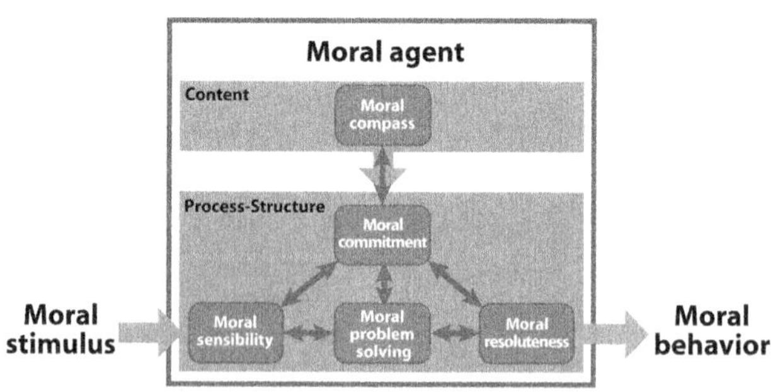

Fig. 2 The five building blocks representing competencies of moral intelligence in relation to the multistage model of moral functioning (adaptation from Tanner and Christen 2013)

capacity that influences all stages of the process, and in particular provides a motivational force to the semantic content of the moral compass.

- *Moral sensibility*: The ability to recognize morally salient aspects of a particular situation. The relevance of moral sensibility is obvious: If morally relevant aspects of a situation are not recognized, there is no cause to be concerned with the question of right action.
- *Moral problem solving*: The ability to bring the morally salient features of a situation to the decision-making process, and depending on the degree of conflict involved (e.g., if the problem has the structure of a dilemma), to arrive at a decision consistent with the subject's particular moral compass.
- *Moral resoluteness*: The ability to carry out one's own decisions despite, inter alia, external or internal resistance and barriers.

The concept of moral intelligence integrates the findings of (moral) psychological research into a unified model. As such, it enters an area with a rather long tradition. What distinguishes our model from other approaches is the central role of moral commitment, i.e., the capacity to uphold the demands of morality throughout this entire process and to align one's cognitions, decisions, and actions with one's moral ends. Moral commitment is to some extent the bridge between the moral compass and the other competencies of moral intelligence, and expresses the will to apply the contents of the moral compass.

It is unlikely that the building blocks of moral intelligence are related to distinct and clearly separable neuronal modules, because it is generally questionable whether mental processes can be defined and separated in a way that permits them to be associated with particular brain regions (Uttal 2001). The competencies also differ in their degree of exactness: Whereas moral sensibility is conceptually the simplest component, moral problem solving and moral resoluteness are more rich theoretical constructs. In addition, the psychological literature on the five competencies is not equally well developed. Moral problem solving or decision-making and, to a lesser degree, moral commitment within motivation psychology have been the subject of research for decades, in particular within developmental moral psychology advanced, among others, by Jean Piaget (1932) and Kohlberg (1981). Moral sensibility and moral resoluteness, however, are less well studied. Despite these difficulties, the framework of moral intelligence provides more precise considerations of which competencies may be affected by brain lesions, although one cannot expect that a specified lesion affects only one of those competencies, leaving the others intact.

2.3 Problems in Interrelating the Brain with Moral Agency

Our current knowledge on the neuroscience of ethics supports the expectation that there is no one-to-one correspondence between clearly discernible brain structures and functions on the one hand and the competencies that outline moral intelligence, or even moral agency in general, on the other hand. The so-called moral brain obviously consists of a large functional network including both cortical and subcortical anatomical structures (recent overviews: Mendez 2009; Fumagalli and Priori 2012; Pascual et al. 2013). Because moral agency is based on a complex process, these brain structures share their neural circuits with those controlling other mental processes, such as emotions, motivations, decision-making in general, impulse control, and theory of mind (ToM). A moral brain does not exist per se; rather, many subsystems of the emotional and the cognitive brain systems are engaged in moral processes. These complex processes are influenced by many genetic, endocrine, and environmental factors (Fumagalli and Priori 2012; Pascual et al. 2013).

Among the anatomical structures implicated in moral agency are the frontal, temporal, and cingulate cortices; i.e., considerable parts of the cortical hemispheres: The prefrontal cortex (PFC) regulates activity in subcortical emotional centers and is involved in planning and supervising moral decisions. When its functionality is disturbed, impulsive aggression becomes more probable (Fumagalli and Priori 2012). Patients with prefrontal lesions—especially in the orbito-prefrontal and medial regions—are often significantly impaired in both cognitive and affective empathy (Shamay-Tsoory et al. 2004; Eslinger et al. 2004). Patients with bilateral lesions of the orbitofrontal cortex show impairments in social behavior (Hornak et al. 2003). Furthermore, changes in moral decision-making have been found in prefrontal lesion patients (Koenigs et al. 2007; Ciaramelli et al. 2007). Based on such studies, it has been claimed that the ventromedial prefrontal cortex (VMPFC) attaches moral and emotional value to social events and anticipates their future outcomes. It is involved in ToM and empathy, mediates automatic moral and prosocial reactions, and participates in social emotions, including guilt, embarrassment, and compassion. The temporal lobes are also involved in ToM, and their dysfunction is often implicated in violent psychopathy. The dorsolateral prefrontal cortex (DLPFC) is involved in cognitive empathy and in the application of reasoned analysis to moral situations (Mendez 2009). The cingulate cortex mediates conflicts, particularly conflicts between emotional and rational components of moral reasoning (Fumagalli and Priori 2012). Current research suggests that the cortical structures most directly involved in abnormal moral behavior are the right medial orbitofrontal cortex and the right ventromedial prefrontal cortex (Fumagalli and Priori 2012). Subcortical structures are also involved in moral behavior, particularly the amygdalae, the hippocampus and the basal ganglia (Mendez 2009; Fumagalli and Priori 2012; Pascual et al. 2013).

Although the relationship between frontal lobe damage and morally relevant behavior aberrations like aggression has been part of clinical experience for more than 60 years, most of the evidence is case-based. There are only a few larger studies with appropriate design (Hawkins and Trobst 2000). The most known retrospective study examined the relationship between frontal lobe lesions and aggression in 279 veterans who had sustained penetrating head injuries, which were compared to 57 veterans without brain injury matched by age, education, and time served in Vietnam (Grafman et al. 1996). They found that the veterans with brain injury were more aggressive than control veterans, as reported by family and friends. In particular, veterans with ventromedial frontal lobe lesions were reported to be most aggressive, when compared to veterans with lesions elsewhere in the brain. But the authors also reported that "not all patients with these lesions had such behavior, and some patients with lesions elsewhere in the brain, and even normal controls, can show an increased tendency toward aggressive and violent behavior" (Grafman et al. 1996, p. 1237). There are also other case studies of patients with massive frontal lesions that are not compatible with a clear causal link between lesion and lasting behavior changes. For example, some patients do not show aberrant social behavior despite the lesions (Feinstein et al. 2010), in some patients the behavior changes after the lesions are reversible (Frías Ibáñez et al. 2008), and in some patients the behavioral and personality changes are compatible with stable functioning in family, professional, and social settings (Mataró et al. 2001). Therefore, the available evidence does not provide conclusive evidence that frontal lesions inevitably lead to such behavior changes. A too schematic, one-to-one connection between lesions in specific brain areas and specific moral behavior aberrations is misleading.

One reason for this variety found in the literature on the interrelation between the brain and moral behavior refers to the experiments that are used in these studies. Currently, a gross variety of tasks is used for assessing morality in the context of moral psychology or the neuroscience of ethics, which makes it difficult to compare the results of these studies. Furthermore, most moral tasks have intrinsic limitations. For example, they are not ecologically valid in that they reflect environmental and daily experience only poorly, or they request abstract judgments that exclude the complex decisional context. Additionally, task instructions usually forbid the subjects to make additional assumptions not included in the text, even though problem solving automatically intervenes in these situations. Finally, moral items distinctly differ from one another and involve different moral rules, violations, and values such as honesty, money, life, health, probity, or solidarity.

An important methodological limitation of most experimental studies is that they focus on moral judgments; i.e., the researchers account for, predict, or find neural correlates to moral judgments that they use in their experiments. These moral judgments are of a specific kind and have several defining features (Abend 2013). They are made in response to specific stimuli in imaginary situations, and they use only thin ethical concepts such as: okay, appropriate, permissible, acceptable, wrong, etc. In addition, they are fixed, verdict-like, and clear—not conceptually or semantically muddled, incoherent, etc. But moral judgments do not only occur as responses to specific stimuli or eliciting situations. Rather, some moral judgments develop over longer periods and are based on the reflection of many experiences and theoretical deliberation. In addition, morality cannot be reduced to moral judgments. This problem concerns in particular virtue ethics—an ethical approach that evaluates the character of persons in contrast to approaches that evaluate their actions, either in terms of duties or rules (deontology) or their consequences (consequentialism). Moral evaluations of actions are more easily expressed by moral judgments. Abend (2013) argues correctly that the object of study of much recent work on the connection between the brain and morality is not morality per se, but a particular kind of individual moral judgment.

And even within this special sample, complexity remains. Parkinson et al. (2011) investigated moral scenarios that involved disgusting, harmful, and dishonest behavior along with a neutral scenario, and asked subjects to judge the general moral wrongness of the actions within each scenario as well as the degree of disgust, harm or dishonesty while in a fMRI scanner. They found that the latter three statements were subserved by distinct neural systems and these differences were much more robust than differences in wrongness judgments within a moral area. The dorsomedial prefrontal cortex was the only region activated by all scenarios judged to be morally wrong in comparison with neutral scenarios. However, this region was also activated by dishonest and harmful scenarios judged not to be morally wrong. Furthermore, these scenarios were not suggestive of a domain-general role that is neither specific for nor predictive of moral decisions. The results suggest that moral judgment is not a wholly unified faculty in the human brain, but rather, instantiated in dissociable neural systems that are engaged differentially depending on the type of transgression being judged.

In summary, this brief overview suggests that the attempt to find clear-cut connections between a fine-grained understanding of moral agency and defined neuronal structures may lead to a picture that is too complex to be useful in a clinical context. For example, there may be different neuronal systems that are responsible for moral sensibility related to harm versus moral sensibility related to honesty. A focal lesion may thus impair one aspect of moral sensibility more than others—but the relevance of this imbalance will depend on the situation in which this competence is needed. In addition to impairing one aspect of moral sensibility, a focal lesion may also influence other competencies (maybe also nonmoral ones), as the affected brain region serves many basic functions. This complexity, however, is usually not assessed in experimental studies involving lesion patients because it is not feasible to perform a full evaluation of all possible impairments a brain lesion may cause.

The basic problem (see Fig. 3) is that on the side of the phenomenology of moral behavior, one needs a sufficiently elaborated but not too complex set of constructs that describe competencies relevant for moral behavior, such as the moral intelligence model. A rehabilitation specialist can neither work with a too general concept such as moral behavior, nor with a too fine-grained understanding of moral behavior such as impairment of honesty-related moral resoluteness. On the side of the investigation of the lesions, a similar problem emerges: One needs a partitioning that is compatible with the size of regions that can be affected through focal lesions —and with the methods available to actually identify regions and their degree of impairment. It is no coincidence that the current neuroscience of ethics denotes still rather large regions as being relevant for moral behavior, for example, the orbitofrontal cortex, which extends over several square centimeters. Taking all methodological issues of properly identifying such regions aside, it is clear that they are involved in many basic functional networks that may be clearly identified sometime

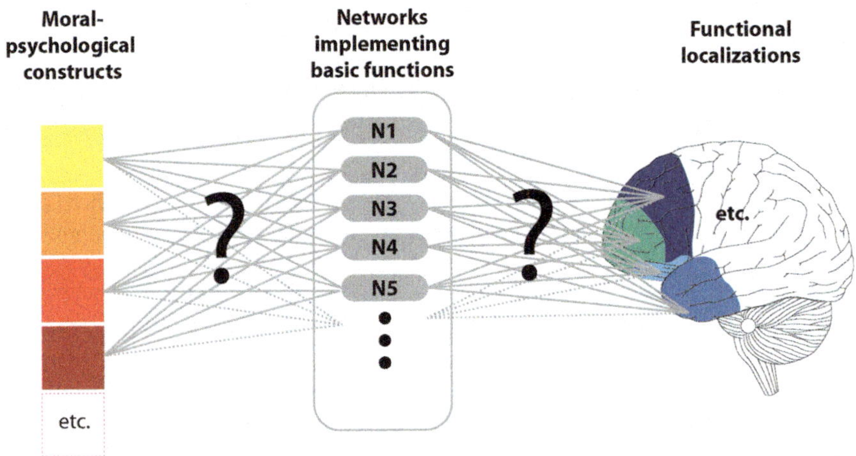

Fig. 3 Illustrating the problem of connecting moral–psychological constructs that describe relevant and usable moral competencies (*left side*) and functional localizations (*right side*) that actually involve many networks (N1, N2, N3, etc.) that may be affected by a lesion

in the future. This will be a challenging endeavor, as it is still rather unclear what constitutes a basic function and what should be the demarcation criterion within the huge cortical networks. Several of those networks that implement basic functions will then be recruited in order to form a defined moral–psychological construct that is useful, for example, in rehabilitation.

A way out of this problem is to resign from an elaborated phenomenology of moral agency and to focus on very few behavior types that seem to have clear moral impact, like violence. Fumagalli and Priori (2012) write: "From a behavioral point of view, the major consequence of moral abnormality is violence," which stands exemplarily for this position. However, as we will outline in Sect. 3.3, this position also raises ethical questions. We now proceed by providing a typology of brain lesions that may affect moral behavior.

3 Effects of Brain Injuries on Moral Behavior

Various types of pathological processes can affect the brain in a way that produces changes in behavior. Some of them occur instantly such as in trauma or stroke; others develop over a longer time scale, for example tumor growth or neurodegeneration. In the following, we use the notion of lesion or damage in a rather general way to indicate any kind of structural damage to brain tissue that have functional consequences. Examples of brain lesions include the following:

- Direct injury of brain tissue (e.g., gunshot)
- Ischemic damage to brain tissue (e.g., stroke, aneurysm rupture)

- Tumor-related damage to brain tissue (e.g., damage due to infiltrating tumor growth or expansion lesion due to increased pressure)
- Neurodegenerative processes (e.g., death of specific cell types as in Parkinsonism)
- Brain inflammation (e.g., encephalitis)

For analyzing differences between types of brain injuries, we classify them along two dimensions: the temporal scale of the brain injury and the plasticity potential of the brain injury.

The first dimension describes the typical temporal course of different types of brain injuries, namely the temporal course of their onset, of the subsequent development of changes in personality and behavior, and of the necessary therapies and rehabilitation processes. For example, the type of brain tumors determines the occurrence of symptoms (suddenly or gradually), the duration of necessary therapies (several hours for tumor resection or life-long for pharmacotherapy) and of necessary rehabilitation (short training course or life-long training). This dimension also influences how other people, particularly from the direct social surrounding of the patient, will react to lesion-related changes. For example, personality changes that develop slowly allow the family a better customization to changed behavior of the patient.

The second dimension is the plasticity of the brain that may allow for a partial or full reversibility or compensation of functional losses. This dimension comprises both healing processes of the affected brain tissue and functional shifts. An example for the latter is the transfer of the language centers from the left to the right hemisphere after resection of the left hemisphere due to Rasmussen encephalitis causing therapy-refractory epilepsy, which has been reported only from children younger than 5–6 years (Varadkar et al. 2014). Several factors influence the plasticity of the brain:

- The patient's age at the time of a brain lesion
- The exact location of the lesioned area and its physiological functions
- Healing processes
- Compensation processes (e.g., shift from certain functions to another than the affected area)
- The therapy and rehabilitation measures used (including medication) and their efficiency

Figure 4 provides an overview on these two dimensions. On the x-axis, brain injuries are sorted according to their temporal scale, that is, whether they develop fast or slowly. On the y-axis, brain injuries are sorted according to their plasticity potential. These two dimensions are relevant for brain damage resulting from pathological processes (white boxes) and from interventions as unintended side effects (gray boxes). Again, ideal types are shown, and the location of each example within each of the four quadrants is not intended to be precise.

Using this classification, we will now provide a case-based overview to outline the diversity of moral behavior changes due to brain lesions.

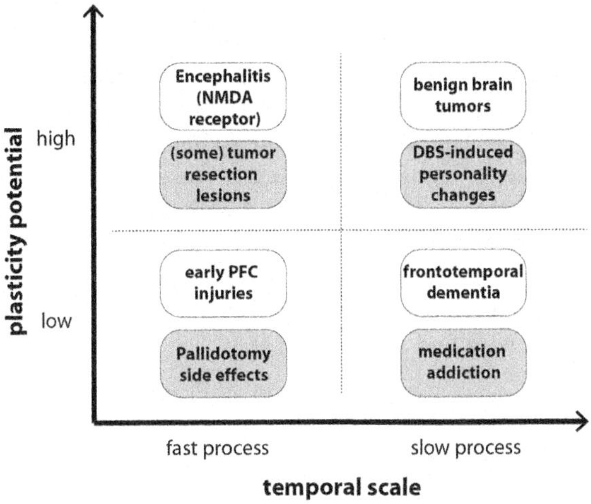

Fig. 4 Examples of pathological processes or interventions side effects influencing the brain. The examples are structured along the dimensions temporal scale of the process and the plasticity potential (potential of functional reversibility of the lesion sequelae). The *white boxes* denote changes due to pathological processes, the *gray boxes* non-intended changes due to interventions. The figure only identifies ideal types in the four quadrants of the scheme (*dotted line*)

3.1 Fast Processes with High Plasticity Potential

If changes in personality and behavior are caused by fast developing brain lesions which are reversible by adequate therapies or by natural healing processes, then for the patient and people in his/her social surrounding it becomes obvious that the changes were caused by a brain lesion, and not by the patient's "evil will." Such cases may be caused either by disease, injury, or therapy. They are particularly interesting since they allow us to study causal relationships between brain lesions and changes in personality and behavior in a bidirectional way.

An impressive example is NMDA receptor antibody encephalitis (NMDA: N-Methyl-D-aspartate), which was first described in 2007 as an autoimmune disease characterized by rapid development of psychosis, paranoia, aggressiveness, and other symptoms which may lead to a misdiagnosis of schizophrenia. Fortunately, the inflammation-caused mental sequelae are mostly reversible with timely administration of an effective therapy consisting of cortisol administration, hemodialysis, and immunotherapy (Dalmau et al. 2007, 2008). But since this disease has not been discovered before 2007, and since new scientific findings need some time for clinical translation, it is likely that many patients suffering from psychosis have been misdiagnosed with schizophrenia and thus have not received an effective therapy.

Sometimes interventions in the brain cause relatively fast changes in personality and behavior that may be reversible. After right pallidotomy for medically treatment-refractory Parkinsonism, a 59-year-old patient developed hypersexuality including pedophilic behavior. Immediately after the pallidotomy, he became markedly hypersexual. He forced his wife to have sex with him, masturbated frequently, propositioned his wife's female friends, hired strippers and prostitutes, and spent hours viewing Internet pornography. The patient was accused of touching his 5-year-old granddaughter inappropriately and asking her to touch his penis. He was ashamed of his behavior, complained of intrusive sexual thoughts and urges that overwhelmed him, and desired to just have his libidinal urges normalized again. The patient had no history of psychiatric illness, unusual sexual behavior, or drug-induced behavioral changes prior to his surgery. A reduction of his dopaminergic drugs resulted in a gradual decrease in sexual behaviors but worsened the symptoms of Parkinsonism (Mendez and Shapira 2011).

3.2 Fast Processes with Low Plasticity Potential

Particularly dramatic are fast processes with a low plasticity potential. In these cases, the personality and the behavior of an individual change rapidly, that is, within minutes, hours, or days. In addition, the changes are not reversible. Such cases can occur both by brain disease and by interventions in the brain. Again, the fast development of the lesion increases the likelihood that the behavioral changes are perceived as externally caused; however, the low reversibility potential also increases the likelihood of stigmatization.

Strokes can suddenly affect personality and behavior, often irreversibly. In one case, a 70-year-old man developed hemiballism, persistent hypersexuality, memory and executive dysfunction, and poor judgment after a small stroke involving the nucleus subthalamicus (Absher et al. 2000). Another example is the resection of brain tumors, which can change personality or behavior directly and often irreversibly. Although there is no evidence-based knowledge on the incidence, direction and extent of personality changes after brain tumor resection, several studies reveal a relationship between brain tumor surgery and changes in personality and (moral) behavior: Patients who had brain surgery for tumors have higher degrees of emotional and social dysfunction compared to extra-cerebral neurosurgery patients and terminally ill cancer patients (Andrewes et al. 2003, $n = 69$). Particularly, tumor resections from the frontal lobes can cause a lack of emotion and problems with decision-making, even in case of intact cognitive functions. In severe cases, psychopathy can develop which is characterized by impulsivity, antisocial behavior, and uncontrollable aggressions (Phineas Gage syndrome; Damasio 1994; Eslinger and Damasio 1985; Eslinger et al. 2004; Meyers et al. 1992; Tranel et al. 2002, patient SB-2046). Different behavioral disorders have also been reported after surgery for frontolimbic tumors. By way of example, a patient developed kleptomania and compulsive gambling after removal of a craniopharyngioma. Besides this, he became circumstantial and logorrheic, and displayed hypergraphia and a preoccupation with religious and moral ideas (Nyffeler and Regard 2001). In children, aggressive microsurgery for craniopharyngiomas has a significant impact on social–emotional and behavioral functioning (Sands et al. 2005). A prospective study reports that the majority of children who had total resections of craniopharyngiomas were more or less severely affected by a hypothalamic syndrome that altered their social integration and caused academic failure (Pierre-Kahn et al. 2005, $n = 14$). High rates of intellectual impairment, poor social adaptation, and emotional lability in craniopharyngioma survivors (30–60 %) might be caused by an impaired frontal lobe function following surgery (Stelling et al. 1986).

In some cases, the resection of brain tumors can cause the onset of new psychiatric symptoms. For example, a patient with no previous mental illness developed major depression with psychosis after resection of a giant middle fossa hemangiopericytoma (Sade et al. 2006). Another patient developed a schizophreniform psychosis after excision and postoperative radiotherapy of an oligodendroglioma (Mace and Trimble 1991, case C).

Tumor resections from brain areas, which have recently been considered irrelevant for cognitive capacities, personality, and behavior, can cause a wide spectrum of neuropsychological and behavioral abnormalities. Behavioral deficits or attention deficit problems were detected in 33 % or 12.5 % of patients, respectively, who were operated for benign cerebellar tumors during childhood (Steinlin et al. 2003, $n = 24$). Some demonstrated psychiatric symptoms such as mutism, addiction problems, anorexia, uncontrolled temper tantrums and phobia. Patients had difficulties in selective and, more notably, sustained attention, which resemble dysfunctions seen in patients with frontal lesions (Steinlin et al. 2003). There is evidence that cerebellar dysfunction includes a mild frontal dysfunction due to destroyed cerebello-frontal connections (Steinlin et al. 2003). The resection of benign cerebellar tumors causes the posterior fossa syndrome in 28 % of children, which is characterized by mutism, oropharyngeal dyspraxia, emotional lability, different neuropsychiatric symptoms, and autistic behavior (Catsman-Berrevoets and Aarsen 2010, $n = 148$).

Hypersexuality following brain surgery has also been reported. Two patients developed hypersexuality with inappropriate sexual behavior following the placement of ventriculoperitoneal shunts for the treatment of hydrocephalus (Gorman and Cummings 1992). The sexual disorder was likely caused by septal damage due to the shunt placement. Both lesions and stimulation of the septum have caused hypersexuality in animals and humans. The septum is presumably one locus of a circuit of structures mediating sexual behavior (Gorman and Cummings 1992). Further loci of this circuit include the inferior frontal cortex, the hypothalamus, and the amygdaloid nuclei. Lesions in any of these regions have a major, site-specific impact on sexual behavior. Bilateral lesions of the amygdaloid nuclei produce hypersexuality (Klüver Bucy syndrome); whereas lesions in the hypothalamus reduce sexual activity (Gorman and Cummings 1992). A few long-lasting cases of disinhibition and inappropriate sexual behavior following pallidotomy (lesion of parts of the globus pallidus) for treating medically treatment-resistant Parkinsonism have been published (Shannon et al. 1998, $n = 3$, persistent for at least 6 months).

Fast negative changes in personality and behavior can also occur after psychiatric neurosurgery, which is used very rarely for the treatment of severe therapy-resistant cases. By way of example, possible sequelae of capsulotomy include aggressiveness, dysexecutive function deterioration, and sexual disinhibition (Cosgrove and Rauch 1995; D'Astous et al. 2013; Dougherty et al. 2002; Feldman et al. 2001; Rück et al. 2008). Following subcaudatetractotomy, the development of undesirable personality traits has been reported in some patients (Feldman et al. 2001). After ventromedial frontal leukotomy, most patients with lesions in the ventral striatum (8 out of 11) developed substance dependence (Irle et al. 1998). The reports indicated no potential of reversibility of these sequelae.

3.3 Slow Processes with High Plasticity Potential

When changes in personality and behavior occur gradually due to slowly developing brain lesions, they allow for better adaptation to the changes, both for the patient and for people in the social surrounding. Slow processes could be considered as less dramatic than fast processes. On the other hand, slow development can conceal the fact that problematic changes in personality and behavior are caused by a disease and not by the patients' "evil will." Particularly, if the disease is not yet diagnosed, or if the patient's significant others do not understand its effects on the patient's personality and behavior, the patient may be blamed for aberrant behavior. However, this risk is diminished if these slowly developing changes in personality and behavior are reversible by adequate medical or neurosurgical therapies. Notable examples can be found particularly in studies about the neurosurgical treatment of epilepsy, Parkinson's disease, and brain tumors.

In patients suffering from epilepsy, changes in personality and behavior mostly develop over several years. However, in many cases, they are reversible after neurosurgical resection of the epileptic focus. Patients with epilepsy have a higher prevalence of lifetime psychiatric disorders (35 %) than the general population (20.7 %; Téllez-Zenteno et al. 2007), and particularly high are the rates in patients with temporal lobe epilepsy (Foong and Flugel 2007). Following surgery for epilepsy, depression, anxiety, behavioral disorders, and severe obsessive–compulsive disorders—which are often, but not always comorbidities of the disease—are often improved (Devinsky et al. 2005; Guangming et al. 2009; Guarnieri et al. 2005; Hannan et al. 2009; Lendt et al. 2000; Witt et al. 2008; review: Foong and Flugel 2007). In many patients, increased warmth in social relationships and reduced egotism have been described (Hill et al. 1957). Improvements in aggressive behavior in children following surgery for temporal lobe epilepsy have been reported in several papers (review: Foong and Flugel 2007). In children, the most notable improvements after surgery for epilepsy include decreased hyperactivity, greater emotional well being, and improved socialization (review: Spencer and Huh 2008).

In patients suffering from Parkinson's disease, some disease-related changes in personality are sometimes reversed after deep brain stimulation (DBS) of the nucleus subthalamicus (Schneider et al. 2003). Changes in behavior caused by

Parkinsonism drug therapy, such as impulse control disorders, pathological gambling, addiction to levodopa, and hypersexuality, can disappear after DBS since it allows for the reduction of the dopaminergic drugs (Demetriades et al. 2011).

Whether patients suffering from brain tumors develop changes in personality and behavior depends on tumor location, tumor size, and tumor type. In many cases, these changes are reversible after resection of the tumor, irradiation, or chemotherapy. The reversibility of tumor-related personality changes depends both on the healing processes in the damaged brain tissue and on the amount of brain tissue that is further damaged by treatment. A notable case in the discussion to follow illustrates how detrimental personality changes can result from a brain tumor and how brain surgery can restore the personality. A 40-year-old married schoolteacher became obsessed with child pornography and started to solicit prostitutes and to molest his stepdaughter. His wife evicted him from the family home after discovering his sexual advances to her daughter. He was accused and found guilty of molesting children. He had to enter a treatment program for convicted sexual offenders where he continued asking women for sex and was expelled from the program. One day before the start of his prison sentence, he was admitted to the hospital for headaches and an indomitable sex drive. An MRI scan revealed that he had an egg-sized brain tumor in the frontal lobe, a brain area essential for judgment, social behavior, and self-control. The tumor had already infiltrated the hypothalamus, which also controls sex drive. After tumor resection, the pedophiliac drive vanished completely, and the patient went home to his family. But several months later, he secretly started to collect pornography again. An MRI scan showed that the tumor had regrown. It was removed once more, and his abnormal sexual drive vanished again (Burns and Swerdlow 2003).

3.4 Slow Processes with Low Plasticity Potential

When changes in personality and behavior occur gradually due to slowly developing brain lesions that in addition have low plasticity potential, the risk that these changes are attributed to the "evil will" of the patient resurfaces, and thus, the problem of misunderstanding, blame, and stigmatization will be aggravated.

Many neurodegenerative brain diseases affect an individual's personality and behavior deeply and enduringly, particularly frontotemporal or vascular dementia, Parkinson's disease and Huntington's disease. Frontotemporal dementia (FTD) is a notable example of sociopathic behavior caused by an acquired frontal brain disorder. FTD is a progressive neurodegenerative disorder that affects mainly the ventromedial prefrontal cortex, anterior temporal regions, and/or the orbitofrontal cortex. The behavioral changes of FTD patients seem to correlate with a decreased metabolism or perfusion in these regions as measured with PET or SPECT (Mendez 2010). FTD patients show marked impairments in moral reasoning despite knowledge of moral and social rules, emotion recognition particularly for anger and disgust, empathy as rated by caregivers, and executive functions (Lough et al. 2006).

FTD patients undergo a change in personality and behavior which is characterized by the following core features: transgression of social norms, sociopathic behavior, altered moral feelings, loss of emotional empathy, and disinhibited, compulsive acts. Although cognition remains largely intact, knowledge of moral behavior and of potential consequences of rule violations is preserved (Mendez 2010). Although they can make reasoned moral judgments, the emotional morality of FTD patients is altered, so that they respond to moral dilemmas in a calculated rather than an emotional fashion (Mendez and Shapira 2009). Typical examples of behavioral problems of FTD patients are loss of social tact and propriety, unacceptable physical contact, neglect of personal hygiene, and compulsive eating or hoarding. More than half of patients fail to conform to lawful behavior. Several reports describe stealing, unethical job conduct, indecent exposure, and inappropriate sexual behavior such as child molestation, illegal driving acts, and physical assaults or violence (Mendez 2010). According to Mendez, FTD patients have a "specific, brain-based impairment in moral reasoning"; their "sociopathic behavior is consistent with decreased emotional moral judgment plus a lack of empathy and disinhibited, compulsive drives" (Mendez 2010, p. 322).

In addition, the growth of brain tumors can affect personality and behavior on a longer timescale. Both tumor type and tumor location are significant influential factors for emotional and social dysfunctions such as anger, helplessness, fatigue, emotional dyscontrol, indifference, and maladaptive behavior (Andrewes et al. 2003, $n = 69$). Tumors in the temporal lobes can be associated with behavioral problems, including aggression and rage attacks (Nakaji et al. 2003). The behavior of patients with lesions in the orbitofrontal and ventromedial PFC has been described as aggressive, lacking responsibility, and concern for social and moral rules (Eslinger and Damasio 1985; Damasio 1994).

4 Ethical Pitfalls in Investigating Changes in Moral Behavior After Brain Lesions

So far, we have outlined the methodological problems when relating moral agency with the brain and provided a scheme to describe the complexity of brain lesion types and moral behavior changes that may result from these lesions. We now discuss the extent to which the investigation of changes in moral behavior after brain lesions poses ethical problems.

Why should it be ethically problematic to relate brain lesions with, in most cases, unwanted changes in moral behavior? This may sound like an odd question, because such findings might contribute to novel therapies that prevent or reverse behavior changes. However, the impetus to find a neuronal cause for disturbed moral behavior aligns with a general tendency in popular culture to find brain-based explanations for behavior (Frazzetto and Anker 2009). Researchers and clinicians who investigate how changes in the brain lead to changes in moral behavior propose to use their findings for interventions. For example, Fumagalli and Priori (2012, p. 2017) write: "From a clinical point of view, subjects manifesting abnormal moral behavior should be screened for neurological disorders to promote an early diagnosis. A potentially important issue arises when clinicians treat patients whose social position makes them responsible for others (including state leaders and politicians) with abnormalities of moral behavior or with other conditions (or treatment) that could influence their "moral brain." In these cases, an early diagnosis and, whenever possible, effective treatment is important both for the patient and for the welfare of society."

We think that the following questions should be considered in order to assess the ethics of moral behavior interventions based on findings in neuroscience:

1. On the grounds of which ethical theory should the borders between still tolerable behavior and morally blameful behavior be defined?
2. Does it make a difference when a change in moral behavior is caused by a brain lesion due to a pathological process, compared to a change in moral behavior that results as unwanted but maybe inevitable side effect of brain interventions to treat neurological disorders?
3. What do we owe persons who display unpleasant or even immoral behavior due to brain lesions?
4. Does tightening the link between brain damage and behavior aberration increase or decrease the stigmatization of these persons?
5. Given that there is a relation between the brain and moral behavior, what knowledge do we need to better understand this relation?
6. Is there a danger that we pathologize unwanted but legitimate moral standpoints?

In the following, we will briefly discuss these questions and outline some ethical risks that are associated with them.

4.1 Which Ethical Theory?

The first question relates to the basic problem that there is disagreement on what counts as moral behavior. Although we do not support moral relativism, we agree with its observation that moral issues are evaluated very controversially and that the controversies depend significantly on cultural and societal differences. A notable example is the field of sexual morality, where the controversy is enormous. Practices such as prostitution, child marriage, intermarriage, homosexuality, premarital sex, extramarital sex, promiscuity, divorce, polygamy, etc., are socially fully accepted in some cultures and condemned or even illegal in others. This variability may explain the large prevalence differences of hypersexuality, from 2 to 10 % (Chiang et al. 2012) when comparing different countries. The cultural differences in sexual morality probably influence which criteria are used for the diagnosis of hypersexuality and thus the prevalence rates.

For a scientific investigation of brain disorders that cause disorders of moral behavior, a universalistic ethical approach would be optimal, as many ethical theories consider universalizability to be a distinguishing feature of moral judgments and a substantive guide to moral obligation: Moral imperatives should be regarded as equally binding on everyone. However, in philosophy, many arguments have been developed against the feasibility of a universalistic ethic. For example, Beauchamp and Childress' principle-based ethics (2013) that is often considered to be a gold standard in bioethics is exposed to critics from several sides. First, the deductivists (e.g., Clouser and Gert 1990) criticize eclecticism and the lack of a universal, applicable ethical theory. Second, defendants of casuistic ethics (e.g., Jonsen 1995) criticize a too schematic application of principles to particular cases (Harris 2003), and claim that it blocks substantive ethical inquiry (Callahan 2003). Third, the social science critique of bioethics claims that bioethics grounded in philosophy and moral theory gives a dominant role to idealized, rational thought and tends to exclude social and cultural factors, so that it is isolated from practice (overview: Hedgecoe 2004). The dominance of the principle of respect for autonomy in particular has been criticized by many authors from different ideological backgrounds (critical overview: Gillon 2003). In spite of the diversity of these criticisms, they converge in giving collective benefits more weight than individual rights.

In summary, the question of "which morality?" does not only refer to the commonplace, that there is disagreement concerning the morality of certain behaviors. The point is that the relationship of brain lesions with moral behavior tends to blur this variability and that this may happen on a level where this effect is difficult to be discerned, for example, on the level of diagnostic criteria. This impetus to universalize morality is not based on grounds of an elaborated theory of ethical universalism, but is implicitly embedded in the methodology that investigates the phenomenon. Therefore, we identify as the first ethical risk of the neuroscience of ethics that it may suppress legitimate controversies on moral theories.

4.2 What Is Causing the Brain Lesion?

At first glance, there seems to be a fundamental difference between changes of moral behavior caused by pathological processes such as strokes or tumors, and those caused by medical interventions. This distinction may hold when the intervention directly targets the behavior as in psychiatric neurosurgery, but it is less clear for lesions that occur as unintended or unavoidable consequences of therapeutic interventions such as tumor resection.

For dealing with this problem, it is necessary to accumulate knowledge on sequelae that may result from particular interventions. This knowledge then can be used for the shared decision-making process between patient and medical experts. The problem, however, is that changes in moral behavior that are caused by interventions are often hard to measure, whereas their relative life impact is high (Müller and Christen 2011).

However, a fundamental problem remains, namely the large individual variability of human brains (e.g., functional connectivity; Barch et al. 2013) as well as of regeneration processes. Therefore, clinical outcome studies that average across patients to provide a unitary measure of outcome are not sufficient. Because of the large outcome variability, it is necessary to report both good and poor outcomes separately. Cross-sectional group research does not reveal the different individual trajectories and provides only limited clues about which factors are most relevant in effecting positive change for an individual. It is important to study individual outcomes, particularly by identifying subgroup patterns that can become lost in whole-group analyses. To overcome this systematical shortcoming, long-term follow-up studies of outcome, particularly of neuropsychological and socio-psychological outcome, are necessary (Wilson et al. 2005). Particularly cases with unfavorable or unexpected outcome should be investigated, since they offer extraordinary chances for scientific discovery and improving the techniques used (Kubu and Ford 2012). Besides clinical studies, case studies contribute much to clinical experience and to scientific understanding. For example, the knowledge of adverse effects of deep brain stimulation has been spread mainly via case reports (Christen and Müller 2011). The careful documentation and publication of extraordinary single cases are important for scientific progress. This highlights the importance of case studies in addition to knowledge based on statistical evidence.

Given this caveat, an approach in therapy and rehabilitation that focuses on the individual case seems appropriate. However, this approach conflicts with an increasing involvement of knowledge based on statistical evidence and a regulatory or legal structure that more and more relies on such knowledge, for example, in addressing insurance claims. This pinpoints a second ethical risk: Restricting clinical research on the relationship between brain and moral behavior may undermine the value of special case studies involving outliers.

4.3 What Do We Owe Brain Lesion Patients with Socially Aberrant Behavior?

The care of frontal lesion patients is challenging and demonstrates constraints of classical principles of medical ethics like autonomy and beneficence. Disabilities that directly affect social interactions with others pose more challenges to family and caregivers compared to physical disabilities or pure cognitive disabilities. The main reason for this is probably that the disability caused by the brain lesion directly influences morally questionable behaviors like boasting, egocentrism, or obstinacy; or even uncontroversial immoral behaviors like habitual lying, child molestation, or violence. Empathy or even sympathy for these patients is much harder to sustain.

Most people who suffer from brain lesions are not responsible for their lesions, and they are significantly disadvantaged. Therefore, we think that society has the moral duty to support their reintegration. Besides a good medical treatment, rehabilitation programs are necessary, and for some patients a protected environment. Furthermore, we think that research on the responsibility of people with brain lesions for aberrant or even criminal behavior is necessary. This research has to consider both medical and normative issues and therefore requires interdisciplinary programs.

4.4 Will Research Increase the Stigmatization of Patients with Brain Lesions?

It is tempting to believe that a better understanding of how damage to the brain leads to changes in moral behavior will increase the understanding of such patients and enhance their social reintegration. However, we suspect that this hope is misleading for two reasons. First, experimental ethics has shown that the majority of people judge in an incompatibilist way, that is, they believe that determinism excludes moral responsibility. But many people tend toward compatibilism if the cases they have to judge trigger emotions; then most people blame others and hold them morally responsible despite knowing that the person was determined to act in a specific, immoral way (Nichols and Knobe 2007). Second, research on stigmatization has shown that biological explanations of psychiatric disorders have complex effects on stigmatization which depend on several aspects of a given disorder. Biological explanations of psychiatric disorders increase stigmatization particularly if the disorder is explained as irreversible or as genetically based, or if it makes the patients dangerous for third persons (Müller and Heinz 2013). Therefore, we expect that a better biological understanding of how brain lesions cause aberrant behavior and changes in personality might decrease the stigmatization of principally curable or reversible diseases but increase the stigmatization of irreversible brain disorders.

From a theoretical point of view, a deeper understanding of how brain lesions cause social aberrant behavior is likely to support the view that brain lesion patients also deserve help and understanding when they behave immorally. However, it is likely that for many patients the practical effect will be the opposite—namely more stigmatization and discrimination. Thus, the third ethical risk of a neuroscience of ethics is that a better understanding of how brain lesions cause moral behavior changes might undermine people's willingness to support lesion patients to reintegrate in the society.

4.5 What Should We Investigate?

This leads to the problem of deciding on which side of the interrelation between the brain and moral behavior the emphasis of research should be. Here, we are confronted with a fundamental impetus of the scientific method—namely that it aims to generate causal knowledge that is as precise and as deterministic as possible. The neuroscience of ethics proceeds from documenting correlations between brain lesions and social behavior to experimenting by noninvasive interventions with transcranial magnetic stimulation or by drugs. This should allow for finding possible causal relationships underlying the correlations—for example, that inhibition of a certain brain region causes lack in impulse control. However, as our overview in Sect. 3 has shown, there is a surprising lack of knowledge on the reversibility of such behaviors. We propose that future research in the neuroscience of ethics should focus on three issues: the spectrum of behavior aberrations following brain lesions including their interrelation with the social environment, the chances of regeneration, and the development of effective therapies.

4.6 Danger of Pathologizing Ethical Theories

We also pinpoint a fourth ethical risk of the neuroscience of ethics, namely to relate different ethical standpoints or ethical theories like deontology or utilitarianism to specified neuronal differences, or even to certain neurological disorders. It is possible to detect different activation patterns during different forms of moral thinking. But what is the function of such findings in the meta-ethical discourse? We suspect that they are sometimes misused for debunking ethical theories, that is, for discrediting ethical theories not by normative arguments, but by pinpointing inopportune or emotional processes that cause certain types of moral argumentations. An example is the discretization of deontology compared to utilitarianism based on neuroscientific findings (Singer 2005). In extremis, this could subserve a pathologization of ethical theories.

5 Summary and Outlook: Moral Behavior as Target of Therapy

The central aim of our contribution is to outline the complexity of the interrelation between the brain and moral behavior relation when seen from a neuroscientific standpoint. We have argued that dynamic and temporal factors on both sides of the equation—the stability of moral evaluations across time and society or across neuronal plasticity—structure this complexity, but also explain why we cannot expect clear-cut relations between specific brain lesions and specific behavior aberrations. But to achieve feasibility for therapy and rehabilitation, we will need models that have an intermediate degree of complexity, like the model of moral intelligence we have proposed. These models must be complex enough to capture the relevant phenomenon, but simple enough to be understandable for practical purposes. In that way, basic researchers and therapists can approach the question of how to interrelate the brain with moral agency in a more structured way.

However, one has to be aware of ethical risks that accompany this endeavor. The most urgent risk is a lack of knowledge about the spectrum of behavioral aberrations that accompany specified brain lesions, or about the potential of regeneration and its dependence on the social environment. The problem is aggravated by the difficulties in everyday life when dealing with patients that show aberrant social behavior, where one has to balance demands for responsible behavior and lenience due to their disability.

We close by remarking that the complexity outlined in our contribution also calls for caution with respect to recent claims for moral enhancement—the idea that knowledge of the biological foundation of human moral behavior may allow for interventions into the neuronal infrastructure of morality in order to improve the behavior of people or, at least, to diminish some forms of evil (De Ridder et al. 2009; Shook 2012; Persson and Savulescu 2012). Although we are optimistic that more sophisticated and individualized research will certainly help to bear the behavioral burdens caused by some brain lesions, we do not support policies of moral enhancement through brain interventions that disregard the autonomy and dignity of the patients concerned.

Acknowledgments **Funding**: This research has been partly supported by the German Research Foundation (DFG), Germany (MU 3321/1-1).

References

Abend G (2013) What the science of morality doesn't say about morality. Phil Soc Sci 43:157–200

Absher JR, Vogt BA, Clark DG et al (2000) Hypersexuality and hemiballism due to subthalamic infarction. Neuropsych Neuropsychol Behav Neurol 13(3):220–229

Andrewes DG, Kaye A, Murphy M et al (2003) Emotional and social dysfunction in patients following surgical treatment for brain tumour. J Clin Neurosci 10(4):428–433

Appiah KA (2010) The honor code: how moral revolutions happen. W.W. Norton & Company, New York

Barch DM, Burgess GC, Harms MP et al (2013) Function in the human connectome: task-fMRI and individual differences in behavior. NeuroImage 80:169–189

Batts S (2009) Brain lesions and their implications in criminal responsibility. Behav Sci Law 27 (2):261–272

Beauchamp TL, Childress JF (2013) Principles of biomedical ethics, 7th edn. Oxford University Press, Oxford

Björklund F, Björnsson G, Eriksson J et al (2012) Recent work: motivational internalism. Analysis 72:124–137

Burns JM, Swerdlow RH (2003) Right orbitofrontal tumor with pedophilia symptom and constructional apraxia sign. Arch Neurol 60:437–440

Callahan D (2003) Principlism and communitarianism. J Med Ethics 29:287–291
Catsman-Berrevoets C, Aarsen FK (2010) The spectrum of neurobehavioral deficits in the posterior fossa syndrome in children after cerebellar tumour surgery. Cortex 46:933–946
Chiang HL, Huang YS, Chen ST et al (2012) Are there ethnic differences in impulsive/compulsive behaviors in Parkinson's disease? Europ J Neurol 19(3):494–500
Cholbi MJ (2006) Belief attribution and the falsification of motive internalism. Philos Psychol 19:607–616
Christen M, Alfano M (2013) Outlining the field—a research program for empirically informed ethics. In: Christen M, Fischer J, Huppenbauer M, Tanner C, van Schaik C (eds) Empirically informed ethics. Morality between facts and norms. Library of ethics and applied philosophy. Springer, Berlin, pp 3–28
Christen M, Müller S (2011) Single cases promote knowledge transfer in the field of DBS. Front Integr Neurosci 5 (Article 13)
Christen M, Regard M (2012) Der "unmoralische Patient". Eine Analyse der Nutzung hirnverletzter Menschen in der Moralforschung. Nervenheilkunde 31:209–214
Ciaramelli E, Muccioli M, Làdavas E et al (2007) Selective deficit in personal moral judgment following damage to ventromedial prefrontal cortex. Soc Cog Affect Neurosci 2:84–92
Clouser KD, Gert B (1990) A critique of principlism. J Med Philos 15:219–236
Cosgrove GR, Rauch SL (1995) Psychosurgery. Neurosurg Clin North A 6:167–176
D'Astous M, Cottin S, Roy M et al (2013) Bilateral stereotactic anterior capsulotomy for obsessive-compulsive disorder: long-term follow-up. J Neurol Neurosurg Psychiat 84 (11):1208–1213
Dalmau J, Gleichman AJ, Hughes EG et al (2008) Anti-NMDA-receptor encephalitis: case series and analysis of the effects of antibodies. Lancet Neurol 7:1091–1098
Dalmau J, Tuzun E, Wu HY et al (2007) Paraneoplastic anti-N-methyl-D-aspartate receptor encephalitis associated with ovarian teratoma. Ann Neurol 61:25–36
Damasio AR (1994) Descartes' error. Putnam, New York
Demetriades P, Rickards H, Cavanna AE (2011) Impulse control disorders following deep brain stimulation of the subthalamic nucleus in Parkinson's disease: clinical aspects. Parkinson's Disease, doi: 10.4061/2011/658415
De Ridder D, Langguth B, Plazier M et al (2009) Moral Dysfunction and potential treatments. In: Verplaetse J et al (eds) The moral brain. Springer, Berlin, pp 155–183
Devinsky O, Barr WB, Vickrey BG et al (2005) Changes in depression and anxiety after resective surgery for epilepsy. Neurology 65:1744–1749
Dougherty DD, Baer L, Cosgrove GR et al (2002) Prospective long-term follow-up of 44 patients who received cingulotomy for treatment-refractory obsessive-compulsive disorder. Am J Psych 159:269–275
Eslinger PJ, Damasio A (1985) Severe disturbance of higher cognition after bilateral frontal lobe ablation: Patient EVR. Neurology 35:1731–1741
Eslinger PJ, Flaherty-Craig CV, Benton AL (2004) Developmental outcomes after early prefrontal cortex damage. Brain Cog 55:84–103
Feinstein JS, Rudrauf D, Khalsa SS et al (2010) Bilateral limbic system destruction in man. J Clin Exp Neuropsychol 32(1):88–106
Feldman RP, Alterman RL, Goodrich JT (2001) Contemporary psychosurgery and a look to the future. J Neurosurg 95:944–956
Foong J, Flugel D (2007) Psychiatric outcome of surgery for temporal lobe epilepsy and presurgical considerations. Epilepsy Res 75:84–96
Frazzetto G, Anker S (2009) Neuroculture. Nature Rev Neurosci 10(11):815–821
Frías Ibáñez A, Malea Fernández A, Dorado García M et al (2008) Orbitofrontal syndrome after a cerebral aneurysm operation. A case report. Actas España Psiquiatrica 36(6):362–365
Fumagalli M, Priori A (2012) Functional and clinical neuroanatomy of morality. Brain 135 (7):2006–2021
Gillon R (2003) Ethics needs principles—four can encompass the rest—and respect for autonomy should be "first among equals". J Med Ethics 29:307–312

Gorman DG, Cummings JL (1992) Hypersexuality following septal injury. Arch Neurol 49:308–310
Grafman J, Schwab K, Warden D et al (1996) Frontal lobe injuries, violence, and aggression: a report of the Vietnam Head Injury Study. Neurology 46:1231–1238
Guangming Z, Wenjing Z, Guoquiang C et al (2009) Psychiatric symptom changes after corticoamygdalo-hippo-campectomy in patients with medial temporal lobe epilepsy through Symptom Checklist 90 Revised. Surgical Neurol 72:587–591
Guarnieri R, Araújo D, Carlotti CG Jr et al (2005) Suppression of obsessive-compulsive symptoms after epilepsy surgery. Epilepsy Behav 7:316–319
Hannan S, Cross JH, Scott RC et al (2009) The effects of epilepsy surgery on emotions, behavior, and psychosocial impairment in children and adolescents with drug-resistant epilepsy: a prospective study. Epilepsy Behav 15:318–324
Harlow JM (1868) Recovery from the passage of an iron bar through the head. Publ Mass Med Soc 2:327–347
Harris J (2003) In praise of unprincipled ethics. J Med Ethics 29:303–306
Hawkins KA, Trobst KK (2000) Frontal lobe dysfunction and aggression: conceptual issues and research findings. Aggr Viol Behav 5(2):147–157
Hedgecoe AM (2004) Critical bioethics: beyond the social science critique of applied ethics. Bioethics 18(2):120–143
Hill D, Pond DA, Mitschell W et al (1957) Personality changes following temporal lobectomy for epilepsy. J Med Sci 103:18–27
Hornak J, Bramham J, Rolls ET et al (2003) Changes in emotion after circumscribed surgical lesions of the orbitofrontal and cingulate cortices. Brain 126:1691–1712
Irle E, Exner C, Thielen K et al (1998) Obsessive-compulsive disorder and ventromedial frontal lesions: clinical and neuropsychological findings. Am J Psych 155:255–263
Jonsen AR (1995) Casuistry: an alternative or complement to principles? Kennedy Inst Ethics J 5:237–251
Koenigs M, Young L, Adolphs R et al (2007) Damage to the prefrontal cortex increases utilitarian moral judgments. Nature 446:908–911
Kohlberg L (1981/1984) Essays on moral development, vol I/II. Harper & Row, San Francisco
Kubu CS, Ford PJ (2012) Beyond mere symptom relief in deep brain stimulation: an ethical obligation for multi-faceted assessment of outcome. AJOB Neurosci 3(1):44–49
Lendt M, Helmstaedter C, Kuczaty S et al (2000) Behavioral disorders in children with epilepsy: early improvement after surgery. J Neurol Neuro-surg Psych 69:739–744
Lough S, Kipps CM, Treise C et al (2006) Social reasoning, emotion and empathy in frontotemporal dementia. Neuropsychologia 44(6):950–958
Mace CJ, Trimble MR (1991) Psychosis following temporal lobe surgery: a report of six cases. J Neurol Neurosurg Psych 54:639–644
Macmillan M (2000) An odd kind of fame. Stories of Phineas Gage. MIT-Press, Cambridge, Mass
Mataró M, Jurado MA, García-Sánchez C et al (2001) Long-term effects of bilateral frontal brain lesion: 60 years after injury with an iron bar. Arch Neurol 58(7):1139–1142
Mendez MF, Shapira JS (2011) Pedophilic behavior from brain diseases. J Sex Med 8:1092–1100
Mendez MF (2010) The unique predisposition to criminal violations in frontotemporal dementia. J Am Acad Psych Law 38(3):318–323
Mendez MF (2009) The neurobiology of moral behavior: review and neuropsychiatric implications. CNS Spectrum 14(11):608–620
Mendez MF, Shapira JS (2009) Altered emotional morality in frontotemporal dementia. Cogn Neuropsych 14(3):165–179
Meyers CA, Berman SA, Scheibel RS et al (1992) Acquired antisocial personality disorder associated with unilateral left orbital frontal lobe damage. J Psych Neurosci 17(3):121–125
Müller S, Heinz A (2013) Stigmatisierung oder Entstigmatisierung durch Biologisierung psychischer Krankheiten? Nervenheilkunde 12:955–961
Müller S, Christen M (2011) Deep brain stimulation in Parkinsonian patients—ethical evaluation of cognitive, affective, and behavioral sequelae. AJOB Neurosci 2(1):3–13

Nakaji P, Meltzer HS, Singel SA et al (2003) Improvement of aggressive and anti-social behaviour after resection of temporal lobe tumors. Pediatrics 112(5):e430–e433
Narvaez D (2005) The neo-Kohlbergian tradition and beyond: schemas, expertise and character. In: Carlo G, Pope-Edwards C (eds) Moral motivation through the lifespan. University of Nebraska Press, Lincoln, pp 119–163
Nichols S, Knobe J (2007) Moral responsibility and determinism: the cognitive science of folk intuitions. Noûs 41(4):663–685
Nyffeler T, Regard M (2001) Kleptomania in a patient with a right frontolimbic lesion. Neuropsych Neuropsychol Behav Neurol 14(1):73–76
Parkinson C, Sinnott-Armstrong W, Koralus PE et al (2011) Is morality unified? Evidence that distinct neural systems underlie moral judgments of harm, dishonesty, and disgust. J Cog Neurosci 23(10):3162–3180
Pascual L, Rodrigues P, Gallardo-Pujol D (2013) How does morality work in the brain? A functional and structural perspective of moral behavior. Front Integr Neurosci 7:65
Persson I, Savulescu J (2012) Unfit for the future: the need for moral enhancement. Uehiro series in practical ethics. Oxford University Press, Oxford
Piaget J (1932) The moral judgment of the child. Kegan Paul, Trench, Trubner and Co., London
Pierre-Kahn A, Recassens C, Pinto G et al (2005) Social and psycho-intellectual outcome following radical removal of craniopharyngiomas in childhood: a prospective series. Child's Nerv Sys 21:817–824
Rest JR (1986) Moral development: advances in research and theory. Praeger, New York
Roskies A (2006) Patients with ventromedial frontal damage have moral beliefs. Philos Psychol 19:617–627
Rück C, Karlsson A, Steele D et al (2008) Capsulotomy for obsessive-compulsive disorder. Arch Gen Psych 65(8):914–922
Sade B, Budur K, Lee DK et al (2006) Major depression with psychosis after resection of a giant middle fossa hemangiopericytoma. Surg Neurol 65:290–292
Sands SA, Milner JS, Goldberg J et al (2005) Quality of life and behavioral follow-up study of pediatric survivors of craniopharyngioma. J Neurosurg 103(4 suppl):302–311
Schneider F, Habel U, Volkmann J et al (2003) Deep brain stimulation of the subthalamic nucleus enhances emotional processing in Parkinson disease. Arch Gen Psych 60(3):296–302
Shamay-Tsoory SG, Tomer R, Goldsher D et al (2004) Impairment in cognitive and affective empathy in patients with brain lesions. J Clin Exp Neuro-psycho-l 26(8):1113–1127
Shannon KM, Penn RD, Kroin JS et al (1998) Stereotactic pallidotomy for the treatment of Parkinson's disease. Efficacy and adverse effects at 6 months in 26 patients. Neurology 50:434–438
Shook JR (2012) Neuroethics and the possible types of moral enhancement. AJOB Neurosci 3 (4):3–14
Singer P (2005) Ethics and intuitions. J Ethics 9:331–352
Spencer S, Huh L (2008) Outcomes of epilepsy surgery in adults and children. Lancet Neurol 7:525–537
Steinlin M, Imfeld S, Zulauf P et al (2003) Neuropsychological long-term sequelae after posterior fossa tumour resection during childhood. Brain 126:1998–2008
Stelling MW, McKay SE, Carr A et al (1986) Frontal lesions and cognitive function in craniopharyngioma survivors. Am J Dis Child 140:710–714
Tanner C, Christen M (2013) Moral intelligence—a framework for understanding moral competences. In: Christen M, Fischer J, Huppenbauer M, Tanner C, van Schaik C (eds) Empirically informed ethics. Morality between facts and norms. Library of ethics and applied philosophy. Springer, Berlin, pp 119–136
Téllez-Zenteno JF, Dhar R, Hernandez-Ronquillo L et al (2007) Long-term outcomes in epilepsy surgery: antiepileptic drugs, mortality, cognitive and psychosocial aspects. Brain 130:334–345
Tranel D, Bechara A, Denburg NL (2002) Asymmetric functional roles of right and left ventromedial prefrontal cortices in social conduct, decision-making, and emotional processing. Cortex 38:589–612

Uttal WR (2001) The new phrenology. The limits of localizing cognitive processes in the brain. MIT Press, Cambridge

Varadkar S, Bien CG, Kruse CA et al (2014) Rasmussen's encephalitis: clinical features, pathobiology, and treatment advances. Lancet Neurol 13(2):195–205

Welt L (1888) Über Charakterveränderungen des Menschen infolge von Läsionen des Stirnhirns. Deutsches Archiv für klinische Medicin 42:339–390

Wilson SJ, Bladin PF, Saling MM et al (2005) Characterizing psychosocial outcome trajectories following seizure surgery. Epilepsy Behav 6:570–580

Witt JA, Hollmann K, Helmstaedter (2008) The impact of lesions and epilepsy on personality and mood in patients with symptomatic epilepsy: a pre- to postoperative follow-up study. Epilepsy Res 82:139–146

Genetic Testing and Neuroimaging for Youth at Risk for Mental Illness: Trading off Benefit and Risk

Grace Lee, Ania Mizgalewicz, Emily Borgelt and Judy Illes

Abstract *Background* According to the World Health Organization, mental illness is one of the leading causes of disability worldwide. The first onset of mental illness usually occurs during childhood or adolescence, with nearly 12 million diagnosed cases in the United States alone. Neuroimaging and genetic testing have been invaluable in research on behavioral, affective, and attentional disorders, particularly with their potential predictive capabilities, and ability to improve diagnosis and to decrease the associated burdens of disease. The present study focused specifically the perspectives of mental health providers on the role of neuroimaging and genetic testing in clinical practice with children and adolescents. *Methods* We interviewed 38 psychiatrists, psychologists, and allied mental health professionals who work primarily with youth about their receptivity toward either the use of neuroimaging or genetic testing. Interviews probed the role they foresee for these modalities for prediction, diagnosis, treatment planning, and the benefits and risks they anticipate. *Results* Practitioners anticipated three major benefits associated with clinical introduction of imaging and genetic testing in the mental health care for youth: (1) improved understanding of the brain and mental illness, (2) more accurate diagnosis than available through conventional clinical examination, and (3) legitimization of treatment plans. They also perceived three major risks: (1) misuse or misinterpretation of the imaging or genetic data, (2) potential adverse impacts on employment and insurance as adolescents reach adulthood, and (3) infringements on self-esteem or self-motivation. *Limitations* The nature of the

G. Lee · A. Mizgalewicz
National Core for Neuroethics, University of British Columbia, Vancouver, BC, Canada

E. Borgelt
Center for Biomedical Ethics, Stanford University, Stanford, CA, USA

J. Illes (✉)
Faculty of Medicine, University of British Columbia, 2211 Wesbrook Mall, Koerner S124, Vancouver, BC V6T 2B5, Canada
e-mail: jilles@mail.ubc.ca

J. Illes
National Core for Neuroethics, Division of Neurology, Department of Medicine, University of British Columbia, Vancouver, BC, Canada

Curr Topics Behav Neurosci (2015) 19: 189–203
DOI: 10.1007/7854_2014_335

Published Online: 24 July 2014

interview questions focused on the future of neuroimaging and genetic testing research in the context of clinical neuroscience. Therefore, the responses from interview participants are based on anticipated rather than actual experience. *Conclusions* Continued expansion of brain imaging and genetic testing into clinical care will require a delicate balance of brain biology and respect for autonomy in the still-evolving cognitive and affective world of young individuals.

Keywords Mental health disorders · Ethics · Child psychiatry · Neuroimaging · Genetic testing · Qualitative research

Contents

1 Introduction

Mental illness encompasses a set of complex cognitive and affective disorders that impact 10–20% of children and adolescents worldwide (Kieling et al. 2011). The World Health Organization identifies mental and behavior disorders as one of the leading causes of global disability and health-related burden in the first three decades of life (Collins et al. 2011; Gore et al. 2011). The onset and effects of mental illness present challenges to individuals for functioning adequately during daily demands, and to societies for managing pervasive stigmatization (Pescosolido et al. 2007) and rising health care costs (Bloom et al. 2011). Thus, efforts to improve mental health care through research using novel neurotechnologies have garnered tremendous interest and hope. While strategies for early intervention are important, of equal importance are discussions of ethical responsibility to children and adolescents who are most vulnerable to barriers to their full developmental potential.

Research applications of neuroimaging and genetic testing have identified both neurobiological correlates and heritability of mental illness in adults. Various neuroimaging techniques such as positron emission tomography (PET), single photon emission computed tomography (SPECT), and functional magnetic resonance imaging (fMRI) provide measures of the hemodynamic correlates of neural processes in a non-invasive manner and have revealed neurocognitive correlates of mental health disorders and progressive changes associated with illness (Wood et al. 2013; Gogtay et al. 2011; Fusar-Poli et al. 2007). Likewise, genetics studies provide evidence for heritability and validated risk factors (Owen et al. 2007; Chubb et al. 2008; Huang et al. 2010).

Studies in youth are more limited, even while interest in the role of neurotechnology for this population increases (Singh and Rose 2009). Attempts to demonstrate structural or functional abnormalities in at-risk youths are still complicated by the heterogeneity of changes occurring naturally and dynamically (Terry et al. 2009; Pantelis et al. 2007; Velakoulis et al. 2006). Several studies assessing parents' attitudes to pre-symptomatic genetic testing of their own children, for example, suggest a high hypothetical demand (Laegsgaard et al. 2009; Meiser et al. 2008; DeLisi and Bertisch 2006; Jones et al. 2002; Trippitelli et al. 1998). Growing interest has also been expressed in evidence-based predictive models to identify individuals in the At-Risk Mental State (ARMS) who are in the prodromal phase of psychosis (Yung et al. 2003, 2005). While advances in psychiatric research have paved the way for testing and applying these neurotechnologies in youth, they have thus far been unsuccessful at finding consistently reliable and replicable predictors for the onset of mental illness (Paus et al. 2008; Miguel-Hidalgo 2013).

One important mental health area which will very likely integrate neuroimaging and genetic findings is psychiatric research (Tairyan and Illes 2009). Advances in genetic research may lead to an improved understanding of mental health and disease, and support the development of pre-symptomatic and prenatal testing for a more informed diagnosis (Rapoport and Gogtay 2008; Paus et al. 2008; Austin and Honer 2007). When combined, imaging and genetics have three key implications for clinical mental health care: prediction by imaging genetics for early intervention (Pezawas and Meyer-Lindenberg 2010; Meyer-Lindenberg and Weinberger 2006); improved diagnosis using biologically-oriented classification (Adam 2013; Kapur et al. 2012); and tailored interventions as a result of a better understanding of mental disorders (Kemp et al. 2008; Pezawas and Meyer-Lindenberg 2010). These benefits may in turn help foster supportive relationships between patients, providers, and society.

Despite the potential of neuroimaging and genetic testing to help shape clinical care, ethical questions challenge the benefit-risk equation. Prognostication is especially difficult in children and adolescents (Racine et al. 2011), and applications of neuroimaging or genetic tools for prediction oversimplify the complex experiences of living with mental illness (Gillihan and Parens 2011; Huang et al. 2010). Furthermore, a diagnosis based on aberrant imaging or genetic results is comparable to attaching a label that comes with irreversible social consequences (Borgelt et al. 2012a; Austin and Honer 2005). Addressing these and other ethical challenges is essential to guide any translational efforts in clinical neuroscience research.

2 Methods

Participants were recruited through announcements directed at the regional health authority of a Western Canadian metropolitan area as well as from North American professional associations, as previously described (Borgelt et al. 2012c). Inclusion criteria stipulated that respondents work as mental health care providers primarily for children and adolescents and that they be fluent in English.

Respondents participated in semi-structured telephone interviews about their receptivity toward either neuroimaging or genetic testing for the prediction, diagnosis, and treatment of mental illness in children and adolescents. Participants were randomly sorted into the neuroimaging or genetic testing groups.

Interviews were audio recorded and transcribed verbatim and analyzed using NVivo 9 software. Two independent coders (G.L. and A.M.) reviewed and coded the interview transcripts using a constant comparative analytic approach to identify major emergent themes and to establish a consensus list of codes per modality (Boeije 2002; Creswell 2013). This inductive process was iterative and interpretive, revealing new themes that would inform a final coding scheme applied to all interviews. Interviewing stopped when theoretical saturation was achieved in the brain imaging group, which drove the primary research question for the study.

We address themes for each modality separately in describing the results, and contrast but do not necessarily compare them to each other, as is appropriate for qualitative data analyses such as these (Mack et al. 2005). Illustrative quotes highlight major points.

3 Results

Thirty-eight health care providers representing psychiatry, psychology, mental health counseling, nursing, and social work participated in this study. Self-reported sub-specializations in order of increasing frequency were: depression, ADHD, bipolar disorders, autism spectrum disorders, OCD, and posttraumatic stress disorders. Participants' ages ranged from 30 to 75 years, with a mean age of 49 years (Table 1). Sixteen participants were women and 22 of the 37 participants held medical degrees. A total of 21 h of data were collected and analyzed from 28 interviews on brain imaging and 9 interviews on genetic testing.

Overall, three interrelated themes represent the potential benefits that participants attach to neuroimaging and genetic testing. These themes underscore their optimism for eventually including neuroimaging and genetic testing into routine clinical care: (1) improved understanding of the brain and mental health conditions, (2) evidence-based diagnosis to facilitate accuracy and early detection, and (3) legitimization of treatment plans. Respondents also express concerns about risk in terms of: (1) misuse or misinterpretation of results, (2) societal impacts on employment and insurance, and (3) infringements on self-esteem or motivation.

Table 1 Sociodemographic details of participants

Variables	*N* = 38 (%)
Gender	
Male	19 (54)
Female	19 (46)
Marital status	
Married	25 (66)
Single	9 (24)
Common law	1 (3)
Divorced	2 (5)
Widowed	1 (3)
Occupation	
Psychiatrist	24 (63)
Counselor	5 (13)
Registered nurse	3 (8)
Social worker	2 (5)
Psychologist	2 (5)
Mental health clinician	2 (5)
Specialty	
Depression	21 (55)
ADHD	20 (53)
Bipolar disorder	14 (37)
Autism spectrum	12 (32)
OCD	11 (29)
Highest level of education completed	
Baccalaureate	2 (5)
Masters	10 (26)
Medical degree	22 (58)
PhD	4 (11)
Age (Range 30–75 years)	Mean (SD)
	40 (6)

3.1 Benefits

3.1.1 Improved Understanding of the Brain and Mental Health Conditions

Interview respondents acknowledge the potential valuable contributions of neuroimaging for clarifying the patient's and family's understanding of mental illness. Participants regard an improved understanding as a prerequisite for an initial acceptance or admission of the diagnosis.

> I think there would be more compassion for what the patient is enduring. Because with many mental health diagnoses, let's just say depression as an example, I think there's a lack of compassion that people seem to feel you can snap out of it. But having a brain scan to indicate this is not the patient's fault it's something that is going on in the body. And, I think it would be helpful for all of us to have a clearer understanding, to help us appreciate that, you know, these changes are very real (Participant #006, Registered Nurse).

Overall, providers perceive the clinical benefits of neuroimaging as far outweighing the risks, by mitigating conflicts arising from the doctor-patient relationship or offering confidence measures in diagnosis.

> Because the doctors will have some confidence about what they're showing, the modality that they're using is important and worth the time and the money that's involved (Participant #103, Psychiatrist).
> [Brain imaging] would give a clinician a lot of information about the illness and the condition at hand and its response to the treatment. And it will be a very important addition to other kinds of clinical information gathered through other avenues or other techniques (Participant #107, Psychiatrist).

In parallel, providers feel that genetic testing would provide a reliable diagnosis for a young patient's symptoms. Participants' support for genetic testing in the context of improving the current understanding of mental illness relates to their values of having a definitive diagnostic option and mitigating anxieties around insufficient clinical information.

> [Genetic testing] would clarify what they're experiencing. It would just reify it. The disorders in psychiatry and psychology are distorted; [they] are right now, currently, clusters of symptoms . . . So, this would make it more concrete, and that would, actually, really change the nature of the psychiatric diagnostic because it would all of a sudden have something concrete. That could be the defining thing about whether a disorder exists or not (Participant #048, Psychologist).

> Sometimes just having a name to be able to put to what's going on, sometimes just that is a relief. I think it's also a relief because knowing what it is, whether this is true or not, but knowing what it is feels like (Participant #044, Counselor).

> For some clients, having information about their genes may be—they may feel more normalized, they may understand their symptoms better, they might feel, you know, "okay, now I understand why things are the way they are." Versus other people [who] may not believe in that (Participant #055, Counselor).

3.1.2 Evidence-Based Diagnosis to Facilitate Accuracy and Early Detection

Under this general theme, three sub-themes touch on the potential for both modalities to have an impact on mental health diagnosis, diagnostic precision, evidence-based diagnosis, and early detection of mental illness. Diagnostic precision is attributed to the apparently objective nature of brain scans and their perceived ability to differentiate mental disorders with behaviorally indistinguishable phenotypes. In such circumstances, clinical utility of neuroimaging is widely described as a clarification of potentially disputable diagnoses by providing evidence of brain characteristics consistent to a particular disorder.

> So, if there could be some definitive test that says, "Yes, this child really has the brain characteristics of what you see in bipolar disorder." If that were identified, it would be helpful, I think, in treatment for sure—choosing the appropriate medications and kind of overall treatment planning. So that would be terrific to have (Participant #111, Psychiatrist).

> Well I think one of the biggest confounding factors is for us to have a more universal understanding of what diagnosis is. So maybe this would actually help us, by having imaging, because there's so much controversy of how to frame diagnosis . . . So perhaps neuroimaging would actually help resolve that (Participant #125, Psychiatrist).

Providers value early intervention for effective symptom management, and attribute the accomplishment of this goal to the diagnostic accuracy offered by neuroimaging.

> I think it would have good impact in that there would be some public health benefit to early diagnosis, early case findings, and in terms of preventing worsening of disorders (Participant #120, Psychiatrist).

Similarly, participants postulate that genetic testing would improve diagnostic accuracy by providing a scientific basis for validation, and hence minimize the crucial time window between diagnosis and treatment.

> Yeah, if it made the assessment process—like if you could diagnose something, it might make the assessment process go faster, which might mean we could get the treatment sooner (Participant #044, Registered Nurse).

Respondents' views are generally convergent between neuroimaging and genetic testing in the context of diagnosing mental illness. Overall, youth providers describe the merits of both modalities as invaluable supplements to their current diagnostic tools. Providers also emphasize their preference to have access to these modalities for initial clinical assessment and for providing validity to their clinical diagnosis.

3.1.3 Legitimization of Treatment Plans

Providers underscore the desirable outcome of improved patient management through more informed and targeted treatment plans. This theme is often associated with the previous major theme of improving diagnosis.

> Again we would now have some validity, some agreement. Everybody looking at the picture hopefully would see the same thing and would know the implications and treatment would be more streamlined and specific [...] a computer that would sort of use the data and have an ability to match that data and the diagnosis with available treatment. So there would be a greater validity to that as well, more evidence based treatment (Participant #103, Psychiatrist).

Providers suggest that youth at risk would feel empowered by having this insight into their future well-being, and hence take a proactive approach in seeking treatment promptly or making lifestyle changes to possibly prevent or delay symptom onset. In their expressed receptivity to neuroimaging in the context of patient management, participants extrapolate a role for neuroimaging in evaluating current clinical treatment protocols.

> [...] it will also help more specific treatments to be found and explored and discovered for certain illnesses that have some neuroimaging-related findings. So it can have not only diagnostic classification but also developing specific treatment for conditions (Participant #107, Psychiatrist).

> So, if I had a brain imaging scan where I could have someone come back in after they're on medicine, check the brain scan and see how much of that seems to have been corrected in terms of the biology, then I have a better sense that yes, we've got the right kind of medicine, it's doing what it should be doing biologically, and yet we're still having difficulties (Participant #116, Psychiatrist).

Providers affirm that parents would like to know whether their child will have a future without mental illness, especially because the availability of a test that could give either a positive or negative predictive value would help establish an early informed treatment. A positive predictive value would empower health care providers to search for interventions directed toward changing the subsequent course of the disease. On the other hand, a negative predictive value would increase the family's awareness of the possibility of emerging symptoms and dispel any hesitation to seeking mental health care.

3.2 *Risks*

3.2.1 Misuse or Misinterpretation of Results

Participants anticipate that disclosure of neuroimaging findings associated with mental illness might refocus the goal of care toward routinely prescribing treatment for acute symptoms. The focus on prescribing treatment would result in replacement of thorough clinical assessment with neuroimaging to guide medication recommendations. Providers feel that this would be an unjustified use for neuroimaging results. Many providers feel that time may be better spent trying to understand the child's background and experiences to derive a clearer context upon which to make a diagnosis.

> Overall I think it's a positive move, but again it needs to be correlated with the clinical situation, discussed with the patient in an appropriate way and against one of the many tools that we use to evaluate patients. So the larger clinical context, it needs to be put into that context. If it's not, I think it could be misused—or people could come to the wrong conclusion on what a certain set of data may mean that's revealed from the image (Participant #134, Psychiatrist).

Participants also stress the importance of helping the family understand brain scan results in a manner that responsibly incorporates privacy and cultural considerations. Providers urge for better correlation of neuroimaging results with the clinical condition, and for more appropriate discussion of the results in the larger clinical context of brain imaging as one of the many tools used to evaluate patients. They express concerns that misinterpretation of results may cause young individuals to be passive about moral responsibility.

> People may then use that information to try to absolve themselves of some responsibility for their actions and say, "Well, you know, I can't help it. It's just the way my brain is wired" (Participant #133, Psychiatrist).

Similarly, providers express concerns about the careful handling of genetic information given its probabilistic nature, and stress the importance of education and support for both the family and the individual.

> I guess I would be more concerned about what's done with that information afterwards as long as there's education and support for the family or the individual, even though there are say genetic factors that are put into place (Participant #036, Psychiatrist).

Related concerns for medical privacy are raised in the context of access to medical records containing diagnostic and treatment information following neuroimaging or a genetic test. There is an emphasis on the need for establishing safeguards to prevent misuse of these results that could impose limitations on access and cost of health care.

3.2.2 Societal Impacts on Employment and Insurance

Under this theme, respondents address the societal impact of neuroimaging or genetic testing on individuals and their support pillars. Participants expect both increased demands for neuroimaging and genetic testing, and respond to this expectation with concerns about eligibility for employment and insurance.

> [...] someone who is labeled as potentially developing a certain mental illness. There might be employment opportunities, schooling opportunities that would be closed to you if people knew that you were at risk for those conditions (Participant #103, Psychiatrist).

> There is a lot of concern about any kind of prediction of a mental health or a physical health problem that could impact a person's eligibility for insurance coverage and medical care (Participant #109, Psychiatrist).

> And here you have we're again labeling someone as inevitably developing a condition that would perhaps interfere with their job perspective, or relationships. So [brain imaging] would open up, that person for further risk, other sorts of difficulties. It's like any screening tool, now that you know you have it what can you do about it? (Participant #103, Psychiatrist).

> Are people, are employers in the future going to be able to access some kind of [genetic testing] database that says, "You've got the depression gene, I don't want to hire you because you might get depressed in 5 years and miss work." I mean there's all these room for abuses later on (Participant #128, Counselor).

Participants recount how the knowledge of being at risk for mental illness may affect parent hopes and dreams for the child and cause emotional changes to family dynamics. Participants further express concerns about possible detrimental effects of neuroimaging and genetic testing on parents, and in turn, how they relate to the child.

> For some families, knowing that means they are going to change the way they feel about that child. They're going to change their expectations, in a negative way, and perhaps distance themselves. Or more or less reject the child (Participant #127, Psychiatrist).

> I would hate for a parent to either lose hope in their child, you know, because they're thinking, "Oh, this is where we're going to end up anyway. So, what's the point of doing any of these interventions right now, or getting the help that my child needs?" (Participant #036, Registered Nurse).

3.2.3 Infringements on Self-esteem and Motivation

Interview participants also anticipate a possible negative impact on self-esteem and motivation as a risk of positive predictive testing for mental illness. Participants describe a double-edged scenario where brain imaging has the potential to provide clarity on one hand, while condemning individuals into misery associated with morbidity or even mortality in some instances. Several providers express concerns that both neuroimaging and genetic testing would have adverse influences on an individual's attitude or outlook in their private thoughts and outward expressions despite having an informed treatment plan.

> [Brain imaging] is also very limiting and may impact client's motivation levels. And kind of may set them up for a doomed kind of scenario as opposed to a resilient hopeful scenario (Participant #023, Counselor).

> I think no matter what the news is about genetic testing, it would definitely impact an individual's perception of themselves and their, their behavior, their environment, their development; their life, really (Participant #058, Registered Nurse).

To the extent that a predictive diagnosis may impact an individual's behavioral development, practitioners consider the vulnerability to develop affective characteristics such as apathy, depression, and apprehensiveness.

> So you'd worry about suicide, you'd worry about people falling short of their expectations, not pushing themselves (Participant #125, Psychiatrist).

> There may be the sense of, "Hey, this is the way the brain is, and nothing can change." There can be a sense of feeling defeated or feeling that their opportunities in the future are limited (Participant #133, Psychiatrist).

4 Discussion

In this study, we identified perceived benefits and risks of neuroimaging and genetic testing for youth at risk for mental illness, and perspectives from providers who work with children and adolescents in clinical practice. Three major themes emerged about benefits of neuroimaging and genetic testing as diagnostic or predictive tools for mental illness in youth: an improvement in understanding the brain with mental illness and clarity in an emotionally charged time of life; the confirmatory nature of brain scans and acceptance of a diagnosis; and legitimization of treatment recommendations from members of the health care team. The present findings are consistent with other studies suggesting that perceptions of objectivity arising from biological evidence of mental illness through brain imaging replace subjective feelings of "being crazy" (Borgelt et al. 2012b; Buchman et al. 2012; Dumit 2004). Our findings of stakeholder receptivity to neuroimaging as a diagnostic tool for clinical evaluation also parallel other interview studies (Anderson and Illes 2012; Borgelt et al. 2011) and surveys studies (Illes et al. 2008) focused on adults.

The parallel benefits of genetic testing arise from the compassion that participants suggested would accrue to afflicted individuals from others, and a higher personal level of acceptance of a diagnosis (Meiser et al. 2005). Our findings also parallel several studies reporting positive attitudes from family members, parents, and clinicians, and a high hypothetical demand from individuals for genetic testing for mental illness (Smith et al. 1996; Trippitelli et al. 1998; Jones et al. 2002; DeLisi and Bertisch 2006; Meiser et al. 2008). On the other hand, critics have cautioned that genetic testing for mental illness will increase stigma and discrimination, even creating prognostic pessimism, general hopelessness, and self-fulfilling prophecies that hamper recovery (Phelan 2005; Lebowitz et al. 2013; Kvaale et al. 2013). In this regard, it has been argued that in the absence of medical benefit, offering genetic testing to children and adolescents could compromise the child's autonomy as an adult when deciding whether or not to obtain their genetic information (Wertz et al. 1994).

Findings for the benefits of the potential clinical uses for neuroimaging are counterbalanced by weighty potential risks. In particular, providers anticipated that an emphasis on neurobiology might divert attention away from patient history or support networks that are part and parcel of providing comprehensive mental health care. These findings stand apart from other studies that have suggested that neural markers for mental illness would augment clinical decision making rather than replace symptom and behavioral clinical assessments (Linden and Fallgatter 2009). Further research is necessary to understand these conflicting phenomena, alongside the refinement of neuroimaging techniques for seamless integration into the therapeutic process once the issues are resolved (Linden et al. 2012). Although new approaches have already begun to pave the way for translating neuroimaging findings into clinical tools for mental health care, current neuroimaging methodologies face many unmet analytical and clinical challenges (Phillips 2012; Fu and Costafreda 2013).

Providers cautioned that receiving a positive neuroimaging test result for mental illness might also raise the risk of guilt, shame, and stigma that could compromise patient motivation and emotional quality of life. These findings are consistent with previous work reporting public concerns about stigma (Pescosolido et al. 2007) and ceded self-determination or hope to predictive brain imaging (Borgelt et al. 2011).

Perceived stigma by providers and family members was reported to have a negative influence on whether individuals were willing to pursue treatment for mental illness (Raingruber 2002; Wrigley et al. 2005). However, other studies like the present one have suggested that other factors such as confirmative fMRI data could reduce stigma and positively influence treatment compliance and willingness to try medication (Illes et al. 2008; Borgelt et al. 2012a, c; Farah and Gillihan 2012). Future research should examine how integration of imaging and genetics will conspire to affect attitudes about mental health and health care.

The participant sample comprised providers from a range of educational and occupational backgrounds. This heterogeneity is both a benefit and a limitation of the study: it yields a wide range of interdisciplinary views, but limits the transferability of the data to other groups not included in this study, such as community mental health workers, primary educators, or affected youth themselves (Onwuegbuzie and Leech 2007). Future research exploring other groups and perspectives would be insightful. The study reveals projections of future applications of neuroimaging in the mental health care setting, and not actual experiences. This latter limitation is generally accepted in qualitative research (Kvale 2006).

Nevertheless, the results identify the positive potential benefits of neurotechnology, specifically imaging and genetics, for mental health in the youth population and highlight the risk factors associated with such modalities as key challenges to applications in mental health care. Data collection preceded the release of DSM-V, which is aimed at providing more thoroughly delineated classifications of mental disorders than prior versions. Among professionals who disapprove of DSM-V, the most common criticism is that it proliferates diseases by placing a medical disease nomenclature to previously described behaviors, which provides more reasons to administer medications (Frances 2009, 2013; Jabr 2013). Critiques of DSM-V argue that medical classifications are valuable if they are grounded in biological findings, a view that underscores the importance of neurotechnology as studied here. For many, the future application of functional neuroimaging, genetic testing, or both combined, may be key to reaching the goal of understanding disease origin and trajectory.

5 Role of the Funding Source

This study was supported by the NIH/NIMH RO1 #9R01MH084282. The funder had no further role in the study design, data collection, or analysis. Judy Illes is Canada Research Chair in Neuroethics. Grant support from CIHR CNE #85117, the Canadian Foundation for Innovation, the British Columbia Development Fund, and Vancouver Coastal Health Research Institute also enabled this research.

Conflict of Interest The authors declare no conflicts of interest.

Acknowledgments We thank the interview participants for their time and contribution to this study.

References

Adam D (2013) On the spectrum. Nature 496:416–418

Anderson J, Illes J (2012) Neuroimaging and mental health: drowning in a sea of acrimony. AJOB Neurosci 3:42–43. doi:10.1038/mp.2012.105.October

Austin JC, Honer WG (2005) The potential impact of genetic counseling for mental illness. Clin Genet 67:134–142. doi:10.1111/j.1399-0004.2004.00330.x

Austin JC, Honer WG (2007) The genomic era and serious mental illness: a potential application for psychiatric genetic counseling. Psychiatr Serv 58:254–261. doi:10.1176/appi.ps.58.2.254

Bloom D, Cafiero E, Jané-Llopis E et al (2011) The global economic burden of noncommunicable diseases. World Economic Forum and the Harvard School of Public Health, Cologny/Geneva, Switzerland

Boeije H (2002) A purposeful approach to the constant comparative method in the analysis of qualitative interviews. Qual Quant 36:391–409

Borgelt E, Buchman DZ, Illes J (2012a) Practitioners' views on neuroimaging: mental health, patient consent, and choice. In: Richmond S, Rees G, Edwards S (eds) I know what you're thinking: brain imaging and mental privacy. Oxford University Press, Oxford, pp 99–110

Borgelt EL, Buchman DZ, Illes J (2012b) Neuroimaging in mental health care: voices in translation. Front Hum Neurosci 6:1–5. doi:10.3389/fnhum.2012.00293

Borgelt EL, Buchman DZ, Illes J (2011) "This is why you've been suffering": reflections of providers on neuroimaging in mental health care. J Bioeth Inq 8:15–25

Borgelt EL, Buchman DZ, Weiss M, Illes J (2012c) In search of "anything that would help": parent perspectives on emerging neurotechnologies. J Atten Disord 1–7. doi:10.1177/1087054712445781

Buchman DZ, Borgelt EL, Whiteley L, Illes J (2012) Neurobiological narratives: Experiences of mood disorder through the lens of neuroimaging. Sociol Health Illn 35:66–81. doi:10.1111/j.1467-9566.2012.01478.x

Chubb JE, Bradshaw NJ, Soares DC et al (2008) The DISC locus in psychiatric illness. Mol Psychiatry 13:36–64. doi:10.1038/sj.mp.4002106

Collins P, Patel V, Joestl S et al (2011) Grand challenges in global mental health. Nature 475:27–30

Creswell J (2013) Qualitative inquiry and research design: choosing among five approaches (3rd ed.). Sage, Thousand Oaks, p 472

DeLisi LE, Bertisch H (2006) A preliminary comparison of the hopes of researchers, clinicians, and families for the future ethical use of genetic findings on schizophrenia. Am J Med Genet B Neuropsychiatr Genet 141B:110–115. doi:10.1002/ajmg.b.30249

Dumit J (2004) Picturing personhood: brain scans and biomedical identity. Princeton University Press, Princeton

Farah MJ, Gillihan SJ (2012) The puzzle of neuroimaging and psychiatric diagnosis: technology and nosology in an evolving discipline. AJOB Neurosci 3:31–41. doi:10.1080/21507740.2012.713072

Frances A (2009) A warning sign on the radio to DSM-V: beware of its unintended consequences. Psychiatric Times. Available at:http://www.psychiatrictimes.com/articles/warning-sign-road-dsm-v-beware-its-unintended-consequences?verify=0A. Accessed 26 June

Frances A (2013) DSM-5 writing mistakes will cause great confusion. Huffington Post. Available at: http://www.huffingtonpost.com/allen-frances/dsm5-writing-mistakes-wil_b_3419747.html. Accessed 11 June

Fu CHY, Costafreda SG (2013) Neuroimaging-based biomarkers in psychiatry: clinical opportunities of a paradigm shift. Can J Psychiatr 58:499–508

Fusar-Poli P, Perez J, Broome M et al (2007) Neurofunctional correlates of vulnerability to psychosis: a systematic review and meta-analysis. Neurosci Biobehav Rev 31:465–484. doi:10.1016/j.neubiorev.2006.11.006

Gillihan SJ, Parens E (2011) Should we expect "neural signatures" for DSM diagnoses? J Clin Psychiatry 72:1383–1389. doi:10.4088/JCP.10r06332gre

Gogtay N, Vyas NS, Testa R et al (2011) Age of onset of schizophrenia: perspectives from structural neuroimaging studies. Schizophr Bull 37:504–513. doi:10.1093/schbul/sbr030

Gore FM, Bloem PJN, Patton GC et al (2011) Global burden of disease in young people aged 10–24 years: a systematic analysis. Lancet 377:2093–2102. doi:10.1016/S0140-6736(11)60512-6

Huang J, Perlis R, Lee P et al (2010) Cross-disorder genomewide analysis of schizophrenia, bipolar disorder, and depression. Am J Psychiatry 167:1254–1263

Illes J, Lombera S, Rosenberg J, Arnow B (2008) In the mind's eye: provider and patient attitudes on functional brain imaging. J Psychiatr Res 43:107–114. doi:10.1016/j.jpsychires.2008.02.008

Jabr F (2013) New DSV-5 ignores biology of mental illness. Scientific American, Mind and Brain. Available at: http://www.scientificamerican.com/article.cfm?id=new-dsm5-ignores-biology-mental-illness. Accessed 30 April

Jones I, Scourfield J, McCandless F, Craddock N (2002) Attitudes towards future testing for bipolar disorder susceptibility genes: a preliminary investigation. J Affect Disord 71:189–193

Kapur S, Phillips AG, Insel TR (2012) Why has it taken so long for biological psychiatry to develop clinical tests and what to do about it? Mol Psychiatr 17:1174–1179. doi:10.1038/mp.2012.105

Kemp AH, Gordon E, Rush AJ, Williams LM (2008) Improving the prediction of treatment response in depression: integration of clinical, cognitive, psychophysiological, neuroimaging, and genetic measures. CNS Spectr 13:1066–1086 (quiz 1087–1088)

Kieling C, Baker-Henningham H, Belfer M et al (2011) Child and adolescent mental health worldwide: evidence for action. Lancet 378:1515–1525. doi:10.1016/S0140-6736(11)60827-1

Kvaale EP, Haslam N, Gottdiener WH (2013) The "side effects" of medicalization: a meta-analytic review of how biogenetic explanations affect stigma. Clin Psychol Rev 33:782–794. doi:10.1016/j.cpr.2013.06.002

Kvale S (2006) Dominance through interviews and dialogues. Qual Inq 12:480–500. doi:10.1177/1077800406286235

Laegsgaard MM, Kristensen AS, Mors O (2009) Potential consumers' attitudes toward psychiatric genetic research and testing and factors influencing their intentions to test. Genet Test Mol Biomarkers 13:57–65. doi:10.1089/gtmb.2008.0022

Lebowitz MS, Ahn W-K, Nolen-Hoeksema S (2013) Fixable or fate? Perceptions of the biology of depression. J Consult Clin Psychol 81:518–527. doi:10.1037/a0031730

Linden DEJ, Fallgatter AJ (2009) Neuroimaging in psychiatry: from bench to bedside. Front Hum Neurosci 3:1–7. doi:10.3389/neuro.09.049.2009

Linden DEJ, Habes I, Johnston SJ et al (2012) Real-time self-regulation of emotion networks in patients with depression. PLoS One 7:e38115. doi:10.1371/journal.pone.0038115

Mack N, Woodsong C, MacQueen K et al (2005) Qualitative research methods: a data collector's field guide. Family Health International, Research Triangle Park, North Carolina

Meiser B, Kasparian NA, Mitchell PB et al (2008) Attitudes to genetic testing in families with multiple cases of bipolar disorder. Genet Test 12:233–244

Meiser B, Mitchell PB, McGirr H et al (2005) Implications of genetic risk information in families with a high density of bipolar disorder: An exploratory study. Soc Sci Med 60:109–118. doi:10.1016/j.socscimed.2004.04.016

Meyer-Lindenberg A, Weinberger DR (2006) Intermediate phenotypes and genetic mechanisms of psychiatric disorders. Nat Rev Neurosci 7:818–827. doi:10.1038/nrn1993

Miguel-Hidalgo J (2013) Brain structural and functional changes in adolescents with psychiatric disorders. Int J Adolesc Med Health 25:245–256. doi:10.1515/ijamh-2013-0058.BRAIN

Onwuegbuzie AJ, Leech NL (2007) A call for qualitative power analyses. Qual Quant 41:105–121. doi:10.1007/s11135-005-1098-1

Owen MJ, Craddock N, Jablensky A (2007) The genetic deconstruction of psychosis. Schizophr Bull 33:905–911. doi:10.1093/schbul/sbm053

Pantelis C, Velakoulis D, Wood SJ et al (2007) Neuroimaging and emerging psychotic disorders: The Melbourne ultra-high risk studies. Int Rev Psychiatry 19:371–381. doi:10.1080/09540260701512079

Paus T, Keshavan M, Giedd JN (2008) Why do many psychiatric disorders emerge during adolescence? Nat Rev Neurosci 9:947–957. doi:10.1038/nrn2513

Pescosolido BA, Perry BL, Martin JK et al (2007) Stigmatizing attitudes and beliefs about treatment and psychiatric medications for children with mental illness. Psychiatr Serv 58:613–618. doi:10.1176/appi.ps.58.5.613

Pezawas L, Meyer-Lindenberg A (2010) Imaging genetics: progressing by leaps and bounds. Neuroimage 53:801–803. doi:10.1016/j.neuroimage.2010.08.001

Phelan JC (2005) Geneticization of deviant behavior and consequences for stigma: the case of mental illness. J Health Soc Behav 46:307–322. doi:10.1177/002214650504600401

Phillips ML (2012) Neuroimaging in psychiatry: bringing neuroscience into clinical practice. Br J Psychiatry J Ment Sci 201:1–3. doi:10.1192/bjp.bp.112.109587

Racine E, Bell E, Di Pietro NC et al (2011) Evidence-based neuroethics for neurodevelopmental disorders. Semin Pediatr Neurol 18:21–25. doi:10.1016/j.spen.2011.02.002

Raingruber B (2002) Client and provider perspectives regarding the stigma of and nonstigmatizing interventions for depression. Arch Psychiatr Nurs 16:201–207. doi:10.1053/apnu.2002.36230

Rapoport JL, Gogtay N (2008) Brain neuroplasticity in healthy, hyperactive and psychotic children: insights from neuroimaging. Neuropsychopharmacology 33:181–197. doi:10.1038/sj.npp.1301553

Singh I, Rose N (2009) Biomarkers in psychiatry. Nature 460:202–207. doi:10.1038/460202a

Smith L, Sapers B, Reus V, Freimer N (1996) Attitudes towards bipolar disorder and predictive genetic testing among patients and providers. J Med Genet 33:544–549

Tairyan K, Illes J (2009) Imaging genetics and the power of combined technologies: a perspective from neuroethics. Neuroscience 164:7–15. doi:10.1016/j.neuroscience.2009.01.052

Terry J, Lopez-Larson M, Frazier JA (2009) Magnetic resonance imaging studies in early onset bipolar disorder: An updated review. Child Adolesc Psychiatr Clin N Am 18:421–439. doi:10.1016/j.chc.2008.12.004

Trippitelli CL, Jamison KR, Folstein MF et al (1998) Pilot study on patients' and spouses' attitudes toward potential genetic testing for bipolar disorder. Am J Psychiatr 155:899–904

Velakoulis D, Wood SJ, Wong MT et al (2006) Hippocampal and amygdala volumes according to psychosis stage and diagnosis. Arch Gen Psychiatr 63:139–149

Wertz DC, Fanos JH, Reilly PR (1994) Genetic testing for children and adolescents. Who decides? JAMA 272:875–881

Wood SJ, Reniers RLEP, Heinze K (2013) Neuroimaging findings in the at-risk mental state: a review of recent literature. Can J Psychiatry 58:13–18

Wrigley S, Jackson H, Judd F, Komiti A (2005) Role of stigma and attitudes toward help-seeking from a general practitioner for mental health problems in a rural town. Aust N Z J Psychiatry 39:514–521. doi:10.1111/j.1440-1614.2005.01612.x

Yung A, Yuen H, Philips L, et al (2003) Mapping the onset of psychosis: the comprehensive assessment of at risk mental states (CAARMS). Int Congr Schizophr Res 2003

Yung AR, Yuen HP, McGorry PD et al (2005) Mapping the onset of psychosis: the comprehensive assessment of at-risk mental states. Aust N Z J Psychiatr 39:964–971. doi:10.1111/j.1440-1614.2005.01714.x

Externalization of Consciousness. Scientific Possibilities and Clinical Implications

Michele Farisco, Steven Laureys and Kathinka Evers

Abstract The paper starts by analyzing recent advancements in neurotechnological assessment of residual consciousness in patients with disorders of consciousness and in neurotechnology-mediated communication with them. Ethical issues arising from these developments are described, with particular focus on informed consent. Against this background, we argue for the necessity of further scientific efforts and ethical reflection in neurotechnological assessment of consciousness and 'cerebral communication' with verbally non-communicative patients.

Keywords Consciousness · Disorders of consciousness · Neurotechnology · Informed consent

Contents

M. Farisco (✉) · K. Evers
Centre for Research Ethics and Bioethics, Uppsala University, Uppsala, Sweden
e-mail: michele.farisco@crb.uu.se

K. Evers
e-mail: kathinka.evers@crb.uu.se

S. Laureys
Coma Science Group, Cyclotron Research Centre,
University and University Hospital of Liège, Liège, Belgium
e-mail: steven.laureys@ulg.ac.be

Curr Topics Behav Neurosci (2015) 19: 205–222
DOI: 10.1007/7854_2014_338

Published Online: 22 August 2014

1 Introduction

The instrumental investigation of consciousness has witnessed an astonishing progress over the last years. Different neurotechnological tools and methods have been developed in order to assess residual consciousness in patients with disorders of consciousness (DOCs). Functional neuroimaging technologies, such as electroencephalography (EEG), magnetoencephalography (MEG), functional magnetic resonance imaging (fMRI), positron emission tomography (PET), single photon emission tomography (SPECT), event-related potentials (ERPs), magnetoencephalography (MEG), magnetic resonance spectroscopy (MRS), and transcranial magnetic stimulation (TMS) (Laureys et al. 2009), give researchers the possibility to see what happens in the brain during the execution of particular tasks. These emerging neurotechnologies are very promising in regard to the study and the treatment of DOCs. Notably, identification of activated brain areas and real-time observation of cerebral activity potentially allow a new form of technology-based communication in the absence of overt external behavior or speech, thus going beyond the behavioral manifestation of awareness (Evers and Sigman 2013).

It is important to clarify and to assess some issues emerging from this kind of communication. First of all is the relationship between brain activity, which is the specific object of the neuroimaging investigation, and awareness: how to judge when the first implies the second. Another important issue concerns the kind of consciousness that patients with DOCs retain (e.g., can they perceive the same emotional meaning of the provided information?). As a further development of these analyses, the question of how to assess the capacity of patients with DOCs to make an appropriate informed decision will also arise.

In short, the new advancements in neurotechnological assessment of residual consciousness in patients with DOCs raise important ethical issues, such as how to assess residual capacity of self-determination; whether and how much a prospective 'cerebral communication' may be considered as valid for an Informed consent; and whether a prospective direct communication with patients with DOCs through neurotechnology implies the necessity to rethink their clinical management, particularly the role of legal guardians.

According to Laureys and Schiff, the most relevant result of the progress in the neuroimaging investigation of consciousness is the passage from a monolithic way of looking at DOCs to a more graded nosology based on a quantitative assessment of consciousness and on functional neuroimaging technologies. Neurotechnology allowed researchers to detect important neurological differences between patients that are behaviorally classified as equal: As a result, both description and diagnosis of DOCs are more detailed, and new nosographic criteria and categories have been elaborated (Laureys and Schiff 2012). Furthermore, advancements in neuroimaging research have allowed the development of novel investigational paradigms that provide an imaging indication of volition and awareness: This indication may appear but is not unanimously assumed as unambiguous (Laureys and Schiff 2012). One of the earliest studies, conducted by Owen, Laureys and colleagues in 2006 (Owen et al. 2006), is particularly relevant in showing the possible dissociation between the clinical examination based on the behavioral appearance and the results of a neuroimaging assessment (in this case, an fMRI examination). A young woman who survived after a car accident was behaviorally diagnosed as being in a vegetative state (VS) according to the international guidelines. The researchers' team pronounced some sentences (e.g., 'There was milk and sugar in his coffee') and measured through fMRI her neural responses comparing them with responses to acoustically matched noise sequences. Interestingly, the woman's neural reaction to the sentences was equivalent to the control subjects' reactions, yet this result alone is not sufficient to conclude that the woman is aware because there is the possibility of implicit processing: Some aspects of human cognition, as language perception and understanding, can go on without awareness (Fine and Florian Jaeger 2013). For this reason, the research team developed a complementary fMRI study asking the woman to mentally perform two tasks: imagining playing tennis and imagining visiting her house. The relevant result was that the brain activation of the woman was not distinguishable from that of the control subjects, a group of conscious volunteers.

Similar results were obtained in the follow-up study jointly conducted in Liege and Cambridge: 54 patients with severe acquired brain injuries were scanned through fMRI. In response to the request to perform imagery tasks, 5 of them were able to modulate their brain activity by generating blood-oxygenation-level-dependent (BOLD) responses which were judged by the researchers as voluntary, reliable, and repeatable (Monti et al. 2010). Additional tests in one of the 5 responsive subjects revealed his ability to correctly answer yes–no questions through imagery tasks, showing the feasibility of communication. These results are ethically very significant: If new diagnostic tools are available, then it is ethically warranted to use them and to give all the patients the possibility to be rightly diagnosed through them.

Given the possibility that patients with DOCs retain the capacity to communicate and express their own thoughts and preferences, the ethical question of their Informed consent arises.

In this paper, we discuss some technical aspects of fMRI and brain–computer interfaces (BCI) and their prospective use for communicating with patients with DOCs. Furthermore, we analyze the epistemological issue of the relevance of neural activation in the patient for proving or suggesting his/her ability to communicate. Against that background, we analyze emerging ethical issues of Informed consent.

2 The Possibility and Meaning of 'Cerebral Communication'

2.1 fMRI

To date, fMRI is the most commonly used and one of the most promising tools to study DOCs, especially for its noninvasive nature, ever-increasing availability, relatively high spatiotemporal resolution, its capacity to demonstrate the entire network of brain areas activated in particular tasks, and its capacity to provide both anatomical and functional information in the scanned subject. Besides functional data, fMRI techniques also provide other clinically relevant physiological information (e.g., regarding biochemical status, cerebral blood compartment, perfusion, water molecular diffusion, and cerebral microstructure and fiber tracking) (Laureys et al. 2009). There are some limitations to the use of fMRI, for instance in the case of patients who have implanted materials (e.g., metallic implants) that are incompatible with the scanner. In general, the main limitation, or maybe one of the most difficult to assess, especially in case of patients with DOCs, relates to motion artifacts and the duration of the procedure. First, the scanning procedure requires an average time between 15 and 120 min. Second, the methodology used in the fMRI detection of the activated cerebral areas requires repeating the procedure several times in the same subject and/or in different subjects. According to the so-called 'subtraction paradigm,' the brain activation measured before the task (i.e., the control state) is confronted with the brain activation during the task (i.e., the task state), and the difference is assumed to represent the specific brain areas for the task. In order to achieve reliable data, it is necessary to repeat the experiment several times in the same person or in different persons and calculate the average of the results. In this way, it is possible to detect changes in neural activity related to mental activity avoiding the risk of confusing them with false changes resulting from noise (Laureys et al. 2002).

The scientific premise of functional neuroimaging is the functional segregation of the brain. Generally speaking, neuroscientists agree that a cortical area can be specialized for some perceptual or sensorimotor processing and that this specialization is anatomically segregated in the cortex (Laureys et al. 2009). More precisely, it is assumed that the cortical infrastructure of a single function or of a complex behavior can involve different specialized areas combining resources by functional integration between them. As a result, a deep correlation between functional integration and functional segregation is necessary for the brain activities. This coexistence of integration and segregation is the cerebral foundation for functional neuroimaging to be informative about the cerebral activity: Complex behavior can be broken down into more simple and elementary mental operations related to specific cerebral areas.

From a methodological point of view, in the case of the application of fMRI to patients with DOCs, it is important to assess the passive stimulations (i.e., when the patient is not asked to perform any task) and the active paradigms (Boly et al. 2007). Regarding the first point, according to Boly and colleagues, lacking a 'full understanding of the neural correlates of consciousness, even a normal activation in response to passive sensory stimulation cannot be considered as proof of the presence of awareness in patients with DOCs. In contrast, predicted activation in response to the instruction to perform a mental imagery task would provide evidence of voluntary task-dependent brain activity, and hence of consciousness, in non-communicative patients' 'Boly et al. 2007:979'.

We will analyze the issues arising from the assumed 'neural evidence' of consciousness with more details below. What is relevant to note here is that the brain activation in response to passive stimulation is currently not necessarily assumed by the neuroscientific community as proof of consciousness (i.e., awareness). From an ethical point of view, this is relevant, especially regarding Informed consent. The problem is that if brain activation is the only way a patient potentially retains for communicating, but this activation cannot be assumed as proof of conscious activity, then the patient cannot be assumed to be either conscious or able to express a valid Informed consent. For this reason, further technical advancement in the detection of residual consciousness in patients with DOCs is essential in order to resolve the ethical issue of their self-determination (i.e., informed consent).

Regarding the paradigm selection, spatial navigation and motor imagery tasks have been detected as useful mental tasks to identify and assess brain activity and consciousness in patients with DOCs. This new paradigm (i.e., imagery tasks for assessing consciousness through fMRI) could be a useful tool to assess willfulness and consciousness, implement a process of communication with patients with DOCs, and overcome the limitation of the behavioral paradigm based on motor responsiveness.

Neuroimaging in general and fMRI in particular have allowed us to objectively differentiate patterns of cerebral activity in patients with DOCs (Boly et al. 2005). Detection of specific areas yielded by particular tasks is clinically relevant because it potentially gives us the possibility to develop an alternative form of communication with patients lacking the ability to speak and to move (Naci et al. 2013). The aforementioned experiment by Laureys and colleagues, for instance, shows the possibility to communicate with patients by detecting through fMRI the willful activation of specific areas in their brains (Monti et al. 2010). This possibility relies on the identification of the different brain regions and the related mental activities, which have been made possible in recent years.

On the basis of such findings, neuroscientists have defined consciousness as the emergent property of the collective behavior of widespread thalamocortical frontoparietal network connectivity (Laureys 2005a). Moreover, it has been possible to identify the different networks elicited by subjective internal self-related thoughts (self-awareness: midline cortical structures) and by external sensory perceptions (external awareness: lateral frontoparietal structures) (Vanhaudenhuyse et al. 2011). On the basis of this knowledge, an experimental paradigm has been developed in which the brain's response to self-related stimuli such as the patient's own name (Qin et al. 2010), and not to external stimuli, has been measured.

However, as stated above, the activation of a brain area as such is not enough to conclude that the patient is aware, since it could be a case of, for example, passive stimulation reaction or implicit learning (Laureys 2005b). The assumed condition to interpret the neuroimaging data as evidence of consciousness is a time-related condition: The activation of the cerebral area in response to a specific task has to last at least 30 s. In this way, it is possible to disentangle the cerebral activation related to a voluntary (re)action from unconscious reactions that are fleeting (Boly et al. 2005; Greenwald et al. 1996; Naccache et al. 2005). Furthermore, as emerging from the aforementioned experiment by Laureys and colleagues, correct yes–no answers to simple questions such as 'Is your mother's name Yolande?" confirm voluntary origin of the fMRI signal (Monti et al. 2010). Discrimination between voluntary and involuntary brain activity is ethically relevant in regard to the prospective use of neuroimaging for communicating with patients with DOCs and particularly for asking them to give an informed consent.

Research for implementing an fMRI-based communication with patients with DOCs is currently in progress. For instance, a new, noninvasive, relatively fast to apply, and reliable fMRI-based spelling device has recently been proposed as a communication tool, which is potentially promising also for patients with DOCs (Sorger et al. 2012). Yet to date, all these attempts are still at the stage of proofs of concept rather than being practical means to really ensure long-term communication. There are some technical problems in the use of fMRI-based technology to communicate with patients with DOCs. For instance, because of the severe brain damage, the coupling of hemodynamics and neuronal signal, which is at the basis of the fMRI assessment of consciousness, could be very different in patients with DOCs compared to that in healthy people. Moreover, given the plasticity of the brain, the anatomy and functional neuroanatomy could have undergone a functional remapping in patient with DOCs, so that a specific cerebral area could have been functionally replaced by another one.

For the abovementioned difficulties, EEG-based communication devices, the so-called brain–computer interfaces (BCI), are being developed as a potentially more practical, transportable, and cheaper alternative to fMRI for communicating with patients with DOCs (Bruno et al. 2011a; Sorger et al. 2003; Naci et al. 2012; Sellers 2013; Lulé et al. 2013). Other relevant results emerged from a clinical case of complete locked-in syndrome (LIS) showing consciousness via ERP (Schnakers et al. 2009a) and from the measurement of pupil size by a bedside camera to communicate with patients with locked-in syndrome (Stoll et al. 2013).

Another possibility emerging from contemporary neurotechnology is the use of TMS-EEG as a tool to probe consciousness in patients with DOCs (Casali et al. 2013; Jacobo et al. 2013). Furthermore, TMS-EEG potentially gives researchers a tool for developing a communication paradigm with patients with DOCs.

2.2 *Brain–Computer Interface*

BCI is a direct connection between living neuronal tissue and artificial devices that establishes a non-muscular communication pathway between a computer and a brain (Wolpaw et al. 2002). Through BCI, it is possible to detect changes in neuroelectrical activity or brain activity in response to sensory stimulation. The user is then trained to use these changes to select items, words, or letters in communication software or to make choices for neuroprosthesis control (Kübler 2009).

BCI is grounded in a continuous, real-time interaction between living brain and artificial effectors. In this way, a functional hybridization between brain and technology is realized. The operation scheme of a BCI is quite simple: The input is the user's intent coded in the neural activity of her/his brain detected through BOLD response. The output is the device controlled by the user's brain activity.

BCI uses a representation of the subject's mentation as the essential component. The psychological task or the intention of the subject is detected and recorded through invasive or noninvasive methods, mostly EEG using surface or implanted electrodes, but also MEG, fMRI, or functional near infrared spectroscopy (fNIRS). There is a significant difference between these methods regarding the ease of use. For instance, while MEG and fMRI are more demanding, require quite sophisticated instruments, and are quite expensive, EEG, NIRS, and invasive systems are portable and thus suitable for use in daily life (Kübler 2009).

In the particular case of patients with DOCs, EEG offers significant comparative advantages on the aforementioned points. Furthermore, it can be useful to develop EEG-BCI systems that can be used at the bedside to detect volitional brain activity and to enable basic communication.

Thus, to date, EEG-based techniques are the most suitable BCIs for clinical application to patients with DOCs even if other technologies, such as fMRI, allow a more detailed spatial resolution and a more precise allocation of neuronal activity than EEG. Whatever technology is used, the detected and recorded cerebral signals are digitized and differently processed by filtering, amplitude measurement, and spectral analysis (Wolpaw et al. 2002). Specific algorithms then translate the processed signals into commands expressing the users' will. In particular, the subject may communicate choosing the words on a screen moving a cursor through his own mind. In this way, BCI provides subjects with a virtual keyboard where the user can press the keys through the brain activity's modulation.

Importantly, BCI provides the user with real-time feedback on their performance, giving her/him the possibility to improve the use of the BCI over time. BCI thus enables a cerebral communication without motor response. This cerebral communication could give to some behaviorally non-responsive patients, such as patients with DOCs, a new opportunity to communicate.

There are several prerequisites to use BCI for communicating with patients with DOCs. The patient should be able to properly understand verbal commands. The patient should also be able to react to external stimulation and express her/his answers through a minimal form of communication (e.g., a binary yes/no communication) while remaining sufficient cognitive capacities enabling the formulation of a reliable informed decision (Lulé et al. 2013). It is possible that patients retain the ability to partially understand commands, to understand but not to follow commands, or to understand and to follow commands but not well enough to make BCI feasible. In order to use a BCI with patients with DOCs, the understanding of the provided information should be matched with their ability to attend to stimuli, to selectively process the salient ones, and to retain information in working memory (Chatelle et al. 2012).

The results emerging from the aforementioned studies by Laureys, Owen, Schiff and others are relevant and promising also in the direction to use BCI with patients with DOCs. A possible communication protocol through BCI emerges from the experiments by Cruse and colleagues (Cruse et al. 2011). They investigated the capacity of patients with DOCs to perform mental motor tasks that are possible to differentiate in their EEG at the single-trial level. Sixteen patients in VS/UWS were asked to imagine squeezing their right hand or moving all their toes, and in 19 % of the patients a support vector machine predicted the task being executed with an accuracy of between 61–78 %. The same test was performed with MCS patients, where 22 % of them were able to follow commands using motor imagery (Cruse et al. 2012). Starting from these results, it could be possible to implement a binary communication by assuming imagination of right hand as 'yes' and the imagination of toe movement as 'no' (Chatelle et al. 2012).

Another relevant study has been conducted by Lulé and colleagues who tested an EEG-BCI paradigm on 16 healthy subjects and 18 patients in a VS/UWS, in a MCS, and in LIS (Lulé et al. 2013). The results of the study showed that 13 healthy subjects and 1 LIS patient were able to communicate through BCI, and 1 patient in MCS who was unresponsive at the bedside showed command following with the BCI, while all other patients did not show any response to command and could not communicate through BCI. Even if no patients with DOCs were able to functionally communicate through BCI, this study is relevant and promising in showing command following in one patient in MCS.

Further research is needed in order to assess relevant issues limiting the feasibility of BCI with patients with DOCs. Particularly, it is necessary to investigate potential limitations and benefits of multimodal visual–audio–tactile stimulation (Chatelle et al. 2012; Lulé et al. 2013). For instance, a visual-based BCI is sometimes more accurate than an auditory-based BCI (Blankertz et al. 2010). Moreover, auditory stimuli cannot be presented simultaneously and require a longer time to present than visual stimuli (Sellers 2013). Another factor that potentially raises problems for the application of auditory-based BCI to patients with DOCs is that auditory stimulation requires more training (Kübler 2009).

Further studies should investigate the possible long-term mental capabilities potentially retained in patients with DOCs. On the basis of these capabilities, it could be possible to implement a communication that is more complex than a binary yes/no communication (Chatelle et al. 2012).

Because of their critical neurological condition, on the basis of the aforementioned studies, it is reasonable to expect that patients with DOCs will eventually be able to partially understand and execute external requests for mental tasks. The question then arises at what level of accuracy the communication can be considered effective.

Other variables to take into account in the evaluation of the results emerging from experiments with BCI involving patients with DOCs are the possibility of questions too difficult to answer or asked when the patients were sleeping, and the fact that movement, ocular, and respiration artifacts are involuntary and can interfere with the instrumental assessment with false-positive results (Boly et al. 2005). These possibilities are ethically relevant because an inadequate assessment of BCI communication may lead to inappropriate clinical decisions.

In short, three main difficulties emerge from the use of BCI for communicating with patients with DOCs (Chatelle et al. 2012): Patients with DOCs have sensory dysfunction, aphasia, arousal fluctuation, and limited attention span; the suitability of BCI is variable for different patients with DOCs.

Besides the difficulties related to the critical neurological condition of the patients with DOCs, potential limitations to the use of BCI for communication with such patients also arise from the general difficulty to map intended responses to motor imagination, which is a complex task that can be challenging to perform for healthy adults as well (Guger et al. 2003). Given the difficulties summarized above, even though the studies have shown promising results, significant time and effort are needed in order to have a clinical application of BCI with patients with DOCs and to improve their quality of life (Chatelle et al. 2012).

In conclusion, it is clear that further investigations and efforts are essential for developing the communication with patients with DOCs through BCI. However, it is also clear that the clinical application of BCI to detect signs of consciousness in patients with DOCs, particularly in patients with MCS, is already feasible and very promising, especially for solving the major problem of misdiagnosing DOCs and for improving clinical care (Lulé et al. 2013). In fact, if repeated reactions to volitional paradigms are detected, it is reasonable to infer that higher cognitive processing and volition are present in these patients. A different question that still remains open is whether and how these responding patients may be able to use their brain responses for controlling a BCI and how much integrity and connectivity of the brain is necessary for a minimal communication through BCI (Kübler 2009).

3 Informed Consent

As a result of these neuroscientific and neurotechnological developments, the ethically and legally relevant question arises: could a reliable and effective 'cerebral' communication justify an assumption of a right to self-determination of these patients? Could it, for example, be justified to ask them for an informed consent to treatment? With current machine learning technology, the answer currently is negative: Giving an informed consent is not yet realistic for these patients. However, while the actual possibility of DOC patients to give informed consent is absent today, the theoretical possibility is present, and with the further development of these technologies, the situation might come to change. Generally speaking, if communication with patients with DOCs through neurotechnology is feasible, it would be ethically warranted to use and improve it by giving the patients new possibilities to exert their autonomy and self-determination.

The formal condition to have self-determination through informed consent in a medical context is the existence of a relationship between the clinician and the patient. Relationship implies a process of communication, that is, the capacity of the involved subjects to express their thoughts and eventually to answer emerging questions on the basis of those thoughts. A cerebrally communicating patient with DOC is formally able to be in relationship with the clinician, which means that the 'formal condition' for an informed consent could be satisfied.

Yet there is also what could be named a 'substantial condition' for informed consent: The patients have to retain the capacity properly to understand the information provided and to make a choice on this basis between options and the related consequences. This second condition seems problematic if referred to patients with DOCs: We are not sure that they retain the ability to process the provided information and to properly figure out the related consequences.

More specifically, it is generally accepted that to be valid an informed consent has to respect the following requirements (Faden and Beauchamp 1986):

- Disclosure
- Capacity
- Voluntariness

Disclosure implies two requirements for the clinician/researcher: (1) She/he has to give the patient all the needed information for an autonomous decision and (2) she/he has to check the adequate understanding of the information by the patient. Thus, the clinician/researcher has to describe all the possible clinical or experimental options and all the consequent implications for the patient. Furthermore, she/he has to ensure an adequate understanding of the provided information, both choosing an appropriate tool of communication (e.g., written or oral presentation) and a suitable system of checking the understanding of the information (e.g., through pertinent questions asked at different times).

The second requirement for a valid informed consent is the capacity, which entails the actual ability of the patients not only to understand the information provided, but also to make a reasonable judgment concerning the prospective consequences of her/his decision.

The last requirement, voluntariness, refers to the patient's right to decide without undue coercion or influence.

In short, an autonomous decision expressed through an informed consent presupposes that the patient retains four essential components: (1) understanding relevant information concerning treatment or research and related risks and benefits, (2) appreciating different therapeutic or research methods and related consequences, (3) reasoning about different options, and (4) communicating a personal choice (Petrini 2010; American Psychiatric Association 1998).

Tools for assessing these standards have been proposed (Grisso and Appelbaum 1998; Appelbaum 2007), but criticisms have been raised regarding the risk of not taking into account the emotional dimension of the informed consent process (Northoff 2006), and their application to patients with DOCs seems problematic. A conceptual foundation of a mechanistic explanation of capacity has been proposed recently in order to develop the proper tools to empirically detect and assess capacity in patients with DOCs who show responsiveness through BCI or fMRI-BCI (Peterson et al. 2013). The starting point of this proposal is that the inner mental life of some patients behaviorally diagnosed as VS may be richer and more active than assumed so far and that this mental activity could be used to implement a BCI communication in exceptional cases. Even so, to communicate through BCI with patients with DOCs does not guarantee the possibility to involve these patients in medical decision making. The question arises whether or not these BCI-communicating patients retain the ability to make an informed decision regarding their ordinary or experimental treatment. Peterson and colleagues suggest a reductionist approach in order to detect this capacity in patients with DOCs: They start from the MacCAT-T criteria (i.e., understanding, reasoning, justification, appreciation) and try to decompose them in sub-components that can be assessed even in patients with DOCs. This is a robust empirical approach, with prospective relevant clinical implications in the direction of translating BCI-mediated communication from laboratory to clinics. Yet this approach focuses only on cognitive functions without taking into account the emotional dimensions of the informed consent process (Northoff 2006). Autonomy is a complex capacity, which relies not only on cognition, but also on emotion, morality, and sociality (Jox 2013).

It is usually not so easy to implement the communicative process of informed consent described above and particularly tricky to assess the emotional components and the extra-scientific variables that come into play, for example, the sociological and psychological background of the patient that affect and potentially bias his/her understanding of the information (Northoff 2006). These difficulties seem even more relevant in the case of patients with DOCs, even if they are able to cerebrally communicate. It would be very complicated to properly communicate relevant and complex and often specialized information regarding therapy and/or research to these patients. Furthermore, it is reasonably difficult to assess their ability to process this information, to properly understand it, to develop a reasonable judgment regarding the consequences of the prospective choices, and to freely take the better decision for themselves. Thus, even if the experimentally developed system of cerebral communication with patients with DOCs will in due course be translated into clinical practice, its prospective use for a direct consent from the patient remains ethically challenging.

The use of neurotechnology for obtaining informed consent from patients with DOCs is challenging also at the legal level. At present, the role of the legal guardian is not questioned, as this legal instrument is necessary for these patients. Several issues arise concerning the role of guardians, especially concerning the biases they (and clinicians as well) may have regarding quality of life and end-of-life decisions. Empirical results suggest that the personal well-being of chronically ill people is often higher than expected (Albrecht and Devlieger 1999; Bruno et al. 2011b). Even if we have no direct evidence, it is reasonable to assume that a further developed neurotechnology-mediated direct communication can be expected to increase their quality of life, a consideration that strengthens the ethical imperative to support this development. In addition, novel technologies should be further used to better disentangle the different DOCs, whose diagnoses have been shown to have an astonishingly high rate of error estimated between 30 and 40 % (Schnakers et al. 2009b). A better diagnosis would allow a better therapeutic strategy for the patient, for example, by detecting the patient's perception of pain and/or suffering (Demertzi et al. 2013) or her/his inclusion in an appropriate clinical trial. Importantly, even if cerebrally communicative, patients with DOCs remain highly vulnerable and this vulnerability likely affects his/her capacity of self-determination. Because of the aforementioned reasons, the use of neurotechnology to obtain informed consent from patients with DOCs is ethically problematic. A reasonable position could be to distinguish between different kinds of decisions, namely between therapeutic, experimental, and more drastic decisions such as euthanasia, to give a different weight to the patients' answers in relation to the different contexts.

There is also another question related to the 'big issue' of the relationship between neural activity and consciousness, a question pertaining to the meaning of information rather than the processing of information, particularly to its emotional meaning. Even if we must conclude that the correct activation of a particular area to a specific yes–no question suggests the preservation of awareness in the scanned subject, we do not know what kind of awareness she/he preserves. In accordance with the global neuronal workspace model (Dehaene et al. 2011), awareness is the result of functional integration of different areas of the brain. All we can infer from the neuroimaging assessment is that particular areas are processing information, but their interrelation with other cerebral areas remains problematic, so that the significance of their particular activation for awareness also remains problematic. It is theoretically possible, for instance, that the patient is able to functionally process the information but not to meaningfully process the information, in the sense that she/he is not able to relate a specific emotional meaning to the information or to relate the same meaning compared with a healthy person. In that case, we cannot be sure that the meaning that the words have for us is the same meaning that they have for the patient, and while they appear to understand the questions, this apparent understanding remains uncertain and open for interpretation.

Functional responsiveness of the patient, shown to be able to perform specific mental tasks, like moving through a space or playing a sport (Owen et al. 2006), suggests the preservation of semantic capacity. Yet even in this case, it is possible that the capacity to understand the meaning of the information is limited or different from that in healthy people. Particularly, it could be limited to the functional meaning of external information, and the patient could be unable to really understand the meaning of self-related requests concerning, for instance, caring and end-of-life decisions.

The identified conditions for a reliable and effective communication with patients with DOCs for evidence of awareness are robustness, repeatability and correct responses to simple questions. From an ethical point of view, the ability to communicate does not imply the ability to make informed decisions, since capacity is not competence. As stated above, we agree with Peterson and colleagues that, given the present uncertainty regarding the effective capacity of patients with DOCs to make a valid informed decision, they should be allowed to participate in clinical decision making if the capacity threshold for the decision in question is sufficiently low (e.g., for treatments options instead of end-of-life decisions). Notwithstanding the problems summarized above, the involvement of patients with DOCs, who are unable to express an informed autonomous decision, has been gradually justified in clinical practice and research.

4 Discussion

Recent investigations of residual consciousness in patients with DOCs led to new possibilities for communicating with them, i.e., in a 'cerebral' communication without external behavior. This prospective new form of communication potentially raises new ethical challenges, such as the necessity to assess the residual capacity of self-determination in patients with DOCs, the necessity to clarify whether cerebral communication is valid for informed consent, and the necessity to clarify whether cerebral communication implies new forms of management of patients with DOCs.

In the fMRI assessment of consciousness and in the related implementation of communication (with or without BCI), there is an assumption that deserves particular attention. If the BOLD response in the patient is similar to the response in healthy volunteers, it is reasonable to assume that the patient is conscious. Schematically, the logic is the following: (1) Task *t1* implies some signal change (BOLD, EEG or other) which can be identified by classifier (support vector machine learning or other) without a priori assumptions on neuroanatomy or normal patterns; (2) task *t1* implies the yes–no response of the patient as identified by the classifier; and (3) if the yes–no response is correct, then the patient consciously communicated. In active paradigms, a correct communication can be assumed as final evidence of conscious origin. However, as pointed out above, in passive paradigms, the activation of a particular cerebral area per se need not suggest a conscious activity because it could be an 'automatic' processing. Therefore, it is necessary to develop proper clinical and/or neuroimaging protocols in order to assess this risk of false positive. The question of the relationship between brain activation and consciousness however remains open: When is the activation of a cerebral region equivalent to or evidence of the presence of awareness? In the case of patients with DOCs, this issue is ethically highly relevant, particularly regarding the possible neurotechnology-mediated informed consent. This could be required only if it is reasonable to assume that the detected brain signals are evidence of consciousness.

If neurotechnology-mediated communication with patients with DOCs is feasible, it would be ethically warranted to ask them directly for informed consent to ordinary or experimental treatments. This makes it all the more important to clarify the conditions for an effective and reliable communication with the patients through neuroimaging. The theoretical premise for the use of volitional paradigms in the neuroimaging assessment of awareness and volition is that the patient is able to understand instructions, wants and is able to perform what is required (Kübler 2009). In the execution of the investigation and in the interpretation of the emerging data, it is important to assess the risk of false-positive results. A robust and repeated activation of the brain area of interest in response to external instructions has been suggested as proof of the capacity to understand and obey command. In case of evident and repeated brain activation in response to different kinds of tasks at different times, it is reasonable to conclude that a reliable communication is taking place. In other words, we agree that robustness, repeatability, and correct responses to simple questions for communication are essential requirements for assuming neuroimaging evidence of cerebral activation in patients with DOCs as evidence of awareness (Kübler 2009).

Even if a form of cerebral communication with patients with DOCs is possible, assessing their capacity to understand the provided information in order to express a valid informed consent is a challenge. A method commonly used to assess the patient's capacity to understand is to ask her/him to describe with her/his own words what previously communicated by the researcher (Leo 1999; Appelbaum PS-Grisso 1988). This is presently not possible in the case of patients with DOCs, who are able only of a yes–no communication through neuroimaging. It is also problematic to assess the ability of patients with DOCs to appreciate the provided information. Appreciating the information means that the patient is aware that such information is applicable to her/him at a specific time (Gert et al. 2006). In other words, it requires the ability to understand the notions of self and time, and the ability to refer to the self as a dynamic entity shaped through time.

Regarding the reasoning as a requirement for an autonomous decision, a yes–no responsive patient could have impairments in executive function, that is, in the ability to organize, plan, and categorize information. She/he could be able to understand specific information, but not to collect and coordinate different information in order to make a complex decision like withholding or withdrawing life-sustaining treatments. Communication of a personal choice is also problematic in patients with DOCs. Besides possible cognitive impairments, they can present volitional impairments that do not allow them to make and express a personal decision. Finally, the abilities required for an autonomous decision and for a proper informed consent (i.e., understanding, appreciation, reasoning, and choice) are gradable abilities. It is possible that a patient with DOCs retains them only partially. This raises the question of what degree of the aforementioned abilities the subject must retain in order to consider her/him as able to give an informed consent.

The abovementioned difficulties to assess the patient's capacity of proper understanding to give a valid informed consent are technically and ethically relevant. While it can currently be questioned whether a positive cerebral activation in response to particular tasks is evidence of consciousness, it is important to note that a negative result does not imply the absence of willful responsiveness. In fact, different factors could impair the ability to react to external stimulation, even if the patient retains awareness (Boly et al. 2007). Some types of brain damage could impair the patient's ability to understand and/or to perform the selected task. These impairments could lead to a relative reorganization of the brain, involving other areas in the execution of a specific task. Evidence of residual awareness could be flickering and fluctuating. In addition, the patient could decide to not execute the command or be sleeping during the execution of the task. The argument that a negative result cannot be assumed as evidence of absence of consciousness is also ethically relevant for the management of patients with DOCs, especially regarding end-of-life decisions.

Notwithstanding the progress in the use of neurotechnology for diagnosing patients with DOCs and for communicating with them, further studies are needed particularly in the following domains: assessing possible obstacles in the use of neuroimaging for communicating; refining technology in order to disentangle voluntary from involuntary brain activity; training patients on using BCI; defining how much cerebral integrity is necessary for communicating through BCI; investigating possible functional brain remapping affecting patients' capacity to process information; and assessing how to ensure and check an adequate understanding of information by patients. At present, we have the technology, but we need new categories to describe patients who are behaviorally unresponsive but cerebrally communicative (Giacino et al. 2014).

5 Conclusion

The prospective development of a neurotechnology-mediated communication with patients potentially offers new way to exercise the right to self-determination. Specifically, neurotechnology can give clinicians the opportunity to detect covert awareness and facilitate a correct diagnosis or to directly communicate with patients, for example, by asking them if they feel pain. The most important impact of these prospective applications concerns ethical considerations of informed consent. It is theoretically possible to ask patients directly for informed consent by communicating through neurotechnology, but the complexity of the decisions to take in clinical context (which affects the rational as well as the emotional subjective dimensions) urges great precaution. It seems as yet premature to assume that a 'cerebral communication' with patients is enough to assess important ethical issues like informed consent since further investigations are scientifically and ethically appropriate.

In conclusion, to date, the use of neurotechnology to communicate with patients is still at the stage of proof of concepts, but the theoretical possibility and empirical results thus far strongly urge a continued reflection about possible clinical implementations. Particularly, cerebral communication with these patients to express an informed consent raises important theoretical as well as practical issues: the patient's effective ability to understand and process the provided information, her/his ability to integrate the provided information to make a coherent personal decision, and her/his ability to feel the relevance of the clinical options. These issues deserve further reflection at the scientific, technical, legal, and ethical levels. And from an ethical perspective, we should note that even scientifically minor advances could yield important improvements from the patient's point of view.

Acknowledgments This work was done within the European Human Brain Project. The text reflects solely the views of its authors. The European Commission is not liable for any use that may be made of the information contained therein.

References

Albrecht GL, Devlieger PJ (1999) The disability paradox: high quality of life against all odds. Soc Sci Med 48(8):977–988

American Psychiatric Association (1998) Guidelines for assessing the decision-making capacities of potential research subjects with cognitive impairment. Am J Psychiatry 155:1649–1650

Appelbaum PS (2007) Assessment of patients' competence to consent to treatment. New Engl J Med 357:1834–1840

Appelbaum PS-Grisso T (1988) Assessing patients' capacities to consent to treatment. N Engl J Med 319(25):1635–1638

Blankertz B, Sannelli C, Halder S, Hammer EM, Kübler A, Muller KR, Curio G, Dickhaus T (2010) Neurophysiological predictor of SMR-based BCI performance. Neuroimage 51:1303–1309

Boly M, Faymonville ME, Peigneux P, Lambermont B, Damas F, Luxen A, Lamy M, Moonen G, Maquet P, Laureys S (2005) Cerebral processing of auditory and noxious stimuli in severely brain injured patients: differences between VS and MCS. Neuropsychol Rehabil 15 (3–4):283–289

Boly M, Coleman MR, Hampshire A, Bor D, Moonen G, Maquet P, Laureys S, Owen A (2007) When thoughts become action: an fMRI paradigm to study volitional brain activity in non-communicative brain injured patients. Neuroimage 36(3):979–992

Bruno MA, Gosseries O, Ledoux D, Hustinx R, Laureys S (2011a) Assessment of consciousness with electrophysiological and neurological imaging techniques. Curr Opin Crit Care 17 (2):146–151. doi:10.1097/MCC.0b013e328343476d

Bruno M-A, Bernheim JL, Ledoux D, Pellas F, Demertzi A, Laureys S (2011b) A survey on self-assessed well-being in a cohort of chronic locked-in syndrome patients: happy majority, miserable minority. BMJ Open 1:e000039. doi:10.1136/bmjopen-2010-000039

Casali AG, Gosseries O, Rosanova M, Boly M, Sarasso S, Casali KR., Casarotto S, Bruno MA, Laureys S, Tononi G, Massimini M (2013) A theoretically based index of consciousness independent of sensory processing and behavior. Sci Trans Med 5:198ra105. doi:10.1126/scitranslmed.3006294

Chatelle C, Chennu S, Noirhomme Q, Cruse D, Owen AM, Laureys S (2012) Brain–computer interfacing in disorders of consciousness. Brain Inj 26(12):1510–1522. doi:10.3109/02699052.2012.698362

Cruse D, Chennu S, Chatelle C, Bekinschtein T, Fernandez-Espejo D, Junque C, Pickard J, Laureys S, Owen A (2011) Bedside detection of awareness in the vegetative state. Lancet 378: 2088–2094

Cruse D, Chennu S, Chatelle C, Fernandez-Espejo D, Bekinschtein T, Pickard J, Laureys S, Owen A (2012) The relationship between aetiology and covert cognition in the minimally-conscious state. Neurology 78:816–822

Dehaene S, Changeux JP, Naccache L (2011) The global neuronal workspace model of conscious access: from neuronal architectures to clinical applications. In: Dehaene S, Christen Y (eds) Characterizing consciousness: from cognition to the clinic research and perspectives in neurosciences. Springer, Berlin, Heidelberg, pp 55–84

Demertzi A, Racine E, Bruno M-A, Ledoux D, Gosseries O, Vanhaudenhuyse A, Thonnard M, Soddu A, Moonen G, Laureys S (2013) Pain perception in disorders of consciousness: neuroscience, clinical care, and ethics in dialogue. Neuroethics 6(1):37–50

Evers K, Sigman M (2013) Possibilities and limits of mind-reading. A neurophilosophical perspective. Consciousness and Cognition 22(3):887–897. doi:10.1016/j.concog.2013.05.011

Faden RR, Beauchamp TL (1986) A history and theory of informed consent. Oxford University Press, New York

Fine AB, Florian Jaeger T (2013) Evidence for implicit learning in syntactic comprehension. Cogn Sci 37(3):578–591

Gert B, Culver CM, Clouser KD (2006) Bioethics: a systematic approach. Oxford University Press, New York

Giacino JT, Fins JJ, Laureys S, Schiff ND (2014) Disorders of consciousness after acquired brain injury: the state of the science. Nat Rev Neurol 10:99–114

Greenwald AG, Draine SC, Abrams RL (1996) Three cognitive markers of unconscious semantic activation. Science 273(5282):1699–1702

Grisso T, Appelbaum PS (1998) Assessing competence to consent to treatment. A guide for physicians and other health professionals. Oxford University Press, New York, Oxford

Guger C, Edlinger G, Harkam W, Niedermayer I, Pfurtscheller G (2003) How many people are able to operate an EEG-based brain-computer interface (BCI)? IEEE Trans Neural Syst Rehabil Eng 11:145–147

Jacobo DS, King JR, Naccache L, Dehaene S (2013) Ripples of consciousness. Trends Cogn Sci 17(11):552–554

Jox RJ (2013) Interface cannot replace interlocution: why the reductionist concept of neuroimaging-based capacity determination fails. AJOB Neurosci 4(4):15–17

Kübler A (2009) Brain–computer interfaces for communication in paralysed patients and implications for disorders of consciousness. In: Laureys S, Tononi G (eds) The neurology of consciousness. Elsevier, London, pp 217–233

Laureys S (2005a) The neural correlate of (un)awareness: lessons from the vegetative state. Trends Cogn Sci 9:556–559

Laureys S (2005b) Science and society: death, unconsciousness and the brain. Nat Rev Neurosci 6 (11):899–909

Laureys S, Schiff ND (2012) Coma and consciousness: paradigms (re)framed by neuroimaging. NeuroImage 61(2):478–491. doi:10.1016/j.neuroimage.2011.12.041

Laureys S, Peigneux P, Goldman S (2002) Brain imaging. In: D'haenen H, den Boer JA, Willner P (eds) Biological psychiatry, vol 1. Wiley, New York, pp 155–166

Laureys S, Boly M, Tononi G (2009) Functional neuroimaging. In: Laureys S, Tononi G (eds) The neurology of consciousness. Cognitive Neuroscience and Neuropathology. Elsevier, London, pp 31–42

Leo RJ (1999) Competency and the capacity to make treatment decisions: a primer for primary care physicians. Prim Care Companion J Clin Psychiatry 1(5):131–141

Lulé D, Noirhomme Q, Kleih SC, Chatelle C, Halder S, Demertzi A, Bruno MA, Gosseries O, Vanhaudenhuyse A, Schnakers C, Thonnard M, Soddu A, Kübler A, Laureys S (2013) Probing command following in patients with disorders of consciousness using a brain-computer interface. Clin Neurophysiol 124(1):101–106. doi:10.1016/j.clinph.2012.04.030

Monti MM, Vanhaudenhuyse A, Coleman MR, Boly M, Pickard JD, Tshibanda L, Owen AM, Laureys S (2010) Willful modulation of brain activity in disorders of consciousness. N Engl J Med 362(7):579–589. doi:10.1056/NEJMoa0905370

Naccache L, Gaillard R, Adam C, Hasboun D, Clemenceau S, Baulac M, Dehaene S, Cohen L (2005) A direct intracranial record of emotions evoked by subliminal words. Proc Natl Acad Sci USA 102(21):7713–7717

Naci L, Monti MM, Cruse D, Kübler A, Sorger B, Goebel R, Kotchoubey B, Owen AM (2012) Brain–computer interfaces for communication with nonresponsive patients. Ann Neurol 72 (3):312–323. doi:10.1002/ana.23656

Naci L, Cusack R, Jia VZ, Owen AM (2013) The brain's silent messenger: using selective attention to decode human thought for brain-based communication. J Neurosci 33 (22):9385–9393. doi:10.1523/JNEUROSCI.5577-12.2013

Northoff G (2006) Neuroscience of decision making and informed consent: an investigation in neuroethics. J Med Ethics 32(2):70–73

Owen AM, Coleman MR, Boly M, Davis MH, Laureys S, Pickard JD (2006) Detecting awareness in the vegetative state. Science 313(5792):1402

Peterson A, Naci L, Weijer C, Cruse D, Fernández-Espejo Davinia, Graham Makenzie, Owen AM (2013) Assessing decision-making capacity in the behaviorally nonresponsive patient with residual covert awareness. AJOB Neurosci 4(4):3–14

Petrini C (2010) Informed consent in experimentation involving mentally impaired persons: ethical issues. Ann Ist Super Sanita 46(4):411–421. doi:10.4415/ANN_10_04_09

Qin P, Di H, Liu Y, Yu S, Gong Q, Duncan N, Weng X, Laureys S, Northoff G (2010) Anterior cingulate activity and the self in disorders of consciousness. Hum Brain Mapping 31 (12):1993–2002. doi:10.1002/hbm.20989

Schnakers C, Perrin F, Schabus M, Hustinx R, Majerus S, Moonen G, Boly M, Vanhaudenhuyse A, Bruno MA, Laureys S (2009a) Detecting consciousness in a total locked-in syndrome: an active event-related paradigm. Neurocase 15(4):271–277. doi:10.1080/13554790902724904

Schnakers C, Vanhaudenhuyse A, Giacino J, Ventura M, Boly M, Majerus S, Moonen G, Laureys S (2009b) Diagnostic accuracy of the vegetative and minimally conscious state: clinical consensus versus standardized neurobehavioral assessment. BMC Neurol 9:35. doi:10.1186/1471-2377-9-35

Sellers EW (2013) New horizons in brain–computer interface research. Clin Neurophysiol 124 (1):2–4. doi:10.1016/j.clinph.2012.07.012

Sorger B, Dahmen B, Reithler J, Gosseries O, Maudoux A, Laureys S, Goebel R (2003) Another kind of 'BOLD Response': answering multiple-choice questions via online decoded single-trial brain signals. Prog Brain Res 177:275–292. doi:10.1016/S0079-6123(09)17719-1

Sorger B, Reithler J, Dahmen B, Goebel R (2012) A real-time fMRI-based spelling device immediately enabling robust motor-independent communication. Curr Biol 22(14):1333–1338. doi:10.1016/j.cub.2012.05.022

Stoll J, Chatelle C, Carter O, Koch C, Laureys S, Einhäuser W (2013) Pupil responses allow communication in locked-in syndrome patients. Curr Biol 23(15):R647–R648. doi:10.1016/j.cub.2013.06.011

Vanhaudenhuyse A, Demertzi A, Schabus M, Noirhomme Q, Bredart S, Boly M, Phillips C, Soddu A, Luxen A, Moonen G, Laureys S (2011) Two distinct neuronal networks mediate the awareness of environment and of self. J Cogn Neurosci 23(3):570–578. doi:10.1162/jocn.2010.21488

Wolpaw JR, Birbaumer N, McFarland DJ, Pfurtscheller G, Vaughan TM (2002) Brain–computer interfaces for communication and control. Clin Neurophysiol 113(6):767–791

How Does Enhancing Cognition Affect Human Values? How Does This Translate into Social Responsibility?

Laura Y. Cabrera

Abstract The past decade has seen a rise in the use of different technologies aimed at enhancing cognition of normal healthy individuals. While values have been acknowledged to be an important aspect of cognitive enhancement practices, the discussion has predominantly focused on just a few values, such as safety, peer pressure, and authenticity. How are values, in a broader sense, affected by enhancing cognitive abilities? Is this dependent on the type of technology or intervention used to attain the enhancement, or does the cognitive domain targeted play a bigger role in how values are affected? Values are not only likely to be affected by cognitive enhancement practices; they also play a crucial role in defining the type of interventions that are likely to be undertaken. This paper explores the way values affect and are affected by enhancing cognitive abilities. Furthermore, it argues that knowledge of the interplay between values and cognitive enhancement makes a strong case for social responsibility around cognitive enhancement practices.

Keywords Cognitive enhancement · Social responsibility · Values

Contents

L. Y. Cabrera (✉)
National Core for Neuroethics, The University of British Columbia,
2211 Wesbrook Mall, Koerner S124, Vancouver, BC V6T2B5, Canada
e-mail: cabreral@mail.ubc.ca

Institute of Biomedical Ethics, University of Basel, Bernoullistrasse 28,
4056 Basel, Switzerland

Curr Topics Behav Neurosci (2015) 19: 223–241
DOI: 10.1007/7854_2014_334

Published Online: 22 July 2014

1 Introduction

The use of different technological interventions for cognitive enhancement by healthy individuals raises a number of ethical issues, including safety, peer pressure and authenticity, which have become a focus of discussion and analysis in the neuroethics literature. It is also generally acknowledged that the debate on cognitive enhancement touches upon a wide variety of values, perceptions, hopes, and fears, from philosophical and ethical perspectives on desirable human qualities to the proper role of medicine and the equitable distribution of resources (Coenen et al. 2009; Elliott 2003; Farah et al. 2004; President's Council on Bioethics 2003; Wolpe 2002). Even though the role of human values in the ethical debate is acknowledged, an issue that has not yet been widely explored is how enhancing cognitive ability affects human values if at all.

One way in which values can be affected by cognitive enhancement has to do with the organ targeted, namely the human brain. The brain is not only generally considered to be the most dynamic part of our anatomy and physiology and the most sensitive to intervention, but also the organ responsible for affective and cognitive capacities, and reasoning and decision-making. The brain is also considered to be the location and driver of the human being—including consciousness, self, and identity (Glannon 2007; Farah 2010). From these perceptions about the brain, it is reasonable to say that interventions affecting the brain are regarded as having more far-reaching consequences for human behavior, self-perception, and understanding than any other intervention in our body, and consequently likely to affect human values.

However, there is another way in which cognitive enhancement can affect human values, and this has to do with the different technological, political, and cultural changes that cognitive enhancement brings with it. We live in a world that is changing at an accelerated pace driven by rapid technological change (Kurzweil 2005) and globalization. The speed, scale, and depth of the changes that we are part of today mean not only rapid changes in our technological capabilities, which might be outpacing society's and even each individual's capacity to conceive of and agree on new values, but also generate uncertainty, imbalances, and conflicts both socially and personally. It is not that in the past other socio-cultural and technological changes have not affected our values (Gupta et al. 2011); rather, it is the rapid and radical changes that current emerging technological capabilities to enhance human cognition can potentially bring about that make the subject so important (Czerniawski 2010; Sarewitz and Karas 2006). Moreover, in modern democratic societies, which are in many regards more pluralistic than past societies, the diversity of perspectives, aspirations, and capabilities that we encounter can generate strongly divergent views on values among its members. Finally, it can be said that while it is true that technology and its different uses affect human values, it is also the case that values affect the way we use, develop, regulate and perceive technology and the goals that we aim to achieve by its different uses.

Thus, to properly understand the role cognitive enhancement practices play within a society, it is not only necessary to have a good grasp of the values at stake, but also on how our values affect and are affected by enhancement practices. These two aspects are the main focus of this paper.

2 Enhancing Cognition and Human Values

2.1 What Are Human Values and Why Do They Matter?

Values are referred to constantly in almost every important discussion of modern societies, from bioethics to economics and politics. One reason for the constant reference is because human values represent ideals or goals that people in a society strive to achieve. Values orient activities within and between individuals, and as such are a benchmark for human behavior. Thus, it is of no surprise that when discussing topics such as cognitive enhancement, values are an essential theme.

While it is true that needs and attitudes can also be an important part of how we conduct our lives, values can be said to underlie these. In this regard, values not only guide selection and evaluation of behavior, people and events, they also enable groups and individuals to cope with reality as they "cognitively transform the necessities inherent in human existence and express them in the language of specific values about which they can then communicate" (Schwartz 1994). Values then are important because they serve useful social functions, for example, enabling the smooth functioning and survival of groups, as well as enabling members of a given community to share socialization and conventions (Bain et al. 2006).

Values are generally acquired both through socialization to dominant group values and through the unique learning experiences of individuals. As such, they encompass various moral, legal, cultural and religious considerations that are internalized. Values also affect the means that are used to achieve those ideals and goals. Thus, human values are involved in what people want in life, how they feel under certain circumstances and the decisions they will likely make.

There are a variety of value systems, from personal to social, moral and political, economic and cultural, and often the pursuit of each type of value has psychological, practical and social consequences that may conflict or not be compatible with the pursuit of other types of values (Schwartz 1994). For example, seeking personal success for oneself is likely to clash with actions aimed at improving others' situations. As such, conflicting values do not only occur between individuals, they are also common within a single person. Different enhancement interventions can play a role in the reaffirmation, modification, or at times, even the abandonment of values within our own value system. While it might be the case, often people embrace different values that do not add up to an orderly and coherent system, making these values more likely to be abandoned or modified. It is also possible that at least some values are more core to the person and as such less likely to be changed. Baron and colleagues referred to those values that seem unchangeable as "protected values" (Baron and Spranca 1997), that is to say, values that resist trade-offs with other values as they are regarded as highly important to one's self identity. The fact that people engage in violations of their protected values does not mean that the value is less important to them (Baron and Spranca 1997). However, such a violation is likely to create internal conflict. The contingency of our all-too-human values makes it possible that the various pressures, perspectives and compromises that we have to make in today's 24/7 world can often lead to a radical shift in someone's values. Accordingly, for some the benefits promised by cognitive enhancers might not be enough to make them modify or trade off their values, while for others it might.

2.2 Enhancement, Values and Society

Among the different ways in which humans can enhance themselves, cognitive enhancement has been one of the most discussed. There are many reasons for this. Some people are of the view that humans can never have too much cognitive experience. Some scholars take such a position to argue that it is desirable to enhance cognition, as this enables the individual to experience other forms of knowing and being in the world. Other commentators argue that even "a small increase in general cognitive function would likely be sizeable and desirable" (Sandberg and Bostrom 2006).[1] Even if we do not agree that cognitive improvement would have such a significant impact on society, we can still agree that cognition is indeed important for the individual and society. Cognition involves various mental faculties, including perception, attention, representation, memory, learning, and executive functions such as goal setting, planning, decision making, and judgment (Sahakian and Morein-Zamir 2011; Sandberg and Bostrom 2006), which enable us to perceive, understand, and interact with the world. Thus, considering the scope of abilities involved in cognition, it is not hard to see why cognitive ability is generally, particularly in modern societies, regarded as more useful than other abilities from a social perspective (Sandberg and Bostrom 2006), as well as enabling the pursuit of personal goals. Cognition has also become a key faculty in modern societies as it helps us to balance an increasingly complex

[1] Such a view can be contested, as there might be many other things apart from enhanced cognition that would be needed in order to bring the improvement they claim.

society. Considering all this, we can see why cognitive enhancement, understood as the amplification or extension of any of the core cognitive faculties mentioned above, has become such a common topic and among the main goals of the human enhancement movement.

Cognitive enhancement is not a new goal of humankind; for instance, the use of certain herbs and potions with the aim to improve memory and cognition can be traced back to antiquity. However, as we learn more about how the brain works and about new technologies and applications that can alter brain functioning, the possibilities of developing new types of cognitive enhancers take a different scope. Moreover, the world in which we currently live is a more information-rich world compared to the past, a world in which productivity and efficiency have taken priority over other values. This of course shapes the different uses and users of cognitive enhancers. For example, in today's society, domains that are regarded as facilitators of success in life are more likely to be enhanced than those that are not. Given these changes in today's society, it is not far-fetched to think that they have had an impact in the search for mechanisms to help individuals to cope and be able to fulfill the demands imposed on them by modern societies and living styles.

2.3 Interventions

A look at the discussion around cognitive enhancement points towards disagreement in our values as well as polarized views about the prospects of cognitive enhancement. Choices around cognitive enhancement involve different values and expectations about the technologies and their uses. Thus, how these choices are made has important ethical and political dimensions (Nuffield Council on Bioethics 2013). As neuroscience and neurotechnology have advanced, the list of prospective cognitive enhancers has also expanded (Farah et al. 2004). Today, there exists a broad range of interventions that can affect cognition. While there are non-technological driven ways to enhance cognition, such as keeping an appropriate level of nutrients and glucose, proper sleep (Ferrie et al. 2011) and exercise (Vaynman and Gomez-Pinilla 2005), here I will only mention briefly two of the most controversial methods, namely pharmacological-based enhancements and brain stimulation techniques.

2.3.1 Pharmaceutical Interventions

While for many years individuals have tried to enhance their cognitive functions using drugs (President's Council on Bioethics 2003; Savulescu and Bostrom 2009; Savulescu et al. 2011), characteristic of more modern societies is the use of stimulant drugs, ranging from nicotine and caffeine, which are widely used, to medications for attention deficit hyperactivity disorder and wakefulness, which are more controversial (Evans-Brown et al. 2012; Farah et al. 2004; Greely et al. 2008; Turner and Sahakian 2006). This implies that medications are not necessarily used for impaired or at-risk patients, but also for lifestyle uses (Farah 2010; Racine and Forlini 2010). Among the cognitive areas where studies have found evidence of some improvement in normal healthy individuals are: attention, focus, memory, problem solving, and executive function (Elliott et al. 1997). Even though the long-term effects of these kinds of pharmaceutical interventions in healthy individuals remain largely unknown, and their efficacy for healthy individuals is highly contested, these types of interventions are probably among the most widely used and discussed in the literature and among the general public.

2.3.2 Brain Stimulation

Brain stimulation has recently shown potential as a cognitive enhancer. A number of small studies using brain stimulation report improvements in participants' performance in laboratory tasks, for example, in tasks involving memory or language skills, that could be construed as 'enhancements'. Here, I will only mention two minimally invasive techniques, namely transcranial magnetic stimulation (TMS)[2] and transcranial direct current stimulation (tDCS).[3]

Different research groups have been using these technologies to achieve and demonstrate improvements in cognition. In the case of TMS, studies have shown improved performance in various complex motor learning tasks (Kim et al. 2004; Kobayashi et al. 2004), language-related abilities (Mottaghy et al. 1999), visuospatial processing (Hilgetag et al. 2001; Walsh et al. 1998), perceptual abilities (Gallate et al. 2009; Snyder et al. 2003, 2006; Snyder 2009) and in modulating social cognition (Knoch et al. 2006; Lo et al. 2003; Luber et al. 2009; Young et al. 2010). Regarding tDCS, there are studies suggesting that it enhances working (Fregni et al. 2005; Ohn et al. 2008) and declarative (Marshall et al. 2004) memory as well as certain forms of learning (Bullard et al. 2011; Flöel et al. 2008). Evidence for the enhancement of more general complex problem-solving abilities via tDCS is limited, but intriguing. This includes areas such as complex verbal associative thought (Cerruti and Schlaug 2009), planning ability (Dockery et al. 2009), numerical competence (Kadosh et al. 2010), problem solving (Chi and Snyder 2011, 2012), as well as behavior in cases of reward-seeking tasks (Fecteau et al. 2007) and the generation of deceptive responses (Priori et al. 2008).

Given the role that cognition has in our lives and the different faculties associated with cognition, it is not too far-fetched to think that as new capabilities are obtained by humans through cognitive enhancement, their morality and values will change (Hart 1958). Perhaps new values will come to govern society as we move closer to what some have referred as "enhancement societies" (Coenen et al. 2009), where in an era of increasing experimentation with cognitive enhancement technologies and

---

[2] TMS involves a train of magnetic pulses administered by electromagnets in the head.

[3] tDCS involves applying a weak direct current to the scalp via two saline-soaked sponge electrodes.

interventions, taking pills or stimulating our brains with electric currents will be regarded as acceptable, at least for certain cognitive domains. However, this does not mean that cognitive enhancement will always affect human values, nor that different values will be affected equally. For example, it is likely that cognitive enhancers that exert a temporary action might not affect values in the way that more permanent enhancers will. It is also possible that after a certain level of enhanced cognition, the enhanced individuals might develop new values, new perspectives on life and on their relationships, which might conflict with those of unenhanced individuals. Ultimately these are empirical questions, but in the meantime we can explore how changes in different cognitive domains might affect human values.

Consider memory enhancement. Memory is a good example of how complicated it can be to assess how cognitive enhancement affects values. To start with, memory is not a single system since multiple systems are involved. Furthermore, a memory enhancer that enables people to improve working memory might not be as problematic in terms of affecting values as one that targets memories of personal experiences, as these memories help us to build a framework of things we believe to be true about ourselves and the world (Bublitz and Merkel 2009; Elliott 1998; Taylor 1992). It could also be that an increased ability to remember every instance could overflow our capacity to categorize memories, which could impair our selectivity process, our ability to make abstractions from our lived experiences, and our ability to distinguish larger patterns (Borges 1964; Liao and Sandberg 2008; Luria 1987). These are all important aspects, not only of our cognitive faculties, but also of our value system. Regarding learning, it can be said that reducing a learning experience from one that involves engagement and time, to one in which less time and effort is involved due to the use of a cognitive enhancer, will have an impact on our value system.

In the case of attention, certain enhancers might enable people to concentrate better in tasks, but in doing so encourage them to bypass other important goals, such as keeping in contact with friends.[4] It may also be the case that by enhancing attention we would be able to perceive more details about people's behavior, which will enable us to judge their behavior under a different framework.

Probably the areas of enhanced cognition with the greatest impact on our value system will be those affecting social cognition. For instance, Young et al. (2010) found that TMS caused subjects to focus more on the outcome of an act than the intention of the actor when judging permissibility of the act as tested in a short vignettes scenario. Another example of how brain stimulation can affect social cognition comes from Knoch et al. (2006), who found that after stimulation, subjects in an Ultimatum Game were more likely to accept low money offers, even though they still perceived them as being unfair. One last example involves the effect on risk-seeking behavior (Fecteau et al. 2007), which could produce new behaviors that individuals would otherwise not have engaged in.

[4] In particular as the effects of the enhancer in action might not fade away just after a few hours or when someone is done with the task at hand.

Imagine that in the future cognitive enhancers could affect safely and reliably more complex cognitive faculties, such as critical thinking or self control. Such cognitive enhancement interventions could deeply affect human values. For instance, in the case of critical thinking, it is likely that enhancement would enable us to assess in a more critical way many of the biases underlying questionable human decisions. Some of the studies mentioned above claim that there is already evidence for this (Snyder et al. 2003; Snyder 2009).[5] To some extent, some of these types of enhancement interventions are the ones that supporters of moral enhancement have argued for (Douglas 2008; Persson and Savulescu 2008, 2011). However, as some scholars have counter-argued, it is the complexity of the different mechanisms involved (Zarpentine 2013) that raises problems for this type of argument. In the case of human values it is the interplay of social, cultural, and environmental factors, as well as the complexity behind most of our cognitive capacities that pose a problem in assessing the impact of the enhancement of cognition on values.

While it can be argued that these examples have just shown that cognitive enhancers affect behavior, it is plausible that long-term use of cognitive enhancers might result in certain behaviors being more common than others and that the individual might try to adapt his value system accordingly. This is, of course, an empirical question that needs long term follow-up of individuals who engage in different cognitive enhancement interventions. However, the evidence on the increasing number of individuals engaging in cognitive enhancement practices (Hotze et al. 2011; Olfson et al. 2013; Ragan et al. 2013; Smith and Farah 2011) already tells us something about how values have changed compared to previous decades.

2.4 Values at Stake

Another way in which values and cognitive enhancement affect each other is related to the different values underlying individual desires and societal goals. There are a series of values that seem to permeate Western culture and that are likely driving the current cognitive enhancement agenda.

2.4.1 Competitiveness and Success

We are part of a society in which the prime driver of development is competition (Ferrari et al. 2012; Lamkin 2012). Thus, even though at least some competition seems to be desirable, this is not the case when it discourages people from nurturing values such as cooperation or solidarity.

---

[5] The studies argue that by using brain stimulation cognitive conceptual knowledge biases are removed.

In a society dominated by competitiveness, improved cognition will be perceived and evaluated differently than in a society where other types of values are prioritized. Thus, in competitive societies, it is not uncommon to find arguments supporting cognitive enhancement on the grounds that even a small upward shift in cognitive faculties would have a beneficial economic impact. The main argument use here is that cognitive enhancement would enable people to perform better and avoid losses due to inattention in the school or workplace. However, one counter-argument here is that the use of cognitive enhancements under competitive environments can overstretch the natural range of equality, as those who have access or use cognitive enhancement are accrued greater advantages in life compared to those who do not have access or do not use them, to the point where inequality becomes a more salient issue (Brock 1998). But this is not to say that all cognitive enhancements are valuable only on the ground of bringing positional goods. Under societies where competition is a prime drive, this is likely to be one of the main reasons for using them.

Competitiveness reinforces a focus on productivity, efficiency and output (Makridis 2013). For example, students often refer to academic assignments or grades as reasons to take cognitive enhancers (Ragan et al. 2013; Smith and Farah 2011). At the same time, the daily demands of contemporary life, which increasingly moves towards a 24/7 society characterized by processing large amounts of information from different sources, at times simultaneously, often lead to fatigue, sleep deprivation and continuous stress (Morein-Zamir et al. 2009). Thus, cognitive enhancers are regarded as enablers for people to better comply with these demands and remain competitive (Brukamp 2013). Taking this into consideration, it is no wonder why, in places like school or work, individuals use or seek to use cognitive enhancers in order to be better equipped to compete (Farah et al. 2004; Maher 2008; Academy of Medical Sciences 2012). These institutions, through different mechanisms make clear that certain cognitive faculties are valued, and more often than not reward individuals with above average cognitive performance in areas such as memory, concentration and reasoning.

In societies where competition is praised, "declining to use tools that can confer competitive advantages can be viewed as a kind of moral failing" (Lamkin 2012). What is worrisome, as Lamkin argues, is not only that parents, employers, coaches, peers and a whole range of institutional practices are an important source of pressure to compete, but rather it is the individual's internalized value to compete. Moreover, the drive to succeed, inherent in competitive societies, creates a tension between the drive to succeed and the need to be true to who you really are, and this partly explains why feelings of ambivalence regarding enhancement technologies are common among the general public. Similarly, if cognitive enhancement makes life and competition easier, we may lose opportunities to nurture and grow our moral character (Allhoff et al. 2009). Thus, a better understanding of the possible impact that competition and success might have as values driven cognitive enhancement is of great importance.

2.4.2 Continuous Improvement

A related value in this discussion is a desire for continuous improvement. While it can be argued that part of being human is a natural instinct to improve oneself and develop technology to help in this quest (Williams 2006), in today's society it seems this value has been taken to the extreme. In particular, and in connection to the above-mentioned values, many individuals are likely to feel pressure to keep up with the demands of society and lifestyles that are idealized (Morein-Zamir et al. 2009). Some scholars have raised concerns about what would happen if everyone started to feel that they needed to enhance themselves in order not to be left behind. Others have even gone further, suggesting that the current human condition is not the end stage of evolution, seeing the human as work-in-progress and as such in need to overcome what is perceived as current human biological limitations (Bostrom 2003, 2005; Kurzweil 2005).

Presumably, this would not only change ideas of what is perceived as normal, but is likely to create, at least for some individuals, a conflict of values as they feel the pressure to compete and succeed in life. A further problem with this is that it might create environments in which individuals end up engaging in activities or interventions that have not been shown to be reliable or safe, so they risk their own health and safety thinking that this is the best option to 'fit in' or 'succeed' (Evans-Brown et al. 2012). In addition, any success that is already enjoyed is constantly threatened by the possibility that others might catch up or pull ahead, so there is constant pressure to catch up, and once someone has caught up there is the push to make sure to stay in the lead or above average.

2.4.3 Consumerism

Another common value in modern societies is consumerism. Andy Miah has put forward the idea that enhancements can be regarded as an act of consumption. While certain forms of consumerism have been an integral part of many cultures (Miah 2013) and are often regarded as acts of differentiation, not all forms and degrees of consumerism are regarded as acceptable. For instance, consumer acts can also be a function of conformity to the social norm. Thus, it would be wrong to think that consumption practices driven by the exploitation of people's insecurities and environments, where media and popular culture reinforce idealized ways of success via consumption, should be regarded as valuable ways of consumption.

An interesting point that Miah raises is that different people, based on what they regard as important in their lives, will chose different cognitive enhancements. For some, the focus might be concentration, while for others memory might be more important. Thus, the fact that the literature on cognitive enhancement keeps mentioning certain cognitive domains as sensible domains to be enhanced does not mean that these are good enhancements for all kinds of people. This is one significant reason why exploring other values that people might hold is so important.

Moreover, it is important to keep exploring which cognitive domains are more likely to be seen as consumables and why certain individuals are more likely to consume a certain type of cognitive enhancement and not others.

2.4.4 Individualism and Freedom

Other values that seem to be prevalent in contemporary societies are individualism and freedom. Under these values, the individual takes precedence. In connection to freedom, it is what we prioritized as important for freedom, and not the value of freedom per se, which can be damaging to societies, for example, if freedom is valued mostly because it enables individuals to do as they wish without taking into consideration the larger picture of the consequences of their actions. In Western societies, it is common to hear the claim that as long as we do not directly harm others or impinge on their rights we should be able to do whatever we want with our bodies and minds. A clear example in relation to the enhancement agenda has to do with claims regarding "cognitive liberty", which is regarded as the freedom individuals have over decisions concerning whether or how to change their thought processes (Center for Cognitive Liberty and Ethics 2006). Thus, individual freedom and individual choice are emphasized as if individuals were monads who are free to make decisions that would only affect[6] themselves.

By emphasizing the primacy of individual freedom and choice, collective values are neglected, and a distorted view about individuals is reinforced. A view that does not acknowledge that individuals are not isolated, self-sufficient entities; as much as some like to think of themselves as self-sufficient (Sandel 1998; Held 2006; Cabrera 2011). The discourse on cognitive enhancement has focused on the possible benefits for the individual in question. However, this individualistic perspective neglects the views and challenges of those that are also part of that individual's life, as well as the community at large.

2.4.5 Hidden Values in the Cognitive Enhancement Agenda

There are other values that are less explicit and that have to do with the ways in which we value human life and deal with biological limitations. For example, the value we give to hard work and outcomes can create a contradiction between our personal and societal value systems, as more and more cognitive enhancement is regarded as a way to accomplish ends that would once have been attained via slower yet equally efficient methods (Pasquale 2007). Happiness is another value

[6] The classical liberal position doesn't deny that our actions as individuals might have all sorts of effects on others, but it distinguishes between effects and harms, with the latter being a subset of the former. According to the classical liberal, only harms are an issue for morality.

that carries nearly the weight of a moral imperative in modern society (Frances 2013; Sharpe 2012). This creates a vicious link between competitiveness, success and the pursuit of happiness.

Another hidden aspect that enhancement technologies bring is the reconceptualization of certain conditions as medical problems (Elliott 1999, 2003), in which the reinforcement of certain cultural norms leads some conditions to be regarded as something we need to change (Little 1998). A clear example of this involves views on playing and being easily distracted, where what was once seen as an integral part of childhood is now often viewed as part of an attention disorder. Or if in the past, memory decline was a normal part of the aging process, now this is conceptualized as a medical problem in need of therapy. So once more the enhancement agenda cuts two ways: it can make us healthier or (at least in relative terms) sicker. As Gregor Wolbring has pointed out: "as much as human enhancement technology will become an enabling technology for the few, it will become a disabling technology for the many" (Wolbring 2006). Similarly, enhancement interventions could eventually impose a one-sided, tyrannical pattern of the development of human capacities and abilities, leading to ableism—a set of practices, beliefs, and processes that produces, based on the favoritism for certain abilities that are seen as essential, a particular kind of understanding of one's body, oneself and one's relationship with others, including humans, other species and the environment (Wolbring 2005, 2008). Thus, it can be said that the hidden values behind current cognitive enhancement practices coerce and manipulate individuals to accept a very distorted and narrow conception of good life and well-being (Habermas 2003; Held 2006).

To add complexity into this already complicated equation, current modern social movements, such as transhumanism, promote values in which the human condition is not only regarded as a work-in-progress, but also in which common accepted human limitations, for example in regard to energy, will-power, and the ability to shape their own character in accordance with their ideals (Bostrom 2003), are limitations that should be overcome through technological enhancements. Furthermore, it is assumed that other ways of living, thinking, relating, and feeling that are not yet accessible to humans are valuable and desirable for one's own personal well-being and for the development of the species as a whole. However, the fact that certain kinds of experiences are not accessible or not comprehensible to us already shapes human values and is the basis for shared understanding (Cabrera and Weckert 2013). Thus, changing significantly the basis for a shared understanding of values, such as the possibility of radically enhanced humans, might create a clash of *human* values with *post*human values.

Having augmented and extended cognitive faculties does not necessarily translate into making wiser choices about where we are going, nor does it necessarily lead to happier lives. In connection to the latter, for instance, there is evidence that people holding materialistic values are less happy and less satisfied with their life overall than those who do not (Ryan and Dziurawiec 2001). To a great extent, our prevailing cultural values motivate people to use cognitive enhancers in problematic ways. Certain enhancement practices involve an image of humanity which tacitly or openly redirects the basic values of the individual in ways that conflict with other values that are also perceived as important. Reinforcing the values that may be threatened by enhancement would require the harder work of reforming the broader values that drive demand for this type of enhancement intervention. Thus, while we should question the technologies that are helping people to attain enhancement, we would be better served by challenging the underlying values that make these practices attractive. More importantly, it can be argued that knowledge of how human values and cognitive enhancement interventions interact must translate into a stronger focus on social responsibility. In the next and final section, I will elaborate on this point.

3 What About Social Responsibility?

As we have seen above, different values are both challenged and affirmed by cognitive enhancement (Farah et al. 2004). Moreover, values permeate many, if not all, of our social institutions; as such, the ways in which we set up our institutions and prioritize research funding will have a serious influence on the future of cognitive enhancement (Makridis 2013). Key issues include the different ways in which we can potentially enhance our cognitive abilities, our ability to create a variety of views regarding whether this type of change should be embraced or not, as well as the types of interventions that are regarded as acceptable against those ones that are not. Even the way in which the enhancement debate has been framed creates clashes of different individual and communal values.

Considering all this, it can be argued that the knowledge we have about the brain, technology and values give us a strong case for social responsibility regarding cognitive enhancement. While scientists have a responsibility to educate the public about their new discoveries, engineers also have a responsibility to design new technologies taking into account ethical principles. Furthermore, potential consumers of the technology, policy makers who might decide on the regulations under which these technologies will be allowed or not, and ethicists including philosophers and cultural studies experts should help society reflect on the impact of these cognitive enhancement technologies on values and culture.

Thus, social responsibility can be seen as a framework under which "care for the future of mankind is the overruling duty of collective human action in the age of a technical civilization" (Jonas 1985, 136). A social responsibility framework, then, could bridge the tensions in the cognitive debate and allow a more enriched evaluation of the values that are currently driving the cognitive enhancement agenda. We have a social responsibility to engage in meaningful conversations about enhancement technologies in terms of their efficiency in reaching certain goals, underlying values, the values that they reinforce, as well as their contribution to the things that we value and those that we do not. Failing to attend to the values at stake or to address the elements of culture and social institutions that drive demand for cognitive enhancement means that policies in place are unlikely to address what is truly troubling about them (Lamkin 2012). It could also mean that without proper acknowledgement of the values at stake, technologies used for cognitive enhancement might end up being rejected or misused by society (Nordmann 2004).

It is all-too-human that our values are open to change and contingent on context and new circumstances, and as such, we have a social and moral responsibility to question and challenge not only the interventions that are taking place in society, such as a more accepted use of cognitive enhancement, but also the values that are driving this. Furthermore, we also have a social responsibility to start promoting more social values such as cooperation, caring and empathy, as well as more sustainable, responsible and even ethical ways to enhance cognition—for example, social and community-based interventions rather than interventions that are focused on changing directly the biological reality of individuals (Cabrera 2011). Cognitive enhancement under a social responsibility framework would promote responsible research and innovation as well as more thoughtful ways to make use of new technological capacities, in particular uses that "fulfill a valuable social benefit and [that] do not threaten to undermine other social values" (Nuffield Council on Bioethics 2012). Only then can we start shaping more meaningful, ethical and sustainable lives and promote healthier communities.

4 Conclusion

To think about the future of cognitive enhancement means to consider and examine our values, concerns, goals, and perspectives, and to promote an open discussion about how we can justify them. This will help us make more informed choices about the future of cognitive enhancement and the goals that we might cherish as part of humanity. Similarly, the proper governance of enhancement technologies involves an engagement between different values, understandings, and visions that are likely to come into tension creating dilemmas that will have to be constantly confronted. That is why beyond shared values it might be even more important to keep questioning and challenging our current values, and as such, enable a more open discussion around cognitive enhancement. We should be cautious in thinking that we have a responsibility to use cognitive enhancement in order to be able to appreciate other sets of values. Rather, our social responsibility lies in challenging current ways of thinking and practicing enhancement, so we can promote more inclusive and sustainable ways to improve our cognitive faculties. We have a social responsibility to promote social development and improvement through an open acknowledgement of our relation and dependence onsocial infrastructure and support for our individual flourishing. Failing to do so could create a mismatch between what we value now and our future needs. While more empirical research is needed to have a better understanding of how cognitive enhancement affects human values, we can start setting a framework that ensures that if we engage in cognitive enhancement, we do so within a framework of social responsibility.

Acknowledgments I thank Kieran Donoghue and Madeline Kilty for their insightful comments and for helping with proof reading this chapter. I also thank the editors for their insightful comments and suggestions.

References

Academy of Medical Sciences (2012) Human enhancement and the future of work

Allhoff F, Lin P, Moor J, Weckert J (2009) Ethics of human enhancement: 25 questions and answers. Stud Ethics Law Technol 4(1):39

Bain P, Kashima Y, Haslam N (2006) Conceptual beliefs about human values and their implications: human nature beliefs predict value importance, value trade-offs, and responses to value-laden rhetoric. J Pers Soc Psychol 91(2):351–367

Baron J, Spranca M (1997) Protected values. Organ Behav Hum Dec 70(1):1–16

Borges JL (1964) Labyrinths, selected stories and other writings. New Directions Pub. Corp, New York

Bostrom N (2003) Transhumanist values In: Adams F (ed) Ethical issues for the 21st century. Philosophical documentation center press, Charlottesville

Bostrom N (2005) A history of transhumanist thought. J Evol Technol 14(1):1–25

Brock DW (1998) Enhancement of human function: some distinctions for policymakers. In: Parens E (ed) Enhancing human traits: ethical and social implications. Georgetown University Press, Washington DC, pp 48–69

Brukamp K (2013) Better brains or bitter brains? The ethics of neuroenhancement. In: Hildt E, Franke AG (eds) Cognitive enhancement. Trends in augmentation of human performance, vol 1. Springer, The Netherlands, pp 99–112

Bublitz JC, Merkel R (2009) Autonomy and authenticity of enhanced personality traits. Bioethics 23(6):360–374

Bullard LM, Browning ES, Clark VP, Coffman BA, Garcia CM, Jung RE, van der Merwe AJ, Paulson KM, Vakhtin AA, Wootton CL, Weisend MP (2011) Transcranial direct current stimulation's effect on novice versus experienced learning. Exp Brain Res 213(1):9–14

Cabrera L (2011) Between different human enhancement paradigms: the role of nano and neurotechnology. Charles Sturt University, Canberra

Cabrera L, Weckert J (2013) Human enhancement and communication: on meaning and shared understanding. Sci Eng Ethics 19(3):1039–1056

Center for Cognitive Liberty and Ethics (2006) Mission statement

Cerruti C, Schlaug G (2009) Anodal transcranial direct current stimulation of the prefrontal cortex enhances complex verbal associative thought. J Cogn Neurosci 21(10):1980–1987

Coenen C, Schuiff M, Smits M, Klaassen P, Hennen L, Rader M, Wolbring G (2009) Human enhancement. Study commissioned by the science and technology options assessment and carried out by the European Technology Assessment Group

Czerniawski M (2010) Human enhancement and values. Available at http://dx.doi.org/10.2139/ssrn.1633938

Chi RP, Snyder AW (2011) Facilitate insight by non-invasive brain stimulation. PloS ONE 6(2):e16655

Chi RP, Snyder AW (2012) Brain stimulation enables the solution of an inherently difficult problem. Neurosci Lett 515(2):121–124

Dockery CA, Hueckel-Weng R, Birbaumer N, Plewnia C (2009) Enhancement of planning ability by transcranial direct current stimulation. J Neurosci Official J Soc Neurosci 29(22):7271–7277

Douglas T (2008) Moral enhancement. J Appl Philos 25(3):228–245

Elliott C (1998) What's wrong with enhancement. Paper presented at the CHIPS Public Lecture University of Minnesota

Elliott C (1999) A philosophical disease: bioethics, culture, and identity. Routledge Chapman & Hall, New York
Elliott C (2003) Better than well: American medicine meets the American dream. W.W. Norton, New York
Elliott R, Sahakian BJ, Matthews K, Bannerjea A, Rimmer J, Robbins TW (1997) Effects of methylphenidate on spatial working memory and planning in healthy young adults. Psychopharmacology 131(2):196–206
Evans-Brown M, McVeigh JC, Perkins C, Bellis MA (2012) Human enhancement drugs the emerging challenges to public health. North West Public Health Observatory, Liverpool
Farah MJ (2010) Neuroethics: an introduction with readings. Basic Bioethics. MIT Press, Cambridge
Farah MJ, Illes J, Cook-Deegan R, Gardner H, Kandel E, King P, Parens E, Sahakian B, Wolpe PR (2004) Neurocognitive enhancement: what can we do and what should we do? Nat Rev Neurosci 5(5):421–425
Fecteau S, Knoch D, Fregni F, Sultani N, Boggio P, Pascual-Leone A (2007) Diminishing risk-taking behavior by modulating activity in the prefrontal cortex: a direct current stimulation study. J Neurosci Official J Soc Neurosci 27(46):12500–12505
Ferrari A, Coenen C, Grunwald A (2012) Visions and ethics in current discourse on human enhancement. Nanoethics 6(3):215–229
Ferrie JE, Shipley MJ, Akbaraly TN, Marmot MG, Kivimaki M, Singh-Manoux A (2011) Change in sleep duration and cognitive function: findings from the Whitehall II study. Sleep 34(5):565–573
Flöel A, Rösser N, Michka O, Knecht S, Breitenstein C (2008) Noninvasive brain stimulation improves language learning. J Cogn Neurosci 20(8):1415–1422
Frances A (2013) Saving normal: an insider's revolt against out-of-control psychiatric diagnosis, DSM-5, big pharma, and the medicalization of ordinary life. HarperCollins, New York
Fregni F, Boggio PS, Nitsche M, Bermpohl F, Antal A, Feredoes E, Marcolin MA, Rigonatti SP, Silva MT, Paulus W, Pascual-Leone A (2005) Anodal transcranial direct current stimulation of prefrontal cortex enhances working memory. Exp Brain Res 166(1):23–30
Gallate J, Chi R, Ellwood S, Snyder A (2009) Reducing false memories by magnetic pulse stimulation. Neurosci Lett 449(3):151–154
Glannon W (2007) Bioethics and the brain. Oxford University Press, Oxford
Greely H, Sahakian B, Harris J, Kessler RC, Gazzaniga M, Campbell P, Farah MJ (2008) Towards responsible use of cognitive-enhancing drugs by the healthy. Nature 456(7224):872
Gupta N, Fischer ARH, Frewer LJ (2011) Socio-psychological determinants of public acceptance of technologies: a review. Pub Understanding Sci 21(7):782–795
Habermas J (2003) The future of human nature. Cambridge, Polity
Hart HLA (1958) Positivism and the separation of law and morals. Harvard Law Rev 71(4):623
Held V (2006) The ethics of care: personal, political, and global. Oxford University Press, Oxford
Hilgetag CC, Theoret H, Pascual-Leone A (2001) Enhanced visual spatial attention ipsilateral to rTMS-induced 'virtual lesions' of human parietal cortex. Nat Neurosci 4(9):953–957
Hotze TD, Shah K, Anderson EE, Wynia MK (2011) "Doctor, would you prescribe a pill to help me … ?" A national survey of physicians on using medicine for human enhancement. Am J Bioethics 11(1):3–13
Jonas H (1985) The imperative of responsibility: in search of an ethics for the technological age. University of Chicago Press, Chicago
Kadosh CR, Soskic S, Iuculano T, Kanai R, Walsh V (2010) Modulating neuronal activity produces specific and long-lasting changes in numerical competence. Curr Biol (CB) 20(22):2016–2020
Kim YH, Park JW, Ko MH, Jang SH, Lee PK (2004) Facilitative effect of high frequency subthreshold repetitive transcranial magnetic stimulation on complex sequential motor learning in humans. Neurosci Lett 367(2):181–185
Knoch D, Pascual-Leone A, Meyer K, Treyer V, Fehr E (2006) Diminishing reciprocal fairness by disrupting the right prefrontal cortex. Science 314(5800):829–832

Kobayashi M, Hutchinson S, Theoret H, Schlaug G, Pascual-Leone A (2004) Repetitive TMS of the motor cortex improves ipsilateral sequential simple finger movements. Neurology 62(1):91–98

Kurzweil R (2005) The singularity is near: when humans transcend biology. Viking, New York

Lamkin M (2012) Cognitive enhancements and the values of higher education. Health Care Anal 20(4):347–355

Liao SM, Sandberg A (2008) The normativity of memory modification. Neuroethics 1(2):85–99

Little M (1998) Cosmetic surgery, suspect norms, and the ethics of complicity. In: Parens E (ed) Enhancing human traits: ethical and social implications. Georgetown University Press, Washington DC, pp 162–176

Lo YL, Fook-Chong S, Tan EK (2003) Increased cortical excitability in human deception. NeuroReport 14(7):1021–1024

Luber B, Fisher C, Appelbaum PS, Ploesser M, Lisanby SH (2009) Non-invasive brain stimulation in the detection of deception: scientific challenges and ethical consequences. Behav Sci Law 27(2):191–208

Luria AR (1987) The mind of a Mnemonist: a little book about a vast memory. Harvard University Press, Cambridge

Maher B (2008) Poll results: look who's doping. Nature 452(7188):674–675

Makridis C (2013) Converging technologies: a critical analysis of cognitive enhancement for public policy application. Sci Eng Ethics 19(3):1017–1038

Marshall L, Molle M, Hallschmid M, Born J (2004) Transcranial direct current stimulation during sleep improves declarative memory. J Neurosci Official J Soc Neurosci 24(44):9985–9992

Miah A (2013) Justifying human enhancement: The accumulation of biocultural capital. In: More M, Vita-More N (eds) The Transhumanist reader: classical and contemporary essays on the science, technology and philosophy of the human future. Wiley-Blackwell, Oxford

Morein-Zamir S, Robbins T, Turner DC, Sahakian B (2009) Pharmacological cognitive enhancement. In: Cooper CL, Goswami U, Sahakian BJ (eds) Mental capital and wellbeing. Wiley, Oxford

Mottaghy FM, Hungs M, Brugmann M, Sparing R, Boroojerdi B, Foltys H, Huber W, Topper R (1999) Facilitation of picture naming after repetitive transcranial magnetic stimulation. Neurology 53(8):1806–1812

Nordmann A (2004) Converging technologies: Shaping the future of European societies. Report from the high level expert group on "Foresighting the New Technology Wave". Luxemburg

Nuffield Council on Bioethics (2012) Emerging biotechnologies: technology, choice and the public good. Nuffield Council on Bioethics, London

Nuffield Council on Bioethics (2013) Novel Neurotechnologies: Intervening in the brain

Ohn SH, Park CI, Yoo WK, Ko MH, Choi KP, Kim GM, Lee YT, Kim YH (2008) Time-dependent effect of transcranial direct current stimulation on the enhancement of working memory. NeuroReport 19(1):43–47

Olfson M, Blanco C, Wang S, Greenhill LL (2013) Trends in office-based treatment of adults with stimulants in the United States. J Clin Psychiatry 74:43–50

Pasquale F (2007) Technology, competition and values. Minnesota Journal of Law, Science and Technology. Seton Hall public law research paper no. 1002463

Persson I, Savulescu J (2008) The perils of cognitive enhancement and the urgent imperative to enhance the moral character of humanity. J Appl Philos 25(3):162–177

Persson I, Savulescu J (2011) Unfit for the future? Human nature, scientific progress, and the need for moral enhancement. In: Savulescu J, Meulen R, Kahane G (eds) Enhancing human capacities. Wiley-Blackwell, Malden, pp 486–500

President's Council on Bioethics (2003) Beyond therapy: biotechnology and the pursuit of happiness

Priori A, Mameli F, Cogiamanian F, Marceglia S, Tiriticco M, Mrakic-Sposta S, Ferrucci R, Zago S, Polezzi D, Sartori G (2008) Lie-specific involvement of dorsolateral prefrontal cortex in deception. Cereb Cortex 18(2):451–455

Racine E, Forlini C (2010) Cognitive enhancement, lifestyle choice or misuse of prescription drugs? Neuroethics 3(1):1–4
Ragan CI, Bard I, Singh I (2013) What should we do about student use of cognitive enhancers? An analysis of current evidence. Neuropharmacology 64:588–595
Ryan L, Dziurawiec S (2001) Materialism and its relationship to life satisfaction. Soc Indic Res 55(2):185–197
Sahakian BJ, Morein-Zamir S (2011) Neuroethical issues in cognitive enhancement. J Psychopharmacol 25(2):197–204
Sandberg A, Bostrom N (2006) Converging cognitive enhancements. Prog Convergence Technol Hum Wellbeing 1093:201–227
Sandel MJ (1998) Liberalism and the limits of justice. Cambridge University Press, Cambridge
Sarewitz D, Karas T (2006) Policy implications of technologies for cognitive enhancement
Savulescu J, Bostrom N (2009) Human enhancement. Oxford University Press, Oxford
Savulescu J, Meulen RHJt, Kahane G (2011) Enhancing human capacities. Wiley-Blackwell, Oxford
Schwartz SH (1994) Are there universal aspects in the structure and contents of human values? J Soc Issues 50(4):19–45
Sharpe K (2012) Coming of age on Zoloft: how antidepressants cheered us up, let us down, and changed who we are. HarperCollins, New York
Smith E, Farah M (2011) Are prescription stimulants smart pills? The epidemiology and cognitive neuroscience of prescription stimulant use by healthy individuals. Psychol Bull 137:717–741
Snyder A (2009) Explaining and inducing savant skills: privileged access to lower level, less-processed information. Philos Trans R Soc Lond B Biol Sci 364(1522):1399–1405
Snyder A, Bahramali H, Hawker T, Mitchell DJ (2006) Savant-like numerosity skills revealed in normal people by magnetic pulses. Perception 35(6):837–845
Snyder AW, Mulcahy E, Taylor JL, Mitchell DJ, Sachdev P, Gandevia SC (2003) Savant-like skills exposed in normal people by suppressing the left fronto-temporal lobe. J Integr Neurosci 2(2):149–158
Taylor C (1992) The ethics of authenticity. Harvard University Press, Cambridge
Turner DC, Sahakian BJ (2006) Neuroethics of cognitive enhancement. Biosocieties 1(01):113–123
Vaynman S, Gomez-Pinilla F (2005) License to run: exercise impacts functional plasticity in the intact and injured central nervous system by using neurotrophins. Neurorehabilitation Neural Repair 19(4):283–295
Walsh V, Ellison A, Battelli L, Cowey A (1998) Task-specific impairments and enhancements induced by magnetic stimulation of human visual area V5. Proc Biol sci Roy Soc 265(1395):537–543
Williams E (2006) Good, better, best: the human quest for enhancement. Summary report of an invitational workshop convened by the scientific freedom, responsibility and law program American association for the advancement of science, Washington, DC
Wolbring G (2005) The triangle of enhancement medicine, disabled people, and the concept of health: a new challenge for HTA, health research, and health policy (Health Technology Assessment Initiative #23). Alberta heritage foundation for medical research, Edmonton. Retrieved from http://www.ihe.ca/documents/HTA-FR23.pdf
Wolbring G (2006) The unenhanced underclass. In: Miller P, Wilsdon J (eds) Better Humans The politics of human enhancement and life extension. DEMOS, London
Wolbring G (2008) 'Able' is there an end to out-able? Is there an end to the rat race for abilities? M/C J 11(3). Retrieved from http://journal.media-culture.org.au/index.php/mcjournal/article/viewArticle/57
Wolpe PR (2002) Treatment, enhancement, and the ethics of neurotherapeutics. Brain Cogn 50(3):387–395

Young L, Camprodon JA, Hauser M, Pascual-Leone A, Saxe R (2010) Disruption of the right temporoparietal junction with transcranial magnetic stimulation reduces the role of beliefs in moral judgments. Proc Natl Acad Sci 107(15):6753–6758

Zarpentine C (2013) 'The thorny and arduous path of moral progress': moral psychology and moral enhancement. Neuroethics 6(1):141–153

Deep Brain Stimulation: A Principled and Pragmatic Approach to Understanding the Ethical and Clinical Challenges of an Evolving Technology

Eric Racine, Emily Bell and Natalie Zizzo

Abstract DBS has emerged in the past few decades as a powerful clinical tool in the treatment of movement disorders such as dystonia and Parkinson's disease. As a result of its striking effects, the therapeutic utility of DBS has been investigated in a number of different neurological and neuropsychiatric conditions. Ethical discussion has accompanied this evolution of DBS and has led to the identification of a number of important ethical challenges. In this chapter, we review these challenges based on three of the key principles of biomedical ethics (autonomy, justice, and non-maleficence). Specifically, we adopt a pragmatic perspective by reviewing the ethical issues as they emerge within the context of Parkinson's disease, as this can serve to guide further ethical thinking on the future of DBS. Through this contextualization, we enrich the meaning of the Ethical principles and increase their specificity. We hope that this contribution will inform readers and also

For: *Current Topics in Behavioral Neuroscience* Frauke Ohl, Grace Lee and Judy Illes, editors, Ethics in Behavioral Neuroscience

E. Racine (✉) · E. Bell · N. Zizzo
Neuroethics Research Unit, Institut de recherches cliniques de Montréal (IRCM), 110 Avenue des Pins Ouest, Montréal, QC H2W lR7, Canada
e-mail: eric.racine@ircm.qc.ca

E. Racine
Department of Medicine and Department of Social and Preventive Medicine, Université de Montréal, Montréal, Canada

E. Racine
Departments of Neurology and Neurosurgery, Experimental Medicine and Biomedical Ethics Unit, McGill University, Montréal, Canada

E. Bell
Departments of Neurology and Neurosurgery, McGill University, Montréal, Canada

N. Zizzo
Experimental Medicine and Biomedical Ethics Unit, McGill University, Montréal, Canada

Curr Topics Behav Neurosci (2015) 19: 243–263
DOI: 10.1007/7854_2014_336

Published Online: 22 July 2014

stimulate discussion related to areas where important questions remain unanswered and where further research would need to be undertaken to understand and enact ethical principles.

Keywords Parkinson's disease · Deep brain stimulation · Ethics · Pragmatism

Contents

Deep brain stimulation (DBS) is a fascinating procedure in terms of its history, its evolution in health care, and the questions it brings forth to medicine, ethics, and philosophy. From a historical perspective, DBS is a rather old technique, and its exploration began in the nineteen sixties (Talan 2009). At that time, DBS was trialed for Parkinson's disease and pain treatment, but never gained in popularity given unconvincing results (Talan 2009; Miocinovic et al. 2013). In the nineteen eighties, interest was reignited when the French neurologist Alim-Louis Benabid and his team undertook investigations of neurostimulation for replicating the therapeutic effects of surgical lesions in the treatment of Parkinson's disease, i.e., thalamotomy of the ventral intermediate thalamus. DBS gained new life with Benabid's initial publication (Benabid et al. 1987) and subsequent publication of trial results (Benabid et al. 1993). Some of the core effects of DBS in Parkinson's disease, such as rapid tremor reduction, were striking. This feature created lasting impressions that DBS has tremendous therapeutic power in new targets and conditions (Ford 2009). Unsurprisingly, this aspect of DBS was broadly captured in public discourse on DBS (Racine et al. 2007b). Promptly accompanying the evolution of clinical DBS, ethical discussion has led to the identification of important ethical challenges for the appropriate use of DBS in neurological and neuropsychiatric patients (e.g., in terms of patient screening procedures, informed consent)

(Fins 2000; Farris and Gianola 2009; Bell et al. 2009; Comité Consultatif National d'Ethique Pour les sciences de la vie et de la santé 2002).

Building on previous work of ours and others, we review in this chapter three key ethical principles at the core of contemporary biomedical ethics[1] (Beauchamp and Childress 2009) and explain how conceptual and empirical research has helped to specify the meaning and scope of their application for DBS clinical practice. In doing so, we adopt a pragmatic perspective where ethical principles constitute hypotheses to test in light of their real-world implications and outcomes (Racine 2013). Further, ethical principles are part of an iterative (pragmatic) cycle of issue identification, deliberative creative moral thinking, and real-world testing (Racine 2013). In this respect, DBS for Parkinson's disease presents a compelling case where ethical principles need to be considered and duly specified to capture how they can guide ethical research and clinical innovation. In return, through specification, we can enrich the meaning of ethical principles and increase their specificity. We hope that this contribution will inform readers and also stimulate discussion related to areas where important questions remain unanswered and where further research would need to be undertaken to understand and enact ethical principles.

1 Respect for Autonomy and Challenges Associated with the Profile of DBS

Respect for patient autonomy is a central component of contemporary medical ethics and bioethics (Beauchamp and Childress 2009). Much more than the simple application of the doctrine of "informed consent" and related concepts of assent and dissent, patient autonomy entails broader respect for the person able to make decisions on his or her own behalf (Beauchamp and Childress 2009). Patient autonomy has been fostered broadly and encompasses many aspects including informing the patient (e.g., health information, treatment options), engaging the patient in discussions and decisions about medical care, and supporting the patient in developing healthcare preferences and values. Hence, respect for autonomy offers a substantive paradigm of deep commitment to individual values and individual rights. However, in the context of DBS for Parkinson's disease, challenges in the exercise and respect of patient autonomy can arise, for instance, because of the cognitive effects of Parkinson's disease on decision-making, the effects of patient vulnerability and enthusiastic media depictions of DBS on patient expectations of treatment, and the challenges clinicians[2] face in managing patient expectations.

---

[1] We have focused on the three ethical principles of respect for autonomy, justice, and non-maleficence due to space constraints and ability to relate previous work to those three principles.

[2] We use the term "clinician" to capture all healthcare providers (e.g., physicians, nurses, social workers).

1.1 Impacts of Cognitive Effects on Decision-Making

Since Parkinson's disease can entail cognitive and psychiatric comorbidities, a number of ethical considerations for the informed consent process and autonomy surface. These considerations are not necessarily unique to Parkinson's disease patients as they can extend to other neurodegenerative or neuropsychiatric disorders. Parkinson's disease can impact cognition and mood, while in both Parkinson's disease and some neurodegenerative or neuropsychiatric disorders, there can be deficits in executive function, attention, verbal fluency, and working memory (Kim 2004). However, it is important to note that a diagnosis of a neurodegenerative or neuropsychiatric disorder does not mean that patients necessarily relinquish their decision-making capacity. Empirical research has shown that, in some neurodegenerative conditions, a diagnosis should not rule out the capacity for patients to express healthcare preferences and participate in decision-making. Patients may prove to have a good understanding of the information regarding the procedure or understand related risks, benefits, and potential complications (Kim 2004). Notably, in patients with Parkinson's disease suffering from cognitive deficits, empirical evidence has demonstrated impairments in the ability to efficiently encode and organize new medical information, but reasoning and personal appreciation of that information was not affected (Martin et al. 2008). Thus, in some instances, patients may be found to have impaired decisional capacity, but this does not necessarily negate their ability to be involved in the decision-making process. Clinicians ought to assess patients' cognitive ability and bear potential challenges to decisional capacity in mind during the informed consent process, ensuring adequate understanding of potential risks, benefits, and outcomes of treatments.

1.2 Managing Patient Expectations

Risks associated with DBS which need to be disclosed includes those related to the actual surgical procedure as well as those related to the device and its operation. Additionally, discussion about the schedule of follow-up appointments for device programming should be stressed, as well as the time and commitment involved, to support an informed decision by patients. Discussion of benefits and outcomes can provide an additional challenge for clinicians, as patient expectations of DBS may be affected by patient vulnerability and/or media discourse on DBS (Racine et al. 2007b). Unrealistic expectations may also compromise patient autonomy and the informed consent process if patients do not fully appreciate the spectrum of possible risks and benefits.

1.3 The Effects of Patient Vulnerability

Since DBS is typically offered uniquely to patients at a late stage of Parkinson's disease, they may have a heightened vulnerability that influences their willingness to consent. Although there have been calls for earlier DBS intervention (Woopen et al. 2013; Susman 2001), clinical guidelines, such as the 2012 Canadian Guidelines on Parkinson's disease, consider surgical treatment to be an option only for advanced patients, when treatment of motor symptoms by other medical interventions fails (Grimes et al. 2012). Some bioethicists agree that, due to the severity of risks involved in DBS, the intervention should be offered only when no other treatment proves effective (Glannon 2010). However, since DBS is offered as a last option to control the symptoms of the disease, this can create a challenging context where patients may over-invest DBS with therapeutic outcomes it cannot deliver (e.g., tremors will disappear in all patients). This vulnerability could be easily exacerbated and could result in patients being overly willing to consent to DBS surgery due to desperation and hope for recovery. Studies examining the consent process in Parkinson's disease for novel interventions have encountered a similar phenomenon of "hyped hope" in spite of unknown risks (Miller and Fins 2006), which, at least in the case of neural transplantation, can lead to highly problematic tendencies such as a willingness to "risk everything" for a treatment procedure (Lopes et al. 2003).

1.4 The Influence of Media on Patient Expectations

Media depictions of DBS can also contribute to unrealistic patient expectations. In previous research, we have shown that public discourse has portrayed DBS enthusiastically (Racine et al. 2007b). An increasing coverage of neurostimulation has suggested marked enthusiasm for the clinical translation of DBS. Many articles have described "miracle stories" where patients were literally cured (Racine et al. 2007b), while other print media headlines proclaimed broad clinical benefits beyond the treatment of Parkinson's disease. Risk information and discussion of the ethical challenges of DBS was limited, resulting in the depiction of DBS as a promising therapy for neurodegenerative conditions.

The impact of enthusiastic media coverage is often hard to assess for methodological reasons. However, in a multi-site study of Canadian neurosurgical units with DBS programs, we found evidence that enthusiastic media reports were interacting with the prior vulnerability of patients (Bell et al. 2010). Clinicians interviewed in this study (e.g., neurosurgeons, neurologists, psychiatrists, and other clinical staff involved in DBS programs) reported a clear effect of enthusiastic media coverage on patient expectations. The unfortunate impact may then be a perceived failure of DBS to meet patient's expectations, resulting in disappointment that can lead to negative consequences for the patient–provider relationship (see Fig. 1) (Bell et al. 2010).

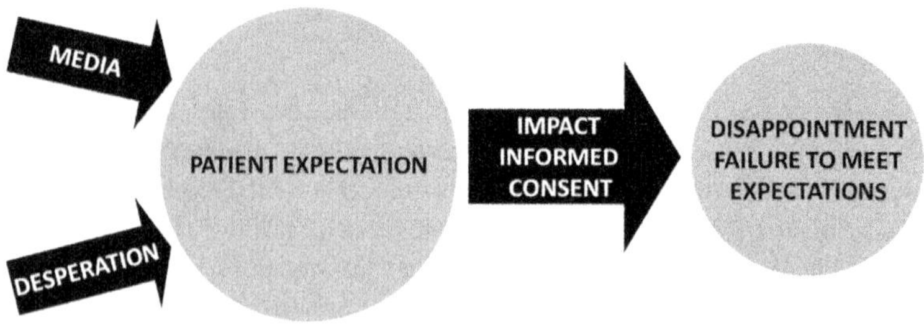

Fig. 1 Desperation and media enthusiasm compound patient expectations—in the clinical context, hope and expectation in DBS for movement disorders challenge informed consent and can lead to a failure to meet pre-defined outcomes

Patients should thus be prompted about the common pitfalls of media depictions of DBS (Racine et al. 2007a). However, the effects of enthusiastic public discourse on DBS are not entirely negative, as increased public awareness about DBS could inform patients and families about this treatment option and bring broader support to healthcare services for neurodegenerative conditions (though these positive features could be augmented) (Bell et al. 2009).

1.5 The Role of Clinicians in Managing Patient Expectations

Patient outcomes in DBS are complex: Some symptoms may improve, others not. DBS can also be transformative in its effects. We have demonstrated that patient expectations of outcomes can be unrealistic, which poses a challenge for clinicians who struggle to manage patient expectations (Bell et al. 2010) (see Table 1). In the literature, measures and practices to address unfounded hope and expectations have been put forth (see Table 2). At this time, more evidence should be gathered on these recommended measures, and dialogue with other clinical areas with similar features (e.g., clinical oncology research) could help tackle these challenges and lay out more comprehensive counseling approaches.

Table 1 High patient expectations of DBS candidates (section A) and clinicians' reactions to expectations (section B) (adapted from Bell et al. 2009)

A. *High expectations of patients reported by clinicians*
"it will be exactly as I was when I was twenty"
"I'm gonna be all better"
"I'll be really good, I'll be really good"
"it is going to fix all the problem"
"I'll be more normal"
B. *Clinician reactions to high patient expectations*
"you are not actually going to cure them"
"this was (is) not a miracle"
"it is not a quick fix"

Table 2 Core ethical principles and their specification as responsive practices within a pragmatic knowledge cycle—inspired by Bell et al. (2009, 2011b) (see text for other references)

Principle and specification	Description of challenging aspects of context	Responsive ethical practices
Principle of respect for autonomy		
Respecting the patient's full or limited ability to engage (or assent) in decision-making. Responsibility to inform patient of risks and commitments involved in DBS. Involves a shared decision-making process where patients may need to be actively empowered to make decisions	PD can lead to cognitive deficits and compromised decision-making capacity with an impact on informed consent	Clinicians probe patient expectations and non-motor expectations; foster reasonable patient expectations
		Multidisciplinary evaluation of candidates
	Patients can be overly optimistic about the expected outcomes of DBS, notably because the media features enthusiastic portrayals of DBS	Social support for patients to deal with perceived failures in outcome
	Patient vulnerability due to last resort nature of DBS can augment desperation or hope for treatment outcomes, challenging balanced understanding of information	Patients accurately communicate expectations and remain aware that chronic illness may change their perception of health information
Principle of justice[a]		
Principle of justice implies considerations of fair access, transparency in principles of distribution of resources, equitable distribution of resources, and technological opportunities, and assessments of global costs and burdens (including opportunity costs)	Not all DBS programs have the same capacity to perform DBS	Committing to an ideal of fairness and equity in allocating resources nationally and regionally (including human resources) and better document national and regional variations to access to DBS
		Maintaining transparency about resource allocation challenges and variations in access when dealing with health policy administrators and patients and their families
	Regional centers approach budgeting for DBS in different ways and are faced with varying levels of challenges related to resource constraints	Advocacy by professional associations to develop national standards of care and access to care commitment to gathering data on DBS use nationally and internationally

(continued)

Table 2 (continued)

Principle and specification	Description of challenging aspects of context	Responsive ethical practices
	Budgets may have an impact on referrals to surgical DBS programs	Identifying strategies to balance the common use of health system resources for research and approved indications of DBS to ensure that approved indications are valued in light of research for emerging indications
Principle of non-maleficence		
Not causing or provoking harm by understanding how physical and psychosocial factors can lead to unsuspected harms; minimizing these harms	There are reports of psychosocial adjustment challenges for PD patients who have undergone DBS	A better assessment of how psychosocial factors are currently taken into account in the informed consent process
	Psychosocial challenges may impact globally the patient's life after DBS and impact relationships to proximate and distal others	More comprehensive follow-up care to respond to psychosocial challenges
	Different hypotheses coexist about these difficulties experienced by patients	Improve the understanding of the mechanism by which DBS exerts psychosocial effects (e.g., using patient-reported outcome measures)

[a] As illustrated and specified in the context of Canada where data have been acquired so far and for which the inherently contextualized aspects of justice, an ethical practice related to social conditions, can be unpacked (Bell et al. 2011a)

While the context of DBS for Parkinson's disease can raise many challenges in the exercise of patient autonomy, it is important to note that these challenges can also provide an opportunity to enrich the exercise of patient autonomy and to enhance the patient–clinician relationship. Challenges to autonomy should encourage clinicians to engage patients in discourse about their illness, the effects of different therapies, and an evaluation of how they are making their choices (including what their goals for treatment are, how they understand and value the different effects of interventions, and what considerations inform their choices). In the age of accessible online information, patients should also be directed to reputable sources of patient-oriented information (Racine et al. 2007a). Common patient-oriented documents available online include those by the National Institute of Neurological Disorders and Stroke's patient page on DBS for Parkinson's disease (National Institute of Neurological Disorders and Stroke 2012), the National Parkinson Foundation (National Parkinson Foundation 2012), as well as those published by academic centers such as the Cleveland Clinic (Cleveland Clinic 2013), and the University of Florida (University of Florida Center for Movement Disorders and Neurorestoration 2013)

2 The Principle of Justice and How DBS, as an Expensive Intervention, Generates Questions for Equity and Fairness

In contemporary discussions, the principle of justice is often overshadowed by considerations related to benefit, risk, and the exercise of patient autonomy where individual rights and freedom anchor the dialogue. Concerned with the social exchange of goods within a population, the principle of justice may sometimes be trumped by other considerations, or altogether overlooked, when the primary deontological obligation of clinicians is focused on the individual patient's care (World Medical Association 1948, revised May 2006). In addition, the principle of justice is complex, notably, because it can borrow different and sometimes radically opposing meanings such as (1) to each person an equal share; (2) to each person according to need; (3) to each person according to effort; (4) to each person according to contribution; (5) to each person according to merit; or (6) to each person according to free-market exchanges (Beauchamp and Childress 2009). In spite of this complexity or the primary obligation of physicians to act in the best interests of the individual patient, clinicians have arguably broader societal responsibilities related to resource allocation, use of health services, and access to health care. For example, the code of ethics of the Canadian Medical Association states that physicians should recognize the need to "promote equitable access to healthcare resources" and "use healthcare resources prudently" (Canadian Medical Association 2004).

2.1 DBS and Issues of Justice

In the context of DBS—a high cost (estimated in one US study at $69,329 in 2006 (Lad et al. 2010)) and increasingly routine procedure for Parkinson's disease patients—issues of distributive justice including the equitable and fair allocation of resources must be considered. Despite the high costs of the surgery, device, and battery replacement, evidence supports the medical management of Parkinson's disease patients with DBS to improve motor function and reduce motor fluctuations and medication side effects (Grimes et al. 2012). Moreover, cost-effectiveness of research recommends funding DBS for eligible patients, where not already funded or reimbursed, because of the overall reasonable incremental cost-utility (€6,700 per quality-adjusted life year) over best medical therapy (Dams et al. 2013) (see Dams et al. 2013 also for a review of the European literature on cost-effectiveness of DBS). Other European studies have shown that the total costs of caring for a patient with Parkinson's disease decrease after surgery (Fraix et al. 2006; Meissner et al. 2005) and that the savings allow return on the procedure investment (approximately 37,000 € when the study was conducted) over 2.2 years (Fraix et al. 2006). Therefore, based on a position where cost-effectiveness forms a basis for assessing what interventions should be available, and where access is influenced by positive assessments of cost-effectiveness, the necessary evidence and impetus are present to make DBS available to eligible patients with Parkinson's disease. Nonetheless, the principle of justice requires broader reflection than is provided by mere estimates of cost-effectiveness, and necessitates deliberation about to whom, how, and to what extent we can afford to, provide health services.

In fact, despite evidence demonstrating improvements in quality of life and motor symptoms and relative cost-effectiveness of DBS in Parkinson's disease, DBS represents an immediate and large investment. As such, regulators, administrators, and insurers may feel obliged to restrict the amount of DBS procedures performed, impacting access to and distribution of resources for DBS. Unfortunately, although practice guidelines support the use of DBS in Parkinson's disease for good candidates, they do little to address issues of potential inequity in access or resource allocation that might be present. Issues of access and resource limitations are sometimes unacknowledged realities of healthcare delivery, but have important ramifications for clinicians, patients, and their families. In a study of Canadian DBS programs, we found widely discrepant approaches to resource allocation, leading to diverging scenarios for access to DBS (See Table 3), and evidence suggesting wide-ranging challenges for access to DBS (Bell et al. 2011a). We discovered that even patient selection can be impacted by the specific resource situation of a hospital. In some cases, clinicians reported that resource limitations affect how quickly patients can be screened (for instance, how quickly neuropsychological assessments can be made) or affect how many patients can even be provided with DBS in any given budgetary year. At the same time, clinicians working at health centers in other regions did not report facing challenges due to resource limitations or in meeting

Table 3 Wide-ranging scenarios of access and resource allocation in Canadian DBS programs (adapted from Bell et al. 2011a)[a]

First scenario Clinicians facing resource allocation challenges (sites A and B)
i.e., more patients requiring DBS than any given budgetary year will fund
Implications of these resource limitations:
Surgeon and operating room time are not used for DBS
Wait times are increased and unpredictable and may cause patients to be re-evaluated before surgery
Referrals are not proactively sought because this would overwhelm surgical capacity
Second scenario Clinicians facing resource limits but viewing these as less problematic (site C), i.e., a short or reasonable waiting time for patients may be viewed as acceptable or even positive, allowing informed reflection of patients
Third scenario Clinicians reporting they do not face resource challenges (sites D and E), i.e., fixed budget but clinicians not concerned about exceeding that budget
Reasons that resource challenges are not experienced include (1) the number of potential candidates lower because of decrease in referral rate and lack of awareness of the program among referring physicians and (2) the limited number of patients because of strict screening criteria

[a] The different scenarios were described by clinicians interviewed at different Canadian DBS programs (sites A, B, C, D, E)

demand for DBS. Another center reported that despite overall limitations set on the resources that can be expended (i.e., personnel), they felt the waiting time is appropriate (see Table 3).

2.2 Access to DBS in the Canadian Context

Given that provincial healthcare systems are bound to federal obligations of access and performance through the Canadian Healthcare Act, to "protect, promote and restore the physical and mental well-being" of Canadians by ensuring "free and universal access to publicly insured health care" (Government of Canada 1984), it is striking to find that these variations can exist. We also question whether patients, families, or clinicians are truly aware of potential differences in access across the country for approved health interventions. Ultimately, the federal legislation only requires the provinces to guarantee the comprehensive coverage of insured care, to cover insurance for all residents, and to ensure equal access to insured services (Government of Canada 1984). It does not explicitly guide the provinces, who independently administer healthcare services, on how to provide access to insured services nor does it clearly mandate the level or standard of health care between provinces. Importantly, access is meant to be equitable, but the meaning of equity remains unclear. In the case of DBS, how can this resource be allocated equitably? Suggestions on how each province could be allocated DBS implantations include equally (i.e. the same share to each province), allocation based on the number of cases of Parkinson's disease, or allocation based on the resources already available (e.g., specialized neurosurgeons). Ultimately, the current mechanism by which healthcare resources are administered in Canada may lead to a neglect of the broader national needs, allowing disparities or variation in access to persist.

Multiple factors could explain such an internally discrepant situation. Challenges in access within publically funded systems have been noted, particularly in the field of medical devices. For example, in Canada, variation in access to implantable cardioverter defibrillators (ICDs) has been shown, as some provinces (i.e., British Columbia, Saskatchewan, Manitoba, Prince Edward Island) fall well below the national average of implantation of ICDs in Canada (Canadian Heart Rhythm Task Force MEDEC 2004). In this case, different factors could explain these variations such as a culture of under-referral (e.g., physicians are hesitant to refer patients to programs that they sense are unavailable, or when they think that wait lists are too long), lack of national standards for access to ICDs, lack of a tertiary center in a province, and some clinical programs being held to fixed budgets (Simpson et al. 2005). Variations in implantation rates of ICDs have also been observed between and within some European countries (Ector et al. 2001; Boriani et al. 2010; Cunningham et al. 2005; Mond and Proclemer 2011). Widely different reasons for these variations have been proposed (McComb et al. 2009). In one Canadian province (Ontario), important inequalities in ICD implantations between those living in urban areas and rural regions have been shown (Lee et al. 2008). Similar inequalities may exist in the context of DBS implantation; however, intra-provincial regional variation in this context has been largely unexplored, likely because of the limited centers across the country providing this specialized neurosurgery (approximately one or two academic centers in every province). In light of this, it is likely that patients have to travel from remote communities to larger centers or, if no program is available, travel from one province to another to access DBS. For DBS, further investigation of how rural populations are serviced by larger academic and urban centers is warranted. Other issues impacting access to DBS could relate to manpower and specialty training in relevant fields (e.g., functional neurosurgery); complex tensions exist in Canada between recognized shortages in specialized practitioners in certain fields and a lack of jobs for many postgraduates in the same medical specialties (Woodrow et al. 2006; Vogel 2011). Moreover, a 2005 report by the Institute for Clinical Evaluative Sciences, which examined the Health Human Resources for Neurosurgical Services in Ontario, describes that one of the visible symptoms of stress in the neurosurgical service delivery system is unequal access to appropriate technology and care across the province (Tepper et al. 2005).

The challenges faced by some clinicians with regard to access and resource restrictions for DBS mean there are difficult rationing and resource allocation decisions to be made. These situations become even more challenging when clinicians are not informed about the allocation process, but are nevertheless charged with explaining to patients and their caregivers allocation decisions, the reasons for wait lists, and the variations between regions or provinces (Holloway et al. 2000). In the context of DBS, resource allocation issues can translate into a number of difficult and arguably unethical situations that require transparency when dealing with patients and families. For example, a Parkinson's disease patient who was a good candidate can develop comorbid conditions (e.g., depression or dementia), while long wait lists may lead to necessitate re-assessments (Bell et al. 2009, 2011a). This leads to a deleterious cycle, where a long wait list necessitates frequent re-assessments, and frequent re-assessments add to the long wait list. As a result, resources must be re-allocated for the same patient and there are even more substantial personal implications for patients and families who are on waiting lists. Table 2 presents some responsive practices to respond to challenges of justice.

3 Non-maleficence and Psychosocial Aspects of DBS

The principle of "first, do no harm" captured in the Latin aphorism "primum non nocere" has been an enduring maxim of medical ethics. Although often attributed to Hippocrates who called upon physicians to abstain from doing harm, its origins are debated (Beauchamp and Childress 2009).

Modern descriptions of the principle have focused on avoidance of physical harms to patients (Beauchamp and Childress 2009). However, non-maleficence cannot be interpreted as the absolute avoidance of harm since some side effects are almost always to be expected from treatments. Hence, the principle stresses rather the avoidance of *undue* or *unjustified* harm. Accordingly, harms need to be considered in relationship to expected benefits. Further, harm can be modulated by the clinical context and social or personal circumstances in which an intervention is offered. Therefore, a contextual appreciation of what the principle means for a specific patient in a specific context needs to take place.

The historical descriptions found in the Latin formulation and related statements in Hippocratic writings also capture another aspect, that is, the fundamental or primary nature of the principle that "first" or "above all," clinicians should not engender harm. Therefore, the principle mandates a prudential attitude calling for a circumspect and reflexive evaluation of one's own knowledge about the beneficial effects of a treatment and that, first and foremost, the physician should avoid harming the patient. This prudential attitude is certainly relevant in the context of DBS given its rapid evolution, partial knowledge of its long-term, potential global effects on the patient, and high expectations toward DBS upheld by patients and their families.

Beyond the physical harms (and benefits) associated with DBS (discussed above), another set of harmful (or potentially beneficial) consequences (Ashkan et al. 2013; Wolz et al. 2012; Sevillano-Garcia and Manso-Calderon 2010) concern the psychosocial (or "non-motor") effects of DBS in Parkinson's disease. These consequences of DBS on quality of life and more comprehensive psychosocial domains are still poorly understood (Sandvik et al. 2012). A landmark qualitative study, published in two papers, showed clearly the issue at hand; contrary to studies suggesting increased independent living after surgery (Krack et al. 2003), patients faced difficulties with their familial or marital relationships after surgery and felt conflicted about returning to work (Agid et al. 2006; Schüpbach et al. 2006). In this study, Schüpbach and colleagues (2006) examined 29 patients with Parkinson's disease before and 18 to 24 months after stimulation. Follow-ups and in-depth interviews with patients, their spouses, and families revealed three challenging areas: the patients' perception of themselves and their bodies (the self), the couple (the proximate other), and the social and professional life (the distal other) (Schüpbach et al. 2006). Some patients expressed difficultly accepting lost years of their lives after being relieved by DBS surgery, while others did not adapt to the sudden motor improvement and still associated with their "ill selves" (Agid et al. 2006). Additionally, a qualitative interview study of patients and clinicians performed by Gisquet demonstrated that some patients who have undergone DBS communicate "a loss of control over managing their illness and over their life," and patients felt tied to the medical team to manage their stimulator and their treatment unlike before (Gisquet 2008). Other authors have proposed that DBS may create adaptation challenges for patients because of a discord between the patients' narrative identity before and after DBS (Schechtman 2010), or because of an abrupt alteration created in the patients' experience of chronic illness (Gisquet 2008). The rapid clinical changes caused by DBS could be partly responsible for such consequences. In the context of epilepsy, others have found that "sudden health" subsequent to medical or surgical treatment "may eliminate the patient's disease and the disease label from the patient's identity" (Seaburn and Erba 2003). In our own research across Canadian surgical centers, we have found corroboration for the findings of Agid, Schüpbach, and colleagues (Bell et al. 2011a, b). Table 4 features examples of testimonials illustrating these three domains and their profound impact on the patient and his social networks (Bell et al. 2011b). Readers should note that the prevalence of such experiences needs to be better established.

3.1 Impact on the Self

The influence of DBS on behavior and personality has not clearly been delineated, and there is conflicting evidence that changes in mood and anxiety occur after DBS (Ballanger et al. 2009; Bell et al. 2009; Frank et al. 2007; Halbig et al. 2009). Moreover, Gisquet has suggested that the experience of mood or behavior changes after DBS may be so far-reaching for patients that they "have the feeling that their identity has been affected" (Gisquet 2008). The larger question remains whether these types of changes, or others observed after DBS, are profound alterations in the personality of the patient (Synofzik and Schlaepfer 2008). Importantly, changes in mood or behavior observed after DBS could be unrelated to the procedure itself, but rather due to the stimulation parameters and targets of stimulation as well as the psychological profile of patients (e.g., cognitive decline prior to surgery), age, and prior L-dopa response (Smeding et al. 2011; Soulas et al. 2011). Discussion regarding the site of choice for stimulation in advanced Parkinson's disease patients and the side effects or advantages of these targets continues (Follett and Torres-Russotto 2012). For example, stimulation of the subthalamic nucleus (STN) may result in more mood-related adverse events (e.g., depression, anxiety) than stimulation of the

Table 4 Illustrative examples of DBS implications for self, proximate other, and distal other (from Bell et al. 2011b)

Changes in personality, changes to the "self"
"So for example one man who had ... bilateral subthalamic stimulation, his wife described him after the surgery basically as being like a spontaneous, impetuous, difficult, teenager. They would be out driving ... they lived near to a boarder ... with the United States and he would say: 'Hey let's go see if we can get across the border without our passports.' You know this a man in his sixties. He would come home with an all-terrain vehicle. You know this is a man who previously hiked and enjoyed sort of peaceful serenity in the outdoors and now wanted to drive an all-terrain vehicle through the woods"
Proximate other (relationships with spouse, friends, and family members)
"A wife who has had a husband who has really been I would have thought a great care to her in terms of his Parkinsonian needs and she fulfilled that role, it was doing something for her. (...) At any rate, she got satisfaction on the fact that he was dependent. Where he had previously been the dominant party in the pair, he was now dependent. I don't think that there was abuse in the story, in the particular case, I don't think there was abuse involved but she got satisfaction on the fact that he was now dependent and in need of her. That was satisfying a need with her. [...] There began to be conflict situations between husband and wife because now he was much more independent. He was driving again, so he said 'look, I am going (...) to go down and see some of my friends'. So I must say that I am not sure that if in the past he had gone down with some of his buddies and spent a lot of time away from home etc. etc. that I am not sure about. Anyway, it was a bad situation. So the two of them had a great deal of conflict and we had to deal with that and get some counseling for the two of them because of these new exchanged roles"
Distal other: Employment, vocational opportunities, and disability
"[...] a very striking example of a young woman who developed a pretty bad movement disorder specifically a generalized dystonia at a young age and as a result was disabled enough that she couldn't really work and uh and at age forty, having failed medical treatment over the years and the surgery comes along and now is the treatment option, we treated her and it cured her, and so now all of a sudden you've got a forty year old who's for the first time in her life normal, and uh that was a major problem. You wouldn't think fixing a disorder would be an issue uh in that manner but all of a sudden this person's normal, the social services people are saying: 'well look, you're now well you should go get a job'. She had not had any employment experience at all in her life, her peer group who were other people that were living in at kind of that level of society, all of a sudden says: 'well you kind of don't belong with our group anymore there's nothing wrong for you,' and so there is an issue of when it works really well people not really being prepared for not being disabled anymore, which often is what we talk about"

pallidal target (Vitek 2002). These adverse events to mood may be higher over the long term in patients who undergo STN stimulation than globus pallidus interna (GPi) stimulation (Moro et al. 2010). Okun and colleagues (2009) have observed that stimulating the ventral contacts of both STN and GPi DBS produces negative mood effects, which they suggest is due to the ventral spread of activity to non-motor and limbic circuits (Okun et al. 2009). Improved understanding of stimulation target sites will enable better management of adverse mood events.

Unfortunately, there is only a small but still thought-provoking literature regarding adaptation challenges for Parkinson's disease patients after DBS (Agid et al. 2006; Schupbach and Agid 2008; Schüpbach et al. 2006). In spite of the lack of causal explanations for such changes, in their obligation to not engender harm, clinicians have to bear in mind such consequences. Psychological care and psychosocial education could be relevant components to include in the care of patients to mitigate harms (Cohen et al. 2007). Research should document psychosocial well-being after surgery as well as conditions and interventions for better outcomes and quality of life. Importantly, some recent clinical trials are taking into account the patient's specific goals in choosing target stimulation sites, because this choice may have an impact on the symptoms they correct (Bell and Racine 2013).

3.2 Proximate Other

Data gathered by our group and others show the existence of psychosocial challenges between the patient and the spouse (and proximate others) after DBS (Agid et al. 2006; Bell et al. 2011b; Schüpbach et al. 2006). One scenario involves patients reclaiming the independence they previously lost and, as a result, rejecting, advertently or inadvertently, their spouse as caregiver, causing the spouse to give up the role they had been playing over the length of the illness. Another scenario involves the patient being "rejected by (the) spouse" leading to, for example, marital problems because the spouse's expectations of outcome are not met by the patient's actual real-life abilities (Agid et al. 2006; Schüpbach et al. 2006). Agid et al. report that 65 % of married patients experienced a "conjugal crisis" following DBS (Agid et al. 2006). Moreover, of the couples studied, Schüpbach et al. (2006) reported that 33 % of the spouses suffered depression after their partners underwent DBS. They also report that a greater percentage of patients were "rejected by their spouse," but our own research suggests this could be a minority (Bell et al. 2011b). Similar inter-spouse conflicts have been described by others where caregivers were reluctant to maintain the role of caregiver after surgery, while patients were reluctant to give up the attention and special treatments that they received from others prior to DBS (Perozzo et al. 2001). The failure of DBS to meet spousal or caregiver expectations, much like the failure of reaching patient expectations, risks creating disappointment and conflict (Bell et al. 2010), a phenomenon also observed in other neurosurgical procedures (Bladin 1992).

Based on our work and that of others, it is clear that the factors contributing to marital conflict following DBS and possible ways to manage or alleviate patient and caregiver distress warrant further investigation. Specifically, a better understanding of how spousal and patient expectations of outcomes can influence the marital relationship after DBS may constitute a key area where DBS healthcare teams could intervene to minimize harms and enhance quality of life.

3.3 Distal Other

The topics of employment and occupational disability have not been extensively discussed or investigated in the context of DBS. General literature on Parkinson's disease patients suggests that many Parkinson's disease patients can rapidly loose employment after being diagnosed, or retire early (Schrag et al. 2003), and most Parkinson's disease patients are no longer working within 10 years of diagnosis (Schrag and Banks 2006). Nonetheless, loss of employment might be highly consequential for young (onset before age 50) Parkinson's disease patients, potentially leading to marital conflicts and a greater perceived impact of the disease (Schrag et al. 2003).

Following DBS, Agid et al. have described that a "retrospective disaster" can be experienced by patients. Although their motor symptoms have improved, patients have suffered irreparable consequences of the disorder such as the loss of employment (Agid et al. 2006). For some patients, the loss of opportunities to gain the skills necessary to be employable may pose the specific problem, rather than the loss of current employment. Highlighting this challenge is the regret that some patients may feel, following improvement of symptoms with DBS, due to not having achieved what they considered to be their full potential. In order to minimize harm to Parkinson's disease patients with respect to "distal others," occupation, and social roles, there could be a need for Parkinson's disease patients to plan for the future earlier in their disease progression and for dedicated support to help them remain in the workforce later into the course of their disease (Schrag and Banks 2006). There may also be a role for clinicians to assist patients and employers in finding appropriate new roles in the workplace for patients with DBS. Unfortunately, there are no data, to our knowledge, which captures the challenges directly related to social support programs and access for patients to such services after DBS. Alternatively, earlier DBS could be a promising way to prevent loss of employment and its related financial implications (Woopen et al. 2013). This does not imply that younger patients are all good candidates for DBS nor that DBS will have a (positive or negative) impact on employment or occupational opportunities for every patient, since many patients undergoing DBS for Parkinson's disease may have already left their professional occupation.

The principle of non-maleficence brings forth questions about the known physical and psychosocial harms of DBS. At this time, there are very little data demonstrating or replicating some of the important results on the effects of DBS on proximate and distal others, or to oneself. More empirical work to understand this problem, hopefully feeding into consent processes and practices for patient autonomy, could help better inform patients about these effects. They could also be used to structure programs (e.g., long-term follow-up, psychological support), or simply increase awareness about these effects for patients, families, and clinicians (Mathieu et al. 2011). Table 2 captures some recommendations to promote non-maleficence in DBS programs.

4 General Conclusion

Following the analysis of some of the emerging issues within three key principles of contemporary biomedical ethics, this paper has reviewed some of the basic ethical challenges that DBS raises in the context of Parkinson's disease. We hope to have shown the relevance of ethical analysis to clinical practice, and the potential for ethical research and deliberation to generate concrete ideas for practice changes and clinical innovation. At this time, more precise understandings of the issues would be beneficial. Participatory and action research could lead to the greatest insights into the impact and benefits of tackling ethical challenges explicitly and directly in the context of DBS for Parkinson's disease.

Acknowledgments Support for this work comes from a grant from the Canadian Institutes of Health Research (Eric Racine, PI; Emily Bell, co-PI) as well as a Career Award from the Fonds de recherche du Québec—Santé (Eric Racine).

References

Agid Y, Schüpbach M, Gargiulo M et al (2006) Neurosurgery in Parkinson's disease: the doctor is happy, the patient less so? J Neural Transm Suppl 70:409–414

Ashkan K, Samuel M, Reddy P et al (2013) The impact of deep brain stimulation on the nonmotor symptoms of Parkinson's disease. J Neural Transm 120(4):639–642

Ballanger B, van Eimeren T, Moro E et al (2009) Stimulation of the subthalamic nucleus and impulsivity: release your horses. Ann Neurol 66(6):817–824

Beauchamp T, Childress J (2009) Principles of biomedical ethics, 6th edn. Oxford University Press, Oxford

Bell E, Mathieu G, Racine E (2009) Preparing the ethical future of deep brain stimulation. Surg Neurol 72(6):577–586

Bell E, Maxwell B, McAndrews MP et al (2010) Hope and patients' expectations in deep brain stimulation: healthcare providers' perspectives and approaches. J Clin Ethics 21(2):112–124

Bell E, Maxwell B, McAndrews MP et al (2011a) Deep brain stimulation and ethics: perspectives from a multisite qualitative study of Canadian neurosurgical centers. World Neurosurg 76 (6):537–547

Bell E, Maxwell B, McAndrews MP et al (2011b) A review of social and relational aspects of deep brain stimulation in Parkinson's disease informed by healthcare provider experiences. Parkinson's Dis 2011:871874

Bell E, Racine E (2013) Ethics guidance for neurological and psychiatric deep brain stimulation. Handb Clin Neurol 116C:313–325

Benabid AL, Pollak P, Louveau A et al (1987) Combined (thalamotomy and stimulation) stereotactic surgery of the VIM thalamic nucleus for bilateral Parkinson disease. Appl Neurophysiol 50(1–6):344–346

Benabid AL, Pollak P, Seigneuret E et al (1993) Chronic VIM thalamic stimulation in Parkinson's disease, essential tremor and extra-pyramidal dyskinesias. Acta Neurochir Suppl (Wien) 58:39–44

Bladin PF (1992) Psychosocial difficulties and outcome after temporal lobectomy. Epilepsia 33 (5):898–907

Boriani G, Berti E, Biffi M et al (2010) Implantable electrical devices for prevention of sudden cardiac death: data on implant rates from a 'real world' regional registry. Europace 12(9):1224–1230

Canadian Heart Rhythm Task Force MEDEC (2004) Provincial variations in access to implantable cardioverter defibrillators (ICDs), Toronto, ON. http://www.medec.org/webfm_send/794. Accessed 18 Dec 2013

Canadian Medical Association (2004) CMA code of ethics. Canadian Medical Association, Ottawa, ON. http://policybase.cma.ca/dbtw-wpd/PolicyPDF/PD04-06.pdf. Accessed 18 Dec 2013

Cleveland Clinic (2013) Deep brain stimulation for Parkinson's disease patients. http://my.clevelandclinic.org/services/deep_brain_stimulation_dbs/hic_deep_brain_stimulation_for_parkinsons_disease_patients.aspx. Accessed 18 Dec 2013

Cohen DB, Oh MY, Baser SM et al (2007) Fast-track programming and rehabilitation model: a novel approach to postoperative deep brain stimulation patient care. Arch Phys Med Rehabil 88 (10):1320–1324

Comité Consultatif National d'Ethique pour les sciences de la vie et de la santé (2002) La neurochirurgie fonctionnelle d'affections psychiatriques sévères. http://www.ccne-ethique.fr/sites/default/files/publications/avis071.pdf. Accessed 6 March 2014

Cunningham AD, Plummer CJ, McComb JM et al (2005) The implantable cardioverter-defibrillator: postcode prescribing in the UK 1998–2002. Heart 91(10):1280–1283

Dams J, Siebert U, Bornschein B et al (2013) Cost-effectiveness of deep brain stimulation in patients with Parkinson's disease. Mov Disord 28(6):763–771

Ector H, Rickards AF, Kappenberger L et al (2001) The world survey of cardiac pacing and implantable cardioverter defibrillators: calendar year 1997—Europe. Pacing Clin Electrophysiol 24(5):863–868

Farris SM, Gianola FJ (2009) Ethical issues surrounding deep brain stimulation in Parkinson's disease. JAAPA 22(2):57–58

Fins JJ (2000) A proposed ethical framework for international cognitive neuroscience: a consideration of deep brain stimulation in impaired consciousness. Neurol Res 22:273–278

Follett KA, Torres-Russotto D (2012) Deep brain stimulation of globus pallidus interna, subthalamic nucleus, and pedunculopontine nucleus for Parkinson's disease: which target? Parkinsonism Relat Disord 18(Suppl 1):S165–S167

Ford PJ (2009) Vulnerable brains: research ethics and neurosurgical patients. J Law Med Ethics 37 (1):73–82

Fraix V, Houeto JL, Lagrange C et al (2006) Clinical and economic results of bilateral subthalamic nucleus stimulation in Parkinson's disease. J Neurol Neurosurg Psychiatry 77(4):443–449

Frank MJ, Samanta J, Moustafa AA et al. (2007) Hold your horses: impulsivity, deep brain stimulation, and medication in parkinsonism. Science 318(5854):1309–1312

Gisquet E (2008) Cerebral implants and Parkinson's disease: a unique form of biographical disruption? Soc Sci Med 67 (11):1847–1851

Glannon W (2010) Consent to deep brain stimulation for neurological and psychiatric disorders. J Clin Ethics 21(2):104–111

Government of Canada (1984) Canada Health Act. Revised Statutes of Canada, 1985, c. C-6. Minister of Justice

Grimes D, Gordon J, Snelgrove B et al (2012) Canadian guidelines on Parkinson's disease. Can J Neurol Sci 39(4 Suppl 4):S1

Halbig TD, Tse W, Frisina PG et al (2009) Subthalamic deep brain stimulation and impulse control in Parkinson's disease. Eur J Neurol 16(4):493–497

Holloway RG, Ringel SP, Bernat JL et al (2000) US neurologists: attitudes on rationing. Neurology 55(10):1492–1497

Kim SYH (2004) Evidence-based ethics for neurology and psychiatry research. NeuroRX 1:372–377

Krack P, Batir A, Van Blercom N et al (2003) Five-year follow-up of bilateral stimulation of the subthalamic nucleus in advanced Parkinson's disease. N Engl J Med 349(20):1925–1934

Lad SP, Kalanithi PS, Patil CG et al (2010) Socioeconomic trends in deep brain stimulation (DBS) surgery. Neuromodulation 13(3):182–186

Lee DS, Tu JV, Juurlink DN (2008) Do Ontario patients with heart failure have equal access to life-saving technology? Healthc Q 11(3):20–21

Lopes M, Meningaud JP, Behin A et al (2003) Consent: a Cartesian ideal? Human neural transplantation in Parkinson's disease. Med Law 22(1):63–71

Martin RC, Okonkwo OC, Hill J et al (2008) Medical decision-making capacity in cognitively impaired Parkinson's disease patients without dementia. Mov Disord 23(13):1867–1874

Mathieu G, Bell E, Racine E (2011) Subjective outcomes measurement and regulatory oversight for deep brain stimulation outcomes in Parkinson's disease. AJOB Neurosci 2(1):16–18
McComb JM, Plummer CJ, Cunningham MW et al (2009) Inequity of access to implantable cardioverter defibrillator therapy in England: possible causes of geographical variation in implantation rates. Europace 11(10):1308–1312
Meissner W, Schreiter D, Volkmann J et al (2005) Deep brain stimulation in late stage Parkinson's disease: a retrospective cost analysis in Germany. J Neurol 252(2):218–223
Miller FG, Fins Joseph J (2006) Protecting human subjects in brain research: a pragmatic perspective. In: Illes J (ed) Neuroethics: defining the issues in theory, practice, and policy. Oxford University Press, Oxford, pp 123–140
Miocinovic S, Somayajula S, Chitnis S et al (2013) History, applications, and mechanisms of deep brain stimulation. JAMA Neurol 70(2):163–171
Mond HG, Proclemer A (2011) The 11th world survey of cardiac pacing and implantable cardioverter-defibrillators: calendar year 2009—a world society of arrhythmia's project. Pacing Clin Electrophysiol 34(8):1013–1027
Moro E, Lozano AM, Pollak P et al (2010) Long-term results of a multicenter study on subthalamic and pallidal stimulation in Parkinson's disease. Mov Disord 25(5):578–586
National Institute of Neurological Disorders and Stroke (2012) NINDS deep brain stimulation for Parkinson's disease information page. http://www.ninds.nih.gov/disorders/deep_brain_stimulation/deep_brain_stimulation.htm. Accessed 18 Dec 2013
National Parkinson Foundation (2012) Deep brain stimulation. http://www.parkinson.org/Parkinson-s-Disease/Treatment/Surgical-Treatment-Options/Deep-Brain-Stimulation. Accessed 18 Dec 2013
Okun MS, Fernandez HH, Wu SS et al (2009) Cognition and mood in Parkinson's disease in subthalamic nucleus versus globus pallidus interna deep brain stimulation: the compare trial. Ann Neurol 65(5):586–595
Perozzo P, Rizzone M, Bergamasco B et al (2001) Deep brain stimulation of subthalamic nucleus: behavioural modifications and familiar relations. Neurol Sci 22(1):81–82
Racine E (2013) Pragmatism and the contribution of neuroscience to ethics. Essays Philos Humanism 21(1):13–20
Racine E, Van der Loos HZA, Illes J (2007a) Internet marketing of neuroproducts: new practices and healthcare policy challenges. Camb Q Healthc Ethics 16:181–194
Racine E, Waldman S, Palmour N et al (2007b) Currents of hope: neurostimulation techniques in US and UK print media. Camb Q Healthc Ethics 16(3):314–318
Sandvik U, Hariz GM, Blomstedt P (2012) Quality of life following DBS in the caudal zona incerta in patients with essential tremor. Acta Neurochir (Wien) 154(3):495–499
Schechtman M (2010) Philosophical reflections on narrative and deep brain stimulation. J Clin Ethics 21(2):133–139
Schrag A, Banks P (2006) Time of loss of employment in Parkinson's disease. Mov Disord 21 (11):1839–1843
Schrag A, Hovris A, Morley D et al (2003) Young- versus older-onset Parkinson's disease: impact of disease and psychosocial consequences. Mov Disord 18(11):1250–1256
Schüpbach M, Gargiulo M, Welter ML et al (2006) Neurosurgery in Parkinson disease: a distressed mind in a repaired body? Neurology 66(12):1811–1816
Schüpbach WM, Agid Y (2008) Psychosocial adjustment after deep brain stimulation in Parkinson's disease. Nat Clin Pract Neurol 4(2):58–59
Seaburn DB, Erba G (2003) The family experience of "sudden health": the case of intractable epilepsy. Fam Process 42(4):453–467
Sevillano-Garcia MD, Manso-Calderon R (2010) nonmotor symptoms in Parkinson's disease and deep brain stimulation. Rev Neurol 50(Suppl 2):S95–S104
Simpson CS, O'Neill BJ, Sholdice MM et al (2005) Canadian cardiovascular society commentary on implantable cardioverter defibrillators in Canada: waiting times and access to care issues. Can J Cardiol 21(Suppl A):19A–24A
Smeding HM, Speelman JD, Huizenga HM et al (2011) Predictors of cognitive and psychosocial outcome after STN DBS in Parkinson's disease. J Neurol Neurosurg Psychiatry 82(7):754–760

Soulas T, Sultan S, Gurruchaga JM et al (2011) Depression and coping as predictors of change after deep brain stimulation in Parkinson's disease. World Neurosurg 75(3–4):525–532

Susman E (2001) Deep brain stimulation for Parkinson disease patients should start before medications fail. Neurol Today 1(1):28–32

Synofzik M, Schlaepfer TE (2008) Stimulating personality: ethical criteria for deep brain stimulation in psychiatric patients and for enhancement purposes. Biotechnol J 3(12):1511–1520

Talan J (2009) Deep brain stimulation: a new treatment shows promise in the most difficult cases. Dana Press, New York

Tepper J, Jaigobin C, Wang C (2005) Health human resources for neurosurgical services in Ontario. ICES investigative report. Institute for Clinical Evaluative Sciences, Toronto

University of Florida Center for Movement Disorders and Neurorestoration (2013) Deep brain stimulation surgery. http://mdc.mbi.ufl.edu/surgery/deep-brain-stimulation-surgery. Accessed 18 Dec 2013

Vitek JL (2002) Deep brain stimulation for Parkinson's disease. A critical re-evaluation of STN versus GPi DBS. Stereotact Funct Neurosurg 78(3–4):119–131

Vogel L (2011) Specialty training out-of-sync with job market. CMAJ 183(13):E1016

Wolz M, Hauschild J, Fauser M et al (2012) Immediate effects of deep brain stimulation of the subthalamic nucleus on nonmotor symptoms in Parkinson's disease. Parkinsonism Relat Disord 18(8):994–997

Woodrow SI, O'Kelly C, Hamstra SJ et al (2006) Unemployment in an underserviced specialty?: the need for co-ordinated workforce planning in Canadian neurosurgery. Can J Neurol Sci 33 (2):170–174

Woopen C, Pauls KA, Koy A et al (2013) Early application of deep brain stimulation: clinical and ethical aspects. Prog Neurobiol 110:74–88

World Medical Association (1948, revised May 2006) WMA declaration of Geneva. http://www.wma.net/en/30publications/10policies/g1/. Accessed 18 Dec 2013

Ethical Issues and Ethical Therapy Associated with Anxiety Disorders

Kaylan L. Altis, Lisa S. Elwood and Bunmi O. Olatunji

Abstract The prevalence of anxiety disorders is among the highest of all psychiatric diagnoses, with a lifetime morbidity rate of nearly 30 %. Given this prevalence, it is important to identify effective and ethical treatments. Empirically based treatments considered efficacious for anxiety disorders largely include cognitive behavioral treatments (CBT), and among these, exposure therapy stands out as both useful and potentially concerning. Ethical concerns regarding exposure treatment for anxiety include fears of symptom exacerbation, high treatment dropout rates, client safety concerns, and the blurring of boundary lines between therapists and clients. Although concerns have been raised regarding exposure treatment generally, specific concerns have been raised related to the treatment of post-traumatic stress disorder (PTSD) given the vulnerable nature of the population. Despite these concerns, research largely supports both the efficacy and safety of exposure therapy. The present chapter provides a review of extant literature highlighting potential ethical concerns, research regarding the raised concerns, and suggestions for minimizing risk in treatment.

Keywords Anxiety · Ethics · CBT · Exposure

Contents

K.L. Altis · L.S. Elwood (✉)
School of Psychological Sciences, University of Indianapolis, Indianapolis, IN 46227, USA
e-mail: elwoodl@uindy.edu

B.O. Olatunji
Department of Psychology, Vanderbilt University, Nashville, USA

Curr Topics Behav Neurosci (2015) 19: 265–278

DOI: 10.1007/7854_2014_340

Published Online: 24 August 2014

1 Introduction

The present chapter provides an overview of the current anxiety treatments that are identified as empirically supported, followed by an in-depth review of the ethical concerns that have been raised regarding treatment of anxiety, with a focus on exposure techniques. Criticisms of exposure therapy have frequently included concerns about symptom exacerbation, increased attrition rates, therapist training and beliefs, safety, and boundary issues. The current chapter reviews each of these concerns and the related literature and concludes that the benefits of exposure treatment for anxiety outweigh the risks. Finally, suggestions for minimizing risks while using exposure therapy and other clinical considerations are presented.

1.1 Overview of Anxiety

The Diagnostic and Statistical Manual Fifth Edition (DSM-5; American Psychiatric Association 2013) indicates that anxiety disorders are prevalent in individuals from childhood to adulthood, with most disorders beginning in adolescence or early adulthood, and occurring in both males and females. Furthermore, research has demonstrated that anxiety disorders are widespread, both nationally and internationally. In a study of prevalence rates, Baxter et al. (2013) used data from 87 studies in 44 countries to conduct a systematic review for the international prevalence rates of anxiety disorders. Results of this study revealed that current prevalence estimates range from 0.9 to 28.3 % and past-year prevalence ranges from 2.4 to 29.8 % across cultures (Baxter et al. 2013). A recent study conducted by Kessler et al. (2012) examined the prevalence of lifetime morbidity risk (LMR) and 12-month prevalence rates in the USA. Results revealed the following anxiety disorders listed from the highest LMR to the lowest, with the LMR listed first followed by the 12-month prevalence rate: specific phobia: 18.4/12.1 %; social phobia: 13.0/7.4 %; post-traumatic stress disorder: 10.1/3.7 %; generalized anxiety disorder (GAD): 9.0/2.0 %; separation anxiety disorder: 8.7/1.2 %; panic disorder: 6.8/2.4 %; bipolar disorder: 4.1/1.8 %; agoraphobia: 3.7/1.7 %; and obsessive-compulsive disorder: 2.7/1.2. Highlighting the prominence of anxiety disorders, data indicated that anxiety disorders have the highest overall prevalence rate among psychiatric disorders, with a 12-month rate of 18.1 % and a lifetime rate of 28.8 % (Kessler et al. 2012).

The presence of an anxiety disorder affects both individuals and society. At an individual level, anxiety disorders have been associated with substantial negative impact on quality of life (Mendlowicz and Stein 2000; Olatunji et al. 2010). A long-standing anxiety disorder can lead to both physical and emotional symptoms, and individuals can consequently suffer functional impairment at work and in other areas of life, lasting up to 1 year or longer, depending on the course of the disorder (Hoffman et al. 2008). Economic costs due to the impairment caused by a GAD, for example, include both personnel costs as evidenced by missed work or a lack of productivity at work and direct medical costs associated with seeking medical attention. Individuals with GAD, for example, seek medical attention significantly more times than individuals without GAD or a comorbid disorder (Hoffman et al. 2008). Overall, the estimated costs associated with anxiety disorders have been reported to be between 42 and 45 billion dollars (Kessler and Greenberg 2002). Thus, the effective treatment of anxiety disorders is necessary on both an individual and a societal level.

1.2 Best Practices for Treating Anxiety

When striving to be an ethical clinician, individuals must avoid causing harm and seek to maximize the success achieved in therapy. To facilitate ethical practice through treatment choices, attempts have been made to provide recommendations for treatment approaches by identifying empirically supported treatments (ESTs) for specific disorders. In their efforts to compose a comprehensive list of current empirically validated therapies, Chambless et al. (1998) provide a list of treatments, termed "well-established treatments" and "probably efficacious treatments" used to treat numerous psychological disorders. The specific "well-established treatments" identified for anxiety primarily fall in the cognitive behavioral therapy (CBT) realm, although treatments may vary on their emphasis on either cognitive or behavioral components. CBT is listed as a well-established treatment for panic disorder (with and without agoraphobia) and GAD, while strict exposure therapy is listed as a well-established treatment for agoraphobia, specific phobia, and obsessive-compulsive disorder (OCD). The treatments for anxiety deemed as "probably efficacious" include applied relaxation for panic disorder and GAD; CBT for social phobia and OCD; exposure therapy for post-traumatic stress disorder (PTSD) and social phobia; eye movement desensitization and reprocessing (EMDR) therapy for PTSD; stress inoculation training for PTSD; and systematic desensitization for specific phobia and social exposure/ social phobia (Chambless et al. 1998). As all of these treatments either are variants or include tenets of CBT, CBT is often deemed the gold standard for treatment of anxiety disorders (Rauch et al. 2012). Additionally, CBT has been identified as the most empirically supported treatment for child and adolescent anxiety (James et al. 2012; Sburlati et al. 2011).

CBT utilizes techniques to identify and modify maladaptive thoughts and behaviors (Beck 2011). A key CBT behavioral treatment for anxiety is exposure. Exposure approaches are based on classical conditioning theory. Exposure includes taking a feared, although not actually dangerous, stimulus and providing the individual with opportunities to experience the stimulus without associated negative consequences. The rationale is that repeated exposures without negative experiences other than the fear and anxiety will result in a reduction of fear. Exposure techniques may also include pairing feared stimuli with a positive experience, such as a relaxation exercise. The goal of Exposure is for the client to reach a point of habituation, which is achieved by first igniting the client's fears. The anxiety and fear experienced during an exposure typically increases, reaches a plateau, and then decreases over time (Muller and Schultz 2012). Exposure exercises can take different forms, including either imaginal, in which one imagines the feared stimulus, or in vivo, when one confronts the stimulus directly. Exposure techniques can also vary in the progression of intensity. Systematic desensitization, for example, involves the creation of a hierarchy of feared stimuli and exposes the client gradually. Flooding, on the other hand, is an exposure that starts with the primary feared object. Exposure are a major behavioral technique utilized in CBT for anxiety and are often used to combat avoidance strategies (Beck 2011). Additional behavioral techniques that can be used include skills training and relaxation.

1.3 Current State of the use of Empirically Supported Treatments (EST)

Although CBT techniques have been identified as the predominant treatment for anxiety disorders, the current state of usage for these techniques does not reflect the research that supports their efficacy. A 10-year follow-up survey conducted by Woody et al. (2005) compares the current state of EST usage with the usage in 1993. The study examined the use of ESTs for anxiety and stress disorders in doctoral programs and internship programs. Eight CBT treatments for anxiety and stress disorders were included on both surveys. Of these eight treatments, seven were taught in a supervised way by the majority of doctoral programs in 1993, while only two were taught in a supervised way in the 2003 follow-up. Although only two treatments were taught in a supervised way, most programs provided at least brief instruction in courses for seven of the treatments. A similar pattern can be observed in the percentages of use by internship programs. In 1993, most internship sites provided formal supervision for four of the eight treatment methods, while in 2003, this number dropped slightly to three. Thus, supervised training for ESTs at both the doctoral and internship levels is lacking, despite the evidence supporting these treatments (Woody et al. 2005). Although the frequency of use of ESTs in general clinical practice is unknown, the suspected levels of implementation are low, and research indicates that for CBT specifically, approximately 30 % of patients in therapy receive this treatment (Goisman et al. 1999). Further, exposure therapy is underutilized in clinical practice, largely due to lack of training and effective dissemination (Cahill et al. 2006; van Minnen et al. 2010; McManus et al. 2008).

If one assumes that the majority of clinicians are striving to practice ethically, which would include providing effective and standard of care treatments, it is helpful to consider what factors have resulted in the discrepancy between EST recommendations and the treatments being utilized in care settings. As a result, much debate has emerged related to barriers to dissemination and implementation of ESTs into general clinical practice. Arguments provided as explanations for the lack of implementation include lack of time and resources for training, a shortage of trained supervisors, inappropriateness for certain populations, and philosophical opposition (Woody et al. 2005). A particular area of ethical concern involves the use of exposure therapy for anxiety, especially PTSD. Supporters of the technique suggest that a lack of training leads to misconceptions of the treatment and its effectiveness, and these misconceptions evolve into ethical concerns. The ethical concerns surrounding the use of exposures in clinical practice focus on the possibility of causing harm by exacerbating symptoms, pushing boundaries within the therapeutic relationship, and experiencing hesitation from clients to relive their negative experiences (Olatunji et al. 2009; Wolitzky-Taylor et al. 2012).

2 Review of the Ethical Concerns for Exposure Therapy

exposure therapy is utilized among several anxiety disorders, including specific phobias, OCD, panic disorder, PTSD, social anxiety, and GAD (Gillihan et al. 2012; Olatunji et al. 2009). As noted previously, exposure therapy was deemed a "well-established treatment" or "probably efficacious treatment" for use with numerous anxiety disorders (Chambless et al. 1998). Despite this support, authors have noted that clinicians remain hesitant to utilize this method for various reasons (Foa et al. 2002; Neuner 2012; Olatunji et al. 2009). Some of the most noted ethical concerns surrounding the use of exposure include symptom exacerbation, attrition rates, therapist training and negative beliefs, safety and risk, and boundary issues (Cahill et al. 2006; Foa et al. 2002; Frye and Spates 2012; Olatunji et al. 2009; van Minnen et al. 2010; Wolitzky-Taylor et al. 2012). The following is a review of these concerns and strategies for how to minimize risk during exposure.

2.1 Symptom Exacerbation

The possibility of causing harm to clients through symptom exacerbation is arguably the primary reason for clinician avoidance of exposure treatment (Olatunji et al. 2009). As exposures involve the confrontation of feared stimuli, whether imaginal or in vivo, the critique follows that exposure evokes stress in clients and thereby can cause harm. Further, the argument states that this harm is not worth the outcome and is insensitive to the client's experience (Wolitzky-Taylor et al. 2012). There is some evidence to support the noted concerns for symptom exacerbation within exposure treatment; however, review of the literature suggests that these concerns are minimal and do not interfere with overall treatment. One study compared the use of imaginal exposure with that of cognitive therapy for participants of trauma therapy and found a statistically significant difference between the groups from pre- to post-treatment, with more patients in the prolonged exposure group showing symptom exacerbation (Tarrier et al. 1999). This significant finding was, however, not maintained at the 6-month post-treatment assessment (Tarrier et al. 1999). Foa et al. (2002) conducted a study to determine whether participants experienced an exacerbation in PTSD, anxiety, or depression symptoms due to imaginal exposure. Participants met DSM-IV criteria for chronic PTSD and were divided into two treatment conditions, prolonged exposure and prolonged exposure plus cognitive restructuring. When measuring symptom exacerbation during the implementation of imaginal exposure, the researchers found minimal levels of symptom exacerbation among participants: 10.5 % experienced an increase in PTSD, 21.1 % in anxiety, and 9.2 % in depressive symptoms, overall accounting for 20 participants out of 76 that experienced symptom exacerbation. Further, the researchers found that this temporary symptom exacerbation did not interfere with treatment and that attrition rates and symptom exacerbation were not correlated. Additional reviews on the use of exposure therapy further state that any symptom exacerbation with this treatment is no worse than the exacerbation found with other treatments (van Minnen et al. 2010; Cahill et al. 2006; Neuner 2012). Thus, while there is evidence to support temporary symptom exacerbation during exposure therapy, findings suggest that it does not interfere with treatment outcomes.

2.1.1 Symptom Exacerbation within Group Cognitive Behavioral Therapy

Special ethical consideration concerning symptom exacerbation has been expressed for the use of exposure within group therapy. It is argued that conducting exposure in group settings could lead to increased symptomatology in other group members as they are exposed to a fearful stimulus they previously were not in contact with, for example, in the case of trauma exposure, or develop a negative association with another previously neutral stimulus. Despite the concern that vicarious traumatization can occur during group therapy, a meta-analysis by Barrera et al. (2013) concludes that there is presently no empirical data to support this claim. The meta-analysis determined that the use of exposures in group therapy was neither more nor less effective than group therapy that excluded exposures and further claimed that if symptom exacerbation did occur, it did not affect overall treatment (Barrera et al. 2013). This finding is supported by additional studies evaluating the effectiveness of group CBT (Mott et al. 2013; Castillo et al. 2012).

2.2 Attrition Rates

In addition to the concern that has been raised regarding whether reliving or reencountering the feared stimuli harms clients by causing revictimization, some have argued that the perceived harm or distress associated with the exposure activity prevents clients from engaging in exposure treatment and can ultimately lead to increased rates of dropout (Cahill et al. 2006). Despite the concern, research does not suggest that exposure therapy produces larger dropout rates than other forms of therapy. For example, Hembree et al. (2003) compared dropout rates of twenty-five treatment studies for PTSD and found no statistically significant differences between the dropout rates for exposure therapy and the other treatment modalities evaluated. Specifically, exposure therapy alone produced an average dropout rate of 20.5 %, compared to that of stress inoculation training or cognitive therapy alone (22.1 %), exposure plus cognitive therapy or stress inoculation training (26.9 %), and eye movement desensitization and reprocessing (18.9 %). Additional studies and summaries support this finding that attrition rates are consistent across other treatment modalities for PTSD (Cahill et al. 2006; Olatunji et al. 2009; Wolitzky-Taylor et al. 2012).

Kehle-Forbes et al. (2013) conducted a subsequent study testing whether the addition of non-exposure components with exposure treatment would improve overall outcomes for PTSD, including attrition rates. This study, as part of a larger meta-analytic review, examined the results of eight studies comparing the efficacy of "exposure-only" treatment versus "exposure plus" treatment. Results showed no differences across treatment conditions regarding dropout rates, therefore suggesting that the addition of a non-exposure component does not improve the dropout rate. Similarly, while attrition exists within group exposure treatment, rates are comparable with other non-exposure group treatments (Barrera et al. 2013). Therefore, extant data suggest that exposure, either a stand-alone treatment or a component of another treatment, is not associated with increased attrition compared to other treatments. The failure to find increased dropout for exposure treatments further suggests that even if clients are experiencing some symptom exacerbation, it is not deterring clients from participating in treatment.

2.3 Therapist Training and Negative Beliefs

Another potential ethical concern is the practice of exposure therapy with minimal or inadequate training. Researchers have found that a lack of training is a major reason that exposure therapy is underutilized (Cahill et al. 2006; Olatunji et al. 2009; Schare and Wyatt 2013; van Minnen et al. 2010). Ideally, clinicians should receive substantial and rigorous training by an expert, which involves training in all components of exposure therapy as well as training in the implementation of the treatment across patient populations and presenting problems (Wolitzky-Taylor et al. 2012).

However, as noted above, Woody et al. (2005) found that fewer than half of APA-accredited internship sites provide training and supervision in ESTs, including exposure therapy. Schare and Wyatt (2013) also suggest that the dissemination of exposureExposure has been lacking in the clinical community largely due to a lack of training at both the predoctoral and postdoctoral levels.

Examining rates in practicing clinicians, Cahill et al. (2006) conducted a cross-study analysis of six studies evaluating clinician utilization of exposure treatment to clients with PTSD and found a relationship between therapist's level of training and the utilization of exposure therapy. In their sample, a minimal number of therapists had received training in exposure therapy (28 %), and consistently, the treatment was minimally employed (<20 %). Similarly, researchers have asserted that trauma therapists are undertrained in this treatment (van Minnen et al. 2010) and that exposure therapy is underutilized in settings in which a majority of the cases have PTSD despite the demonstrated efficacy for treating PTSD (Olatunji et al. 2009). The Cahill et al.'s (2006) study also found that even when therapists were both trained and experienced in exposure therapy, the method was only implemented by about half of the clinicians, largely because of clinician concern for client safety.

Even with training, many therapists hold negative beliefs about the treatment, resulting in poor implementation of the technique (Deacon and Farrell 2013; Deacon et al. 2013; Farrell et al. 2013a, b). Farrell et al. (2013b) conducted a study looking at the relationship between therapist attitudes about exposure therapy and the quality of delivery of the treatment. Participants included fifty-three undergraduate students who were divided into two treatment groups, manipulated to hold either positive or negative beliefs about the treatment. Their study concluded that participants in the negative beliefs group created less ambitious fear hierarchies, chose less distressing items on the hierarchy, and allowed more safety behaviors during the exposure. Deacon et al. (2013) found similar results in their study evaluating the psychometric properties and validity of the therapist beliefs about exposure scale (TBES). Negative beliefs appear to be modifiable, however, as didactic training in exposure therapy has been found to successfully decrease negative beliefs (Deacon et al. 2013). Farrell et al. (2013a) additionally suggest a theory-based approach to modify negative beliefs: pairing exposure therapy with the notion that it is safe, tolerable, and ethical; providing both empirical and emotional appeals that support the efficacy of exposure therapy; and instructing how to defend the treatment as safe, tolerable, and ethical. Continued examination of education techniques for clinicians will be important for increasing the implementation of exposure.

2.4 Safety and Risk

The risk of causing physical harm to patients during an exposure poses a safety concern for clinicians, largely related to in vivo exposure therapy, as this includes a real-life confrontation of feared stimuli. As the nature of an exposure is to provoke anxiety while confronting a fear, a certain degree of risk always exists; however, the degree of risk should not exceed that which another individual not participating in an exposure would encounter on a daily basis. Caution should be paid to the potential for dangerous situations (e.g., time of day), and overall, any potential risk should be discussed and, if relevant, reduced or normalized for the client (Gillihan et al. 2012). Olatunji et al. (2009) confirm that a risk-free exposure experience cannot be guaranteed for the client; however, fears can be addressed and risks minimized.

2.5 Boundary Issues

While exposures can be conducted within the therapy office, many are completed outside the office, both via homework and during a session with the therapist. Thus, the blurring of a boundary line between therapist and client as well as the client's confidentiality has been linked with ethical concerns. Although the issue of crossing a therapeutic boundary can occur in various forms of psychotherapy, the boundary of leaving an office appears most relevant to exposure therapists (Olatunji et al. 2009; Wolitzky-Taylor et al. 2012). One risk when leaving the office for an exposure is the potential to form a more casual relationship between the parties. There is, however, an important distinction between crossing versus violating a given therapeutic boundary. While boundary crossings are not standard practice, they are not inherently unethical and can temporarily occur; however, a boundary violation is a deliberate and consistent exploitation of the relationship that results in an inappropriate dual relationship (Olatunji et al. 2009; Wolitzky-Taylor et al. 2012). The secondary issue with out-of-office exposures is the potential for a breach in patient confidentiality. Once the therapy session moves beyond the office walls, there is an increase in the potential for others to identify one of the parties and recognize the nature of the relationship. These concerns can be minimized by developing a cover story of the therapist's role, the therapist maintaining a safe distance from the client if possible, holding the exposures in different neighborhoods, and not holding therapy-like discussions in public. The possibility for a breach in confidentiality should be discussed with the client beforehand, as there are no guarantees that strict confidentiality can be maintained under exposure conditions (Olatunji et al. 2009; Wolitzky-Taylor et al. 2012).

Despite the risks of boundary crossings during an exposure, the out-of-office exposure itself can produce powerful positive outcomes that include strengthening the patient's confidence, increasing the therapeutic alliance, and enhancing therapist support (Wolitzky-Taylor et al. 2012). An additional benefit is the advantage of being able to address feared stimuli that could not have been confronted in a therapy office and to do so with the coaching of a trained professional (Olatunji et al. 2009). Thus, despite the concern for crossing boundaries in the relationship, the out-of-office exposure experience produces several benefits.

2.6 Minimizing Risk During Exposure

Despite the demonstrated safety and efficacy of exposure therapy, clinicians should still practice risk management at the onset of an exposure. Four suggestions when seeking to minimize risk are to conduct an evaluation of the need for an exposure, obtain informed consent, make naturalistic comparisons for the client concerning risk, and manage any unexpected outcomes after an exposure (Olatunji et al. 2009; Wolitzky-Taylor et al. 2012). An evaluation should be conducted at the onset of each exposure, to weigh the need for the exposure with the above-mentioned ethical concerns. Such an evaluation consists of first determining whether the exposure is appropriate, addressing the appropriate fear, and also whether it is consensual, by involving the client in the treatment planning process and assisting them in taking ownership of their treatment. A second determinant is the cost–benefit analysis which concludes whether the benefits of the exposure outweigh the risk (Wolitzky-Taylor et al. 2012). This initial evaluation of the need for an exposure sets up the provision of a rationale as well as the following informed consent process.

Researchers agree that patients should be made fully aware of all of the components of their exposure, as well as the potential risks involved, and should give consent prior to engaging in any exposure. A successful informed consent process should include appropriate written documentation and also place high value on a verbal dialogue between the parties. The process should be collaborative and include a discussion of alternate treatment as well, so that the patient can be fully informed (Goisman and Gutheil 1992; Olatunji et al. 2009; Wolitzky-Taylor et al. 2012). The consent process should be ongoing and allow the patient to make changes and retract their consent at any point, and a clear rationale for the exposure should also be provided to the patient, to increase the likelihood of treatment adherence and participation (Olatunji et al. 2009).

A naturalistic comparison is one that determines whether an exposure poses an acceptable level of risk that is typical for most people in a similar situation, during the course of normal life. Such a comparison is important to make with the client as it again highlights that while the potential for harm might exist, it is no greater than that which another person would typically encounter from day to day. This comparison determines an acceptable level of risk for the client during the exposure (Olatunji et al. 2009; Wolitzky-Taylor et al. 2012). Wolitzky-Taylor et al. (2012) further note that special consideration should be made for PTSD clients, as their traumatic experiences typically are not comparable to one's everyday life, and thus, more caution with regard to safety should be made during such exposures.

With any exposure, clinicians should discuss with their clients the possibility that the exposure will not go as planned and manage any unexpected outcomes throughout the process. A critical shift in the expectation of an exposure can occur when the clinician frames the exposure as an experiment, or a test of probabilities. This intentional shift in the goal can drastically improve the outcome of an exposure, particularly when an unexpected incident occurs (Olatunji et al. 2009; Wolitzky-Taylor et al. 2012). Clients can reap great benefits in learning from an exposure that did not go entirely as planned, and this situation can further serve to improve the effectiveness of the exposure. Wolitzky-Taylor et al. (2012) further assert that client habituation perhaps is not necessary for a successful exposure, but rather improving client tolerance of the situation, due to an exposure, is also a beneficial goal and outcome.

2.7 Clinical Considerations

While the evidence suggests that risks associated with exposure therapy are minimal, there may be times when other treatment options are preferred. Given the goal of increasing distress temporarily, exposure may be best suited for clients with at least a minimal ability to tolerate distress and respond in acceptable ways. While clients do not need to have expert levels of coping skills prior to participating in exposure, clients with demonstrated patterns of harmful coping skills may be better suited for alternate treatments. At the extreme end, clients with active suicidal ideation may benefit from therapy focused on stabilization and the development of coping skills and distress tolerance before beginning exposure therapy. Similarly, clients with regular self-harm behaviors or substance dependence may not be appropriate for exposure. Clinicians working with these clients will often prioritize treatment for substance dependence, emotion regulation, and coping skills as a first response with a possible intention to participate in exposure at a later date. Reviewing potential contraindications for prolonged exposure use with PTSD, van Minnen et al. (2012) reported that clinicians have typically been cautioned against using exposure with patients who presented with serious self-harm behaviors, acute suicidality, active psychosis, and substance dependence. However, the review references recent research examining treatments that combine exposure with techniques that address the comorbid symptoms simultaneously. Another option may be to utilize cognitive approaches to address trauma symptoms with clients while working on improvement in comorbid symptoms. Although concerns are sometimes expressed regarding the use of exposure therapy with other comorbid disorders, van Minnen et al. (2012) concluded that prolonged exposure for PTSD is appropriate for individuals with comorbid dissociation, depression, substance abuse, and mild borderline personality disorder.

3 Conclusion

In the ongoing attempt to practice ethically and improve quality of life while avoiding harm, clinicians regularly consider the pros, cons, and ethical implications of treatment decisions. As a first step, clinicians should consider the ethical implications of treatment choice, particularly the consideration of research evaluating treatment options. Many will argue that selecting treatment options that have consistently demonstrated effectiveness increase the likelihood of improving the quality of the clients' lives through symptom reduction and decrease potential harm to clients. Review of the literature on the EST of anxiety will undoubtedly include recommendations to utilize cognitive behavioral treatments (CBT) of anxiety, including exposure-based treatments. Despite this, clinicians may find themselves considering ethical concerns related to CBT treatment, and exposure in particular, including fears of symptom exacerbation, increased risk of attrition, boundary violations, and risk to the client. However, research suggests that these concerns are generally not empirically supported and those that are partially supported, such as symptom exacerbation, show no interaction with overall treatment and also no significant differences compared to other treatment methods. Currently, the strong evidence supports exposure therapy as one of the most important techniques utilized in CBT. Further highlighting this, the completion of at least one exposure has been identified as the largest predictor of long-term sustained symptom reduction (Glenn et al. 2013).

Ethical concerns about exposure therapy are largely minimized by appropriate clinical training and proper implementation of the technique (Cahill et al. 2006; Olatunji et al. 2009; van Minnen et al. 2010; Wolitzky-Taylor et al. 2012; Woody et al. 2005). Thus, the impetus rests on the clinical community and on individual therapists to insure that proper training and implementation of this technique is possible. This training and implementation should begin in clinical training programs, including graduate training and internships, but should not end once the degree is earned. The clinical community should strive to increase education regarding the effectiveness of CBT treatments for anxiety and opportunities to receive trainings on treatment implementation. While initial CE events and online trainings are meritorious first steps, clinicians would greatly benefit from the opportunity to receive ongoing consultation or supervision during initial attempts to practice new treatments (Karlin et al. 2010). For example, the Veteran's Administration recently implemented a program to disseminate two ESTs of PTSD to VA clinicians (Karlin et al. 2010). Within the training program, participants complete an initial intensive training, followed by ongoing consultation and review of client materials. The opportunity to engage in this type of thorough training post-degree would likely increase the implementation of ESTs by community clinicians. Finally, academic and clinical communities should continue to implement and disseminate findings from empirical studies that evaluate both the effectiveness of CBT treatments for anxiety and studies that support or disclaim ethical considerations in using exposure therapy.

References

American Psychiatric Association (2013) Diagnostic and statistical manual of mental disorders, 5th edn. Arlington, Virginia

Barrera TL, Mott JM, Hofstein RF et al (2013) A meta-analytic review of exposure in group cognitive behavioral therapy for posttraumatic stress disorder. Clin Psychol Rev 33(1):24–32. doi:10.1016/j.cpr.2012.09.005

Baxter AJ, Scott KM, Vos T et al (2013) Global prevalence of anxiety disorders: a systematic review and meta-regression. Psychol Med 43(5):897–910. doi:10.1017/S003329171200147X
Beck JS (2011) Cognitive behavior therapy: basics and beyond, 2nd edn. New York
Cahill SP, Foa EB, Hembree EA et al (2006) Dissemination of exposure therapy in the treatment of posttraumatic stress disorder. J Trauma Stress 19(5):597–610. doi:10.1002/jts
Castillo DT, C'de Baca J, Qualls C et al (2012) Group exposure therapy treatment for posttraumatic stress disorder in female veterans. Mil Med 177(12):1486–1491. doi:10.7205/MILMED-D-12-00186
Chambless DL, Baker MJ, Baucom DH et al (1998) Update on empirically validated therapies. II. Clin Psychologist 51(1):3–16
Deacon BJ, Farrell NR (2013) Therapist barriers to the dissemination of exposure therapy. In: Storch EA, McKay D (eds) Handbook of treating variants and complications in anxiety disorders. New York
Deacon BJ, Farrell NR, Kemp JJ et al (2013) Assessing therapist reservations about exposure therapy for anxiety disorders: The therapist beliefs about exposure scale. J Anxiety Disord 27 (8):772–780. doi:10.1016/j.janxdis.2013.04.006
Farrell NR, Deacon BJ, Dixon LJ et al (2013a) Theory-based training strategies for modifying practitioner concerns about exposure therapy. J Anxiety Disord 27(8):781–787. doi:10.1016/j.janxdis.2013.09.003
Farrell NR, Deacon BJ, Kemp JJ et al (2013b) Do negative beliefs about exposure therapy cause its suboptimal delivery? An experimental investigation. J Anxiety Disord 27(8):764–771. doi:10.1016/j.janxdis.2013.03.007
Foa EB, Zoellner LA, Feeny NC et al (2002) Does imaginal exposure exacerbate PTSD symptoms? J Consult Clin Psych 70(4):1022–1028. doi:10.1037//0022-006X.70.4.1022
Frye LA, Spates CR (2012) Prolonged exposure, mindfulness, and emotion regulation for the treatment of PTSD. Clin Case Stud 11(3):184–200. doi:10.1177/1534650112446850
Gillihan SJ, Hembree EA, Foa EB (2012) Behavior therapy: exposure therapy for anxiety disorders. In: Dewan MJ, Steenbarger BN, Greenberg RP (eds) The art and science of brief psychotherapies: an illustrated guide, 2nd edn. Arlington, Virginia
Glenn D, Golinelli D, Rose RD et al (2013) Who gets the most out of cognitive behavioral therapy for anxiety disorders? The role of treatment dose and patient engagement. J Consult Clin Psych 81(4):639–649. doi:10.1037/a0033403
Goisman RM, Gutheil TG (1992) Risk management in the practice of behavior therapy: boundaries and behavior. Am J Psychother 46(4):532–543
Goisman RM, Warshaw MG, Keller MB (1999) Psychosocial treatment prescriptions for generalized anxiety disorder, panic disorder, and social phobia, 1991–1996. Am J Psychiat 156 (11):1819–1821
Hembree EA, Foa EB, Dorfan NM et al (2003) Do patients drop out prematurely from exposure therapy for PTSD? J Trauma Stress 16(6):555–562. doi:10.1023/B:JOTS.0000004078.93012.7d
Hoffman DL, Dukes EM, Wittchen H-U (2008) Human and economic burden of generalized anxiety disorder. Depress Anxiety 25(1):72–90. doi:10.1002/da
James A, Cowdrey F, James C (2012) Anxiety disorders in children and adolescents. In: Sturmey P, Hersen M (eds) Handbook of evidence-based practice in clinical psychology, Vol 1: child and adolescent disorders. Hoboken, New Jersey
Karlin BE, Ruzek JI, Chard KM et al (2010) Dissemination of evidence-based psychological treatments for posttraumatic stress disorder in the Veterans Health Administration. J Trauma Stress 23(6):663–673. doi:10.1002/jts
Kehle-Forbes SM, Polusny MA, MacDonald R et al (2013) A systematic review of the efficacy of adding nonexposure components to exposure therapy for posttraumatic stress disorder. Psychol Traum: Theor Res Pract Policy 5(4):317–322. doi:10.1037/a0030040
Kessler RC, Greenberg PE (2002) The economic burden of anxiety and stress disorders. Neuropsychopharmacol: Fifth Gener Prog 67:981–992

Kessler RC, Petukhova M, Sampson NA et al (2012) Twelve-month and lifetime prevalence and lifetime morbid risk of anxiety and mood disorders in the United States. Int J Meth Psych Res 21(3):169–184. doi:10.1002/mpr

McManus F, Grey N, Shafran R (2008) Cognitive therapy for anxiety disorders: current status and future challenges. Behav Cogn Psychoth 36(06):695. doi:10.1017/S135246580800475X

Mendlowicz MV, Stein MB (2000) Quality of life in individuals with anxiety disorders. Am J Psychiat 157(5):669–682. doi:10.1176/appi.ajp.157.5.669

Mott JM, Sutherland RJ, Williams W et al (2013) Patient perspectives on the effectiveness and tolerability of group-based exposure therapy for posttraumatic stress disorder: preliminary self-report findings from 20 veterans. Psychol Traum: Theor Res Pract Policy 5(5):453–461

Muller KL, Schultz LT (2012) "Selling" exposure therapy. Prag Case Stud Psychother 8 (4):288–295

Neuner F (2012) Safety first? Trauma exposure in PTSD. In: Neudeck P, Wittchen H-U (eds) Exposure therapy: rethinking the model—refining the method. New York

Olatunji BO, Cisler JM, Tolin DF (2010) A meta-analysis of the influence of comorbidity on treatment outcome in the anxiety disorders. Clin Psychol Rev 30(6):642–654. doi:10.1016/j.cpr.2010.04.008

Olatunji BO, Deacon BJ, Abramowitz JS (2009) The cruelest cure? Ethical issues in the implementation of exposure-based treatments. Cogn Behav Pract 16(2):172–180

Rauch SA, Eftekhari A, Ruzek JI (2012) Review of exposure therapy: a gold standard for PTSD treatment. J Rehabil Res Dev 49(5):679–687

Sburlati ES, Schniering CA, Lyneham HJ et al (2011) A model of therapist competencies for the empirically supported cognitive behavioral treatment of child and adolescent anxiety and depressive disorders. Clin Child Fam Psych 14(1):89–109. doi:10.1007/s10567-011-0083-6

Schare ML, Wyatt KP (2013) On the evolving nature of exposure therapy. Behav Mod 37 (2):243–256. doi:10.1177/0145445513477421

Tarrier N, Pilgrim H, Sommerfield C et al (1999) A randomized trial of cognitive therapy and imaginal exposure in the treatment of chronic posttraumatic stress disorder. J Consult Clin Psych 67(1):13–18

van Minnen A, Harned MS, Zoellner L et al (2012) Examining potential contraindications for prolonged exposure therapy for PTSD. Eur J Psychotraumatol 3:1–14

van Minnen A, Hendriks L, Olff M (2010) When do trauma experts choose exposure therapy for PTSD patients? A controlled study of therapist and patient factors. Behav Res Ther 48 (4):312–320. doi:10.1016/j.brat.2009.12.003

Wolitzky-Taylor KB, Viar-Paxton MA, Olatunji BO (2012) Ethical issues when considering exposure. In: Davis TEI, Ollendick TH, Ost LG (eds) Intensive one-session treatment of specific phobia. New York

Woody SR, Weisz J, McLean C (2005) Empirically supported treatments: 10 years later. Clin Psychologist 58(4):5–11

Just Like a Circus: The Public Consumption of Sex Differences

Donna L. Maney

Abstract The study of sex differences is a rich, productive area of neuroscience, yielding findings that inform our understanding of basic biology and hold promise for clinical applications. There is a tremendous, problematic mismatch, however, between the actual implications of this research and what has generally been communicated to the public. The message communicated by the media, popular press, and in some cases researchers is often inaccurate with respect to what can and cannot be concluded from the data. This misrepresentation of findings has led to a crisis in public education and threatens to do the same in public health. Here, I suggest a number of ways that neuroscientists might address this growing problem. First, we should acknowledge that the term 'sex difference' is usually interpreted by the media and the public as evidence for dichotomous categories that do not actually exist. Because data rarely sort so cleanly into sex-specific categories, clearer presentation of the nature and size of sex differences is warranted. The term 'sex effect' may be preferable to 'sex difference' when the effect is not large. Second, factors that covary with sex, particularly experience, should be considered as causes of sex differences before the idea of "hardwiring" is invoked. Finally, we should be more vigilant about how our own findings are conveyed to policymakers and the public and speak out when they are misrepresented.

Keywords Brain-based learning · Neuromyth · Neurosexism · Pseudoscience · Reverse inference · Sex differences · Single-sex education

Contents

D.L. Maney (✉)
Department of Psychology, Emory University, Atlanta, GA 30322, USA
e-mail: dmaney@emory.edu

Curr Topics Behav Neurosci (2015) 19: 279–296
DOI: 10.1007/7854_2014_339

Published Online: 17 August 2014

1 Brain-Based Learning and the Power of the Crockus

In December 2006, the elementary school where my mother was teaching held a staff development workshop called Boys and How They Learn. In the material distributed for the workshop, the principal noted, 'We decided to concentrate on boys and how they learn because our boys are not performing as well as our girls.' Several pages of pseudoscientific claims were supplied from the popular press, for example, from Gurian and Stevens (2005): Boys need more room on a table than girls do, boys see better in bright light, and boys tend to hear less well than girls. The majority of the material advocated distinct educational methods for boys and girls, implemented in separate classrooms, citing profound dichotomies in learning styles, neuroanatomy, and physiology. The school had already begun offering an all-girls kindergarten class, and one purpose of the workshop was to explore whether single-sex education should be expanded to other grades.

My mother's school was not alone in this endeavor. Earlier that year, the US Department of Education announced a reinterpretation of Title IX, which had previously prohibited segregation of boys and girls in public schools. Federally funded, single-sex classrooms would now be allowed, if necessary to achieve identified, specific educational goals (US Department of Education 2006). Almost immediately, several hundred public schools in the US implemented single-sex classrooms under the pretense that boys and girls are "hardwired" to process information differently and thus require separate learning environments (American Civil Liberties Union 2012). This surge in 'brain-based' single-sex education programs was fueled in large part by misinformation and pseudoscience from the popular press and media, repackaged under the guise of professional development for teachers. Materials for teacher workshops described profound sex differences for which there is no scientific basis—for example, that boys' brains develop from back to front, whereas girls' develop from front to back; that boys use the 'primitive' parts of their brains, whereas girls used the 'advanced' parts; and that boys cannot remember anything that is told to them (Table 1). In a particularly infamous example, an educational consultant addressing a large urban school district asserted that the "detailed area of the brain", a nonexistent structure he called the 'crockus', is four times larger in girls (Hodgins n.d., as cited in Liberman 2007)—allowing girls to see more details in each experience. Proponents of brain-based single-sex programs typically claimed that effective coeducational classrooms are impossible, not only because of profound sex differences in brain development and information processing, but also because optimal temperature, sound, and light levels differ dramatically between boys and girls (American Civil Liberties Union 2012). References to scholarly research supporting these claims were rarely offered.

Table 1 Examples of claims commonly presented to elementary school staff in professional development workshops in the USA

Claim	Source
Boys' brains develop from the back of the head to the front, from the 'doing' part of the brain to the 'thinking' part, whereas girls' brains develop from the front of the brain to the rear. This means that boys are able to act before they are able to think	Hodgins (2011)
The resting female brain is more active than the male brain, which often goes into a pause state after tasks. To break the pause, boys must use loud voices, run, or jump	Gurian and Stevens (2004), Hodgins (2011), McBride (2008)
Boys don't remember what you have told them. Each time an incident happens, it's as if it has never happened before	Hodgins (2007)
Girls' brains experience 15 % more blood flow than boys'	Gurian and Stevens (2004), McBride (2008)
Girls tend to use the more advanced parts of their brains, whereas boys use the more primitive parts	McBride (2008)
Girls develop language 6 years earlier than do boys	McBride (2008)
The corpus callosum is 20 % larger in girls than boys. This means that boys have trouble talking about their emotions, since emotion and language are located on opposite sides of the brain	Gurian and Stevens (2004), Hodgins (2007, 2011), McBride (2008)
Boys have half as much neural tissue devoted to verbal-emotive functioning	McBride (2008)
Boys have less oxytocin than girls, which makes them uncomfortable with eye contact. They should be seated side-by-side, to avoid such	Gurian and Stevens (2004), McBride (2008)
Boys have less serotonin than girls, which makes them more fidgety and impulsive	Gurian and Stevens (2004), Hodgins (2007)
Girls can hear better than boys	Chadwell (2010), McBride (2008)
Girls can see better in dim light	Chadwell (2010)
Boys' visual systems are wired to detect moving objects	Gurian and Stevens (2004), Chadwell (2010)
Girls' visual systems are wired to respond best to the colors red or pink	Chadwell (2010)
Boys are most comfortable at a temperature of 69 °F whereas girls work best at 75 °F	Sax (2006)
Girls are able to see the details of a situation because the detailed area of the brain, called the crockus, is four times larger in girls than boys	Hodgins (2007)

The effect of such presentations on educational policy in the US was stunning. In 2012, the American Civil Liberties Union reported that of the single-sex education programs they investigated, nearly all cited pseudoscientific material from the popular press, not peer-reviewed literature (see Table 1), as justification for separating the sexes (American Civil Liberties Union 2012). In order to accommodate what they believed were scientifically proven sex differences, schools used different colors to decorate the classrooms, set thermostats at different temperatures, and arranged seating with girls face to face to promote social interactions and boys side by side to avoid eye contact. In an all-boys classroom in Idaho, teachers used microphones to adjust their voices to a level they were told is best for boys (Hollingsworth and Bonner 2012). Teachers at a school in West Virginia were told that girls need low light levels; the lighting was so low in a girls' classroom that a visually impaired student could not see well enough to function (Khadaroo 2012). In some cases, parents sued to end mandatory single-sex instruction (e.g., Doe 2012) but were not always successful (e.g., A.N.A. 2011).

The use of pseudoscience to justify these new practices triggered a strong response from scientists and gender studies scholars. Several critical books and articles were published between 2009 and 2011 (Eliot 2009, 2011; Fine 2008, 2010; Halpern et al. 2011; Jordan-Young 2010; Rivers and Barnett 2011). Work remains to be done, however, as proponents of single-sex classrooms continue to perpetuate myths and stereotypes, and school administrators continue to listen. Those myths and assertions (see Table 1) have been thoroughly debunked elsewhere; my goal in this article is instead to consider the ways in which we have failed to adequately communicate the nature of sex differences to the public and to suggest ways in which we might help teachers and parents better evaluate them. First, we need to recognize that after our findings are published in scholarly journals, they are filtered and sensationalized by a series of non-expert translators, such as the popular media and teacher educators (Hardiman et al. 2012). Sex differences are packaged and sold to schools as evidence that boys and girls fall into dichotomous categories with non-overlapping distributions. Certainly, small differences do inform our understanding of the factors that contribute to learning and their value should not be discounted—yet, as scientists, we are obligated to respond to misrepresentation of our findings to promote a social agenda, and to establish a more effective dialog with policymakers. In addition, we need to address our own propensity to draw illogical inferences about the meaning of sex differences, particularly from neuroimaging data. Ultimately, because sex differences are so easily misunderstood and misinformation potentially harmful, we need to hold others and ourselves to a high standard when reporting them.

2 What is a Sex Difference?

Everyone understands intuitively that the sexes are different, because our sex organs are obviously different. With few exceptions, a child is categorized as one sex or the other from the moment he or she is born. Because of the widely recognized differences in genitalia, it is easy to believe that other differences between the sexes could be equally large. MRI technology affords unprecedented views inside areas that historically have been obscured from view. If educational consultants argue that newly discovered sex differences in the brain and behavior are large and meaningful enough to warrant different classrooms for girls and boys, teachers often listen.

Actual sex differences in behavior and the brain, particularly in children, should certainly not compel educators to implement dramatic new policies. For example, consider sex differences in impulsivity or activity, the effect size (Cohen's d) of which is typically about 0.2 (Hyde 2014). A hypothetical effect of this magnitude is plotted in Fig. 1a. Such sex differences, which might instead be called 'sex effects' to avoid the term 'difference' altogether, can be statistically significant, but only when sample sizes far exceed a typical elementary school class. In other words, such an effect would not be detectable within a class, a grade, or possibly even an entire school—an effect of the size depicted in Fig. 1a would require approximately 400 children to detect. Because distributions are almost never plotted in research reports, the degree of overlap between the sexes is usually lost when the finding is communicated to teachers. As a result, effects like the one in Fig. 1a are presented and interpreted as evidence that boys and girls cannot learn optimally in the same classroom. Yet, for every 50 boys above the mean, there are 46 girls also above it. Similarly, for every 50 girls below the mean, there are 46 boys below it. Thus, if it is true that children above and below the mean need different classroom environments, separating them according to their sex would do very little to address that need. Almost as many children would end up in the 'wrong' classroom as in the 'right' one. Such a strategy, which would benefit only the children at the extremes of the distribution, essentially constitutes teaching to the tails; it does not consider the needs of the majority of the children, for whom sex does not predict ability or behavior.

A more concrete example of overlapping distributions appears in Fig. 1b, which depicts a known sex effect on the sizes of two brain regions. According to Neufang et al. (2009), the hippocampus is larger in girls and the amygdala larger in boys. Educational consultants have used such findings to argue for large sex differences in information processing and emotive functioning (Gurian and Stevens 2004, 2005; Sax 2005). A close inspection of the actual data reveals large overlap; if we were to use the median hippocampus or amygdala size to divide the students into groups, the number of boys and girls would be approximately equal in each. The authors of the study found that the surge of testosterone in pubescent boys may explain the larger amygdala; importantly, our class with 'girl-like' amyg-dalae would even contain two older boys that had begun puberty. Thus, although these brain structures are different in that an effect can be detected, sex is a rather poor predictor of their size. Certainly, if a small hippocampus and large amygdala warrant a certain educational approach, dividing students by sex would not be a good strategy by which to implement that approach.

Neuroanatomical and psychological data almost never fall into distinct clusters corresponding to sex (e.g., see Carothers and Reis 2013). For this reason, looking to sex as a source of 'difference' in the brain has been criticized (Jordan-Young and Rumiati 2012). But because most people regard sex as a category rather than as a continuous variable, even the smallest effects are an easy sell. Single-sex education

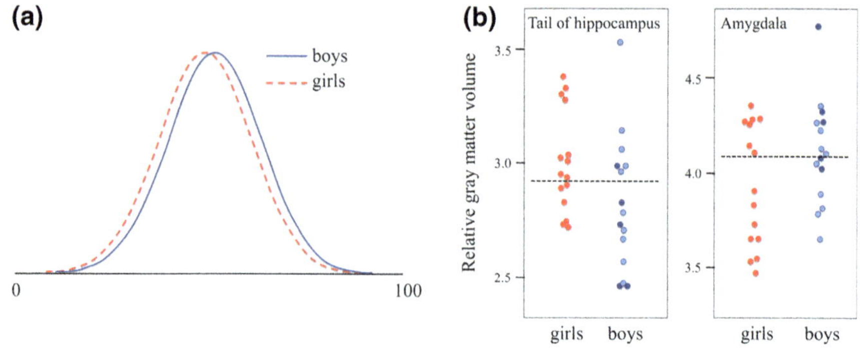

Fig. 1 Sex 'differences' in behavior and the brain typically show large overlap. (**a**) Normal distributions showing a typical sex difference in a behavior or personality trait, such as impulsivity or activity (reviewed by Hyde 2014). The number of boys or girls with scores on a hypothetical scale of 0–100 is plotted. For every 50 boys above the mean, there are also 46 girls above the mean; the sexes overlap by more than 80 %. This difference (effect size $d = 0.2$) is actually much larger than those typically reported for traits such as verbal or mathematic ability. Even for traits with larger sex differences, for example, interest in things vs. people (effect size = 0.93), the overlap is close to 50 %. (**b**) The size of the hippocampus and amygdala varies according to sex in children ages 8–15 (data from Neufang et al. 2009). If this sample of 30 children were split according to the median size of either structure (*dashed lines*), a large proportion of the children would be in the 'wrong' group for their sex. The 'small hippocampus' group would consist of 9 boys and 6 girls; the 'small amygdala' group would consist of 7 boys and 8 girls. Notably, each of the 'girl-like' groups would contain one or two boys with testosterone levels typical of mid-puberty

programs offer a convenient solution to a vexing, urgent problem. Just as Men are from Mars, Women are from Venus (Gray 1992) promised to improve relationships by showing us how to embrace difference, single-sex classrooms promise to save our failing school system. We will need better ways to convey to the public that boys and girls are in fact the same—not from Mars and Venus or even, as some have phrased it, from North and South Dakota (see Eliot 2009). Looking at Fig. 1, I would argue that although a couple of boys may hail from Hoboken and one or two girls from Hackensack, the rest of the children are all from New York City.

3 When a Difference is not a Difference at All

Many sex differences reported in the media and presented to school administrators have uninteresting explanations or no support at all. Here, I will discuss just one example: the amounts of gray and white matter in the brain. Most of the relevant studies suggest that the average amount of gray matter in women is slightly higher than in men (reviewed by Cosgrove et al. 2007). Gray and white matter volumes are closely tied to overall brain volume, which is about 10 % larger on average in men than women. When gray matter volumes are corrected for overall brain size, the sex

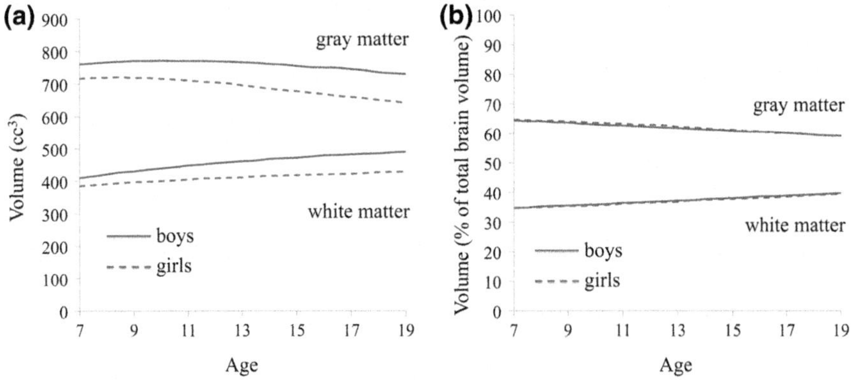

Fig. 2 Some sex differences in the brain may be explained by overall brain size, which is larger in boys. (**a**) Data from Lenroot et al. (2007) suggest that levels of both *gray* and *white* matter are higher in boys. (**b**) When the average values are divided by the average values given for overall brain size, the sex difference appears to be eliminated

effect is substantially lessened (Leonard et al. 2008) or eliminated (Blatter et al. 1995; Courchesne et al. 2000). As an example, data from Lenroot et al. (2007) are shown in Fig. 2a and then redrawn corrected for brain volume in Fig. 2b. The relative amounts of both gray and white matter in children ages 7–19 appear to be exactly the same in boys and girls.

Despite the large literature showing that the sexes have similar amounts of gray and white matter, professional development materials for teachers often paint a starkly different picture. They routinely assert that boys have 6.5 times as much gray matter as girls, and girls have a whopping ten times as much white matter as boys (Box 1). A difference that large would be obvious using techniques available before the dawn of recorded history (e.g., looking at the brain of a deceased person), and would be just as salient as sex differences in reproductive organs. As of this writing, a Google search on ‘women have 10 times more white matter’ produces more than 6,000 hits. The CBS Early Show covered this difference as if it were breaking news (CBS News 2010) but provided no source. With some effort, I traced the myth to media coverage of a paper by Haier et al. (2005). As I expected, these authors did not report sex differences in the amounts of gray or white matter. Rather, they used structural MRI to identify areas of gray and white matter in each participant’s brain and then searched for correlations between the sizes of those areas and scores on an intelligence test. The measures that were 6.5 times larger in men and 10 times larger in women were not the total gray or white matter volumes, but rather the volumes that predicted the score on the test—without regard to whether those correlations were statistically significant. The scores were significantly related to the size of only a few areas of gray matter in men, and one in women. In a press release, the authors commented that human evolution has created two different types of brains designed for equally intelligent behavior (Today@UCI 2005). The media then proceeded to run amok, giving rise to perhaps the most nonsensical neuroscience myth since the one about humans using only 10 % of their brains. If Haier et al. have attempted to address the confusion, their attempts have been swamped by the sheer volume of misrepresentations.

Box 1 An urban legend is born. A 2005 paper by Haier et al. was so grossly misinterpreted by the media that it gave rise to a now-pervasive urban legend: Men have 6.5 times as much gray matter as women, and women have 10 times as much white matter as men. The 2005 paper, which described a structural MRI study relating intelligence to certain voxels of gray and white matter in men and women, contained a version of the figure above. The figure has been reproduced hundreds of times on the internet, sometimes with an accurate caption but more often with a caption such as, "Activity in men and women while taking an IQ test" (no imaging was actually done during the test) or "Men have 6.5 times as much gray matter and women have 10X as much white matter". Figure from Andrew-Sfeir (2012).

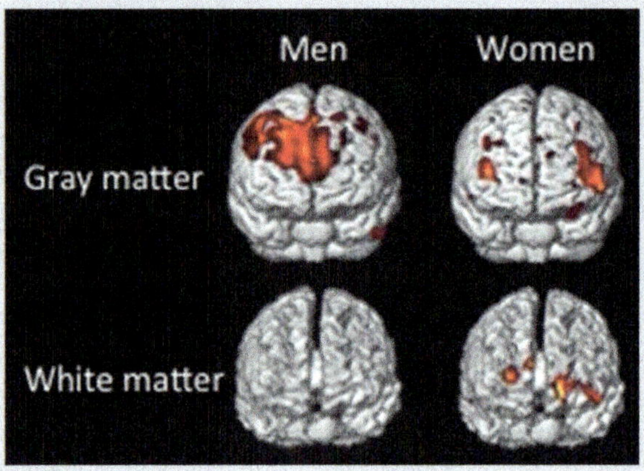

4 Leaps of Logic and the Allure of the Brain Scan

Proponents of brain-based single-sex education have argued that modern imaging techniques have revealed large differences between the brains of boys and girls (Chadwell 2010; Gurian and Stevens 2005). They present images of the BOLD response or white/gray matter distribution, chosen to illustrate differences that may or may not have ever been reported in peer-reviewed literature. Such images are quite powerful; one teacher wrote, 'I was trained in the idea that each student is an individual. But when I saw the PET scans of boys' and girls' brains, I saw how differently those brains are *set up to learn*.' (Gurian and Stevens 2004; italics added). Note that in addition to believing that the images represented a typical boy and girl, the teacher was convinced that they showed something about learning styles. Although that leap of logic is a large one, it is common. Such 'reverse inferences' (Poldrack 2006) rest on the fallacy that if neuroanatomical differences exist, they must explain behavioral differences. The larger hippocampus of girls, according to materials distributed to teachers, endows them with better memory, social skills, and language skills (Gurian and Stevens 2004; McBride 2008). Similarly, the larger amygdala of boys supposedly makes them more aggressive, reduces their attention span, and increases the amount of space they require in the classroom (Gurian and Stevens 2005; Multiplying Connections 2012). The scientific basis for these claims is unclear, but to the non-expert, they apparently seem plausible.

Perhaps the most pervasive of illogical reverse inferences is the attribution of cognitive abilities to the relative volumes of white and gray matter, which is presumed to be related to the degree of interconnectedness among brain regions. The evidence for sex effects on both white matter volume and connectivity has been reviewed elsewhere (e.g., Bishop and Wahlsten 1997; Bruner et al. 2012) and is tangential to my point here: How and whether these factors affect abilities is unknown. Sex differences in connectivity and white matter volume have nonetheless been cited as evidence of either male or female superiority in spatial orienting, language skills, empathizing, map reading, mathematics, and multitasking (reviewed by Fine 2010). To a non-expert, the absence of a known function may not be particularly relevant because a sex difference *implies* function. For example, if women are found to have more white matter and men more gray matter, then white matter is said to be responsible for multitasking and gray matter confers mathematical ability (CBS News 2010). Conversely, if men are found to have more white matter and women more gray matter, then white matter is reported to confer mathematical ability, while gray matter is important for multitasking (Chamberlain 2009). Ignored are the findings that mathematical ability does not vary with sex (reviewed by Hyde 2014), and the only two studies on multitasking showed no female advantage (Hambrick et al. 2010; Mäntylä 2013). A sex difference in the brain appears to be enough to convince many people that a difference in ability must exist, despite the maddening circularity of the logic.

Why are such tenuous arguments so convincing? First, they support stereotypes. The combination of reverse inference and social stereotypes is a dangerous one, as was famously illustrated by Gould in The Mismeasure of Man (1981). Gould debunked nineteenth-century research alleging that intellectual superiority of white men could be explained by their larger cranial capacity, compared with women and men of other races. Gould argued that the conclusions of the researchers were shaped by their own expectations. The stereotypes of that century ensured that the research would be accepted, even embraced. Likewise today, brain-based explanations for effects of sex on achievement capitalize on long-held stereotypes, often triggering aha moments for parents and teachers as they become convinced that their own personal beliefs are validated by science (Kaufmann, n.d.). The more dearly held those beliefs, the harder it is to convince the believer that such arguments are flawed. As one New York Times reader commented, '[Feminists] just love to pretend there are no hard differences between the brains of men and women,' which the reader called a 'brazen denial of what is not only real but thunderingly obvious' (comment on Schott 2010).

Sadly, the use of reverse inference to perpetuate stereotypes is not limited to amateurs. Neuroscientists themselves are guilty (see Bluhm 2012). In a recent imaging study, Ingalhalikar et al. (2014) described effects of sex on the 'connectome' of the brain, arguing that women showed stronger interhemispheric connections, whereas men's brains were better connected within hemisphere. In the discussion section of the paper, the authors inferred that the male-typical pattern would confer an efficient system for coordinated action, whereas the female-typical interhemispheric connections would better integrate the 'analytical' left hemisphere with the 'spatial and intuitive' right hemisphere. The degree of overlap between the sexes was not reported; in the institution's press release, however, the authors described the sex difference as "stark" and "striking" and suggested it might explain why men are better at cycling and women better at socializing and multitasking (Penn Medicine 2013). In an interview, an author remarked, 'I was surprised that it matched a lot of the stereotypes' (Sample 2013). Thus, the authors themselves overinterpreted their own work, providing media-ready sound bytes both in interviews and in the paper itself. If the authors make such claims, can we blame the media, educational consultants, or teachers and parents for doing the same?

Neuroscientists and psychologists worldwide reacted to these statements with disappointment. Dorothy Bishop of the University of Oxford commented, 'The behavior of the scientists doing the study is hard to understand. When they talk of hardwired differences in brains of males and females, and link their results to putative behavioral differences that fit stereotypes but which they haven't measured, they lose credibility among their scientific peers… Do they really believe what they are saying - in which case they are bad scientists? Or are they so swept up in a brief moment of media fame that they don't care about their reputations?' (Bishop 2013). During the days following publication, bloggers posted analyses of the degree of overlap, which they estimated to be large (see Lindeløv 2013; Ridgeway 2013), and raised a multitude of alternative explanations for the results, including experience, head motion in the scanner, and age (e.g., Fine 2013; Scott 2013). The longevity of these online criticisms, compared with the impact of the study itself, remains to be determined, but the authors likely got the message that their interpretations were not universally well received. The swift, scholarly response, which Bishop illustrated online using Storify.com, serves as an excellent example of the kind of post-publication peer review now possible using social media.

Just as flawed arguments can be masked by seeming to support stereotypes, they can also be bolstered by compelling pictures of the brain in action. Weisberg et al. (2008) elegantly showed that faulty explanations for psychological phenomena can be made significantly more convincing, at least to non-experts, by mentioning a brain scan. The addition of 'neuroscience', even if irrelevant, successfully masked flawed logic in a scientific report for everyone but neuroscientists. Since few teachers and school administrators are experts in neuroscience, they are especially vulnerable to the allure of the scan. MRI scans have a troubling tendency to be interpreted by laypersons, and even some researchers, as windows into a brain untouched by experience—a brain in its raw, hardwired state. It should not be so hard to believe that behavior could change the brain—after all, the public generally accepts that exercise changes muscles and eating changes fat. Nonetheless, if a neural structure varies in size or activity between the sexes, that difference is almost invariably communicated to the public as the cause, not an effect, of a behavioral difference. This phenomenon extends far beyond sex differences; brain scans are so convincing of hardwiring that they have led some to question whether criminals are responsible for their actions, or even whether humans have free will (Mobbs et al. 2007; Smith 2011). As human imaging grows out of its adolescence, and as more research emphasizes plasticity, the exquisite sensitivity of the brain to its environment will certainly be recognized. In the meantime, we need to remind ourselves to consider the hypothesis that a sex effect might be explained by experience and speak up when that hypothesis is not considered.

5 Moving Forward

Some authors have argued that because most sex effects in the brain are small and hard to interpret, the benefits of reporting them do not outweigh the tremendous costs (e.g., Jordan-Young and Rumiati 2012). Not all research on the topic is done for the purposes of division or understanding the construct of 'sex'. Such effects can provide important clues about other factors, for example, genes or hormones, that shape the development of the brain. In a sense, sex is simply a natural manipulation of these factors that allows researchers to test their effects conveniently. Discovering a sex effect, particularly in a non-human model for which the effects of experience can be better controlled, provides a valuable starting point for further investigations into causal mechanisms. Rarely, however, does any discovery of an effect provide a valid reason for treating the sexes differently, and should not be used indiscriminately as justification for such (Roy 2012).

Perhaps in response to recent critiques (e.g., ACLU 2012; Halpern et al. 2011; Eliot 2009; Fine 2010), some proponents of single-sex education programs appear to be shifting their focus away from the brain. On the website of the National Association for Single-Sex Public Education, the Gender Differences in the Brain page now redirects to an unrelated page that duplicates material on a different topic. David Chadwell, who for years served as the coordinator of single-sex education programs for the South Carolina Department of Education, now takes a more agnostic view of brain-based approaches. In a 2011 interview, he noted that evidence for brain-based sex differences 'will always be argued by researchers. There are books and research reports that say that there are differences and then there are books and research that say there aren't differences' (Whitmire 2011). His views echo those of teachers at my mother's school, who found the arguments on both sides equally compelling. Ultimately, they decided against expanding their single-sex program because they could not reach a consensus. These signs are encouraging, but of course not satisfying.

Although the number of single-sex classrooms is dwindling, the downturn can be attributed to a fear of lawsuits and lack of funds, not the rejection of pseudoscience (Meder 2012). Single-sex programs remain popular with teachers and parents, and misrepresentations of neuroscience data remain a commonly cited justification.

Before being tempted to cite a 'sex difference' as justification for division, I suggest that educators ask a series of questions about that difference (see also Roy 2012). First, is there actually a difference? As seen in Box 1, careful fact checking is critical. Non-scholarly sources, such as the media and popular press, are not reliable. Second, if there is in fact an effect, what is the degree of overlap between the sexes? Given that the majority of sex 'differences' are no larger than depicted in Fig. 1a, even a statistically significant sex effect is unlikely to justify splitting an entire group by sex. Instead, the only clear separation of boys and girls is likely to occur in the tails of the distribution—students in these tails can be targeted for intervention, but note that such targeting would be best accomplished using something other than sex, for example, test scores, to identify students who might benefit. Finally, in the event that a difference actually does meaningfully distinguish the sexes, does that difference suggest sex-specific needs? Is it possible, for example, to develop teaching materials that actually maximize learning accomplished by a particular part of the brain? Such a program would be ambitious indeed, and likely not in line with sound scientific research (Hardiman et al. 2012).

As neuroscientists, we can improve communication with the public by adopting a few changes in the way we report effects of sex. First, I suggest that unless the effect is quite large, the term 'sex effect' is preferable to 'sex difference.' Second, when possible, data are best presented in a way that allows readers to see the degree of overlap. For example, individual value plots, scatterplots, or distributions can be provided and effect sizes reported. Third, authors should consider factors that covary with sex, for example, experience, hormonal milieu, brain size, etc., as explanations for the effect rather than using terms such as 'hardwired'. Finally, when we see that our results have been misinterpreted or misrepresented, we should publicly take issue. Now, more than ever before, venues are readily available for public commenting and discussion (see Bishop 2013). Of course, even before beginning a study, we should ask ourselves whether the goal is to understand how sex contributes variation to a biologically complex system or simply to divide the sexes. As eloquently pointed out by Roy (2012), research 'primarily driven by an urge to neatly place people into pre-ordained categories… should be accompanied by a warning label' (p. 223).

6 The Future of Neurosexism: Is the Quality of Health Care at Risk?

What has happened in public education should serve as a cautionary tale for medical practitioners, who are under increasing pressure to take the patient's sex into account when considering treatment options. A large number of psychiatric and neurological conditions differ in prevalence or severity between men and women, including depression, anxiety, and pain (Greenspan et al. 2007; Mogil 2012), and some authors have called for more research to understand these effects (Cahill 2006; Beery and Zucker 2011; McCarthy et al. 2012). Some go so far as to suggest we should work toward sex-specific medicines (Gillies and McArthur 2010; Melcangi and Garcia-Segura 2010). In many cases, this strategy will undoubtedly prove to be beneficial, and performing clinical trials with both sexes is critical. As is the case for sex-specific education, however, the size of a sex difference must be considered before implementing any strategy to account for it. If the effect requires a large sample size to detect, or is sometimes not detected, we should proceed toward sex-specific treatment with extreme caution. Sex differences in physiology could be explained by any number of factors that covary with sex, such as body size, fat content, or plasma levels of hormones. In other words, a patient's sex is unlikely to be the best predictor of his or her response to treatment. Our goal should ultimately be to move beyond sex to identify those predictors, particularly if they can be assessed by physicians, in order to make more informed decisions about treatments and maximize their efficacy.

For some disorders, we may not know the true size of a sex difference because the data are colored by gender stereotypes. In a study by Aragonès et al. (2006), physicians made erroneous diagnoses of depression much more often for female than male patients. In other words, overdiagnosis of depression was higher for women than men. Women are more likely to suffer from idiopathic pain and other syndromes with poorly understood etiology (e.g., chronic fatigue syndrome, irritable bowel syndrome), which contributes further to the risk of misdiagnoses of depression and generalized anxiety disorder (Bowman 2001; Meana 1998). A sex effect on the rate of overdiagnosis may contribute to or even explain a sex difference in the prevalence of any illness. Thus, less sexist and more accurate methods of diagnosis may be called for before, or in concert with, the development of sex-specific treatments.

A recent example from cardiovascular medicine illustrates how the reporting of sex effects could potentially cause harm. The past decade has seen many public health campaigns to raise awareness that symptoms of heart attack can be atypical in women. News stories such as Women's Heart Attack Symptoms Are Different from Men's (Longley 2013) and Recognizing the Female Heart Attack (Kam 2009) typically assert that symptoms in men and women are 'drastically different' (Forer 2011). Such articles commonly state that chest pain is a 'man's symptom', whereas 'female heart attack symptoms' are fatigue, anxiety, and indigestion (Tytus 2010). The actual research shows that although symptoms of heart attack do differ somewhat between men and women, those differences are too small to be clinically informative (Gimenez et al. 2014) or to warrant sex-specific public health messages (Canto et al. 2007). Nonetheless, like claims that boys and girls learn differently, the concept of sexually dimorphic heart attacks has been embraced by the public and vigorously defended. For example, when a 2009 study showed no sex difference in rates of chest pain during myocardial ischemia (Mackay et al. 2009; see also Mackay et al. 2011), members of a women's heart health support group posted that the lead author of the study was 'full of garbage' and needed to be 'whacked upside the head' (comment on Tobin 2009). Many websites continue to perpetuate the notion that women's heart attacks do not resemble men's, and cardiologists continue to call for more research on 'fundamental biological differences' between men and women (Maas et al. 2011). In this case, the likely interpretation of those differences by the public could have life-threatening consequences, for example, if women ignore chest pain or men brush off fatigue. A more effective public health strategy would emphasize that anyone, male or female, can present with atypical symptoms (Humphries et al. 2012).

7 Conclusion

If sex-specific educational policies and medical treatments are to be implemented in a fair and logical way, sex must explain an unprecedented proportion of the variability in the relevant trait such as a learning style or response to treatment. Otherwise, sex-specific solutions cannot efficiently address the problem at hand and will only serve to obscure the true source of variability (Jordan-Young and Rumiati 2012). In reality, sex differences in the human brain that meet this criterion—in other words that are absolute (McCarthy et al. 2012) or qualitative (Mogil 2012)—are vanishingly few. Even in other species, absolute sex differences in the brain are relatively rare and relate directly to absolute differences in copulatory or courtship behaviors (McCarthy et al. 2012). The apparent absence of such differences in humans does not mean we will never find them. Mogil and colleagues have shown that the molecular mechanisms underlying inflammatory and neuropathic pain appear to be qualitatively different in male and female mice (Sorge et al. 2011); Woolley and colleagues have shown an interesting dimorphism in the way estradiol modulates synaptic transmission in the hippocampus of rats (Hoang and Woolley 2012). The discovery of both effects was delayed because initial studies were conducted in just one sex, highlighting the need for research that includes both males and females (Cahill 2006; McCarthy et al. 2012; Mogil 2012). Although absolute sex differences do exist, they are not, as many laypersons believe, 'thunderingly obvious' (see Schott 2010), particularly before puberty. Quite the opposite—what is remarkable is not the difference, but the *sameness* (Carothers and Reis 2013; Hyde 2005, 2014)—a concept that has been explored in depth in psychology and gender studies but not yet embraced by neuroscience. It is the sameness we should be communicating to the public, because without that understanding, forthcoming discoveries of large and meaningful sex differences will not rise above the noise.

Acknowledgments I am grateful to Shirley K. Maney, former teacher in the Winston-Salem/Forsyth County School system, for originally calling this controversial topic to my attention, and to Stephan Hamann for his advice and assistance. I also thank Chris Goode, Steve Nowicki, Deboleena Roy, and Kim Wallen for comments on the manuscript.

References

American Civil Liberties Union (2012) Preliminary findings of ACLU "Teach Kids, Not Stereotypes" campaign. http://www.aclu.org/files/assets/doe_ocr_report2_0.pdf. Accessed 15 Jun 2013

A.N.A. et al (2011) Breckinridge County Board of Education, et al. US District Court, Western District of Kentucky at Louisville 3:08-CV-4-S. http://www.chadwellconsulting.com/MSJ%20Opinion.pdf. Accessed 1 Jul 2013

Andrew-Sfeir A (2012) Differences in brain structure between men and women. http://www.verbumsapsat.com/arts-entertainment/2012/02/23/differences-in-brain-structure-between-men-and-women/. Accessed 25 Mar 2014

Aragonès E, Piñol JL, Labad A (2006) The overdiagnosis of depression in non-depressed patients in primary care. Fam Pract 23:363–368

Beery AK, Zucker I (2011) Sex bias in neuroscience and biomedical research. Neurosci Biobehav Rev 35:565–572

Bishop DV (2013) What a difference a day makes: how social media is transforming scientific debate. http://storify.com/deevybee/postpublication-peer-review-on-sex-differences-in. Accessed 12 Dec 2013

Bishop MK, Wahlsten D (1997) Sex differences in the human corpus callosum: myth or reality? Neurosci Behavioral Rev 21:581–601

Blatter DD, Bigler ED, Gale SD, Johnson SC, Anderson CV, Burnett BM, Parker N, Kurth S, Horn DS (1995) Quantitative volumetric analysis of brain MR: normative database spanning 5 decades of life. AJNR Am J Neuroradiol 16:241–251

Bluhm R (2012) Self-fulfilling prophecies: the influence of gender stereotypes on functional neuroimaging research on emotion. Hypatia 28:870–886

Bowman GS (2001) Emotions and illness. J Adv Nursing 34:256–263

Bruner E, Cuetara MJ, Colom R, Martin-Loeches M (2012) Gender-based differences in the shape of the human corpus callosum are associated with allometric variations. J Anatomy 220: 417–421

Cahill L (2006) Why sex matters for neuroscience. Nat Rev Neurosci 7:477–484

Canto JG, Goldberg RJ, Hand MM, Bonow RO, Sopko G, Pepine CJ, Long T (2007) Symptom presentation of women with acute coronary syndromes: myth vs. reality. Arch Intern Med 167:2405–2413

Carothers BJ, Reis HT (2013) Men and women are from earth: examining the latent structure of gender. J Personal Social Psychol 104:385–407

CBS News (2010, September) Size matters: how male, female brains compare. http://www.cbsnews.com/news/size-matters-how-male-female-brains-compare/. Accessed 18 Jun 2013

Chadwell D (2010, February 28) Instructional resources. http://www.chadwellconsulting.com/Instructional.htm. Accessed 15 Dec 2013

Chamberlain LB (2009) The amazing teen brain: what every child advocate needs to know. ABA Child Law Pract 28:17–24

Cosgrove KP, Mazure CM, Staley JK (2007) Evolving knowledge of sex differences in brain structure, function, and chemistry. Biol Psychiatry 62:847–855

Courchesne E, Chizum H, Townsend J, Cowles A, Covington J, Egaas B, Harwood M, Hinds S, Press GA (2000) Normal brain development and aging: quantitative analysis and *in vivo* MR imaging in healthy volunteers. Neuroradiology 216:672–682

Doe V (2012) Wood County Board of Education. US District Court, Southern District of West Virginia 6:12-cv-04355. http://www.wvsd.uscourts.gov/sites/default/files/opinions/612-cv-04355.pdf. Accessed 23 Jun 2014

Eliot L (2009) Pink brain, blue brain. Houghton Mifflin Harcourt, Boston

Eliot L (2011) Single-sex education and the brain. Sex Roles 69:363–381

Fine C (2008) Will working mothers' brains explode? The popular new genre of neurosexism. Neuroethics 1:69–72

Fine C (2010) Delusions of gender. Norton, New York
Fine C (2013) New insights into gendered brain wiring, or a perfect case study in neurosexism? The conversation. https://theconversation.com/new-insights-into-gendered-brain-wiring-or-a-perfect-case-study-in-neurosexism-21083. Accessed 20 Jan 2014
Forer B (2011) Top 5 symptoms of heart disease in women. http://abcnews.go.com/Health/HeartDiseaseNews/top-symptoms-heart-disease-women/story?id=14009993. Accessed 28 Dec 2013
Gillies GE, McArthur S (2010) Estrogen actions in the brain and the basis for differential action in men and women: a case for sex-specific medicines. Pharmacol Rev 62:155–198
Gimenez MR, Reiter M, Twerenbold R, Reichlin T, Wildi K, Haaf P, Wicki K, Zellweger C, Hoeller R, Moehring B, Sou SM, Mueller M, Denhaerynck K, Meller B, Stallone F, Henseler S, Bassetti S, Geigy N, Osswald S, Mueller C (2014) Sex-specific chest pain characteristics in the early diagnosis of acute myocardial infarction. JAMA Int Med 174:241–249
Gould SJ (1981) The mismeasure of man. Norton, New York
Gray J (1992) Men are from mars, women are from Venus: a practical guide for improving communication and getting what you want in your relationships. Harper, New York
Greenspan JD, Craft RM, LeResche L, Arendt-Nielsen L, Berkley KJ, Fillingim RB, Gold MS, Holdcroft A, Lautenbacher S, Mayer EA, Mogil JS, Murphy AZ, Traub RJ (2007) Studying sex and gender differences in pain and analgesia: a consensus report. Pain 132(S1):S26–S45
Gurian M, Stevens K (2004) With boys and girls in mind. Educ Leadersh 61:21–27
Gurian M, Stevens K (2005) The minds of boys: saving our sons from falling behind in school and life. Jossey Bass, San Francisco
Haier RJ, Jung RE, Yeo RA, Head K, Alkire MT (2005) The neuroanatomy of general intelligence: sex matters. NeuroImage 25:320–327
Halpern DF, Eliot L, Bigler RS, Fabes RA, Hanish LD, Hyde J, Liben LS, Martin CL (2011) The pseudoscience of single-sex schooling. Science 333:1706–1707
Hambrick DZ, Oswald FL, Darowski ES, Rench TA, Brou R (2010) Predictors of multitasking performance in a synthetic work paradigm. Cognit Psychol 24:1149–1167
Hardiman M, Rinne L, Gregory E, Yarmolinskaya J (2012) Neuroethics, neuroeducation, and classroom teaching: where the brain sciences meet pedagogy. Neuroethics 5:135–143
Huang GZ, Woolley CS (2012) Estradiol acutely suppresses inhibition in the hippocampus through a sex-specific endocannabinoid and mGluR-dependent mechanism. Neuron 74:801–808
Hodgins DJ (2007) Looking through the eyes of boys and girls. As cited in RationalWiki (2013, October 15). Crockus. Retrieved from http://rationalwiki.org/wiki/Crockus. Accessed 15 Dec 2013
Hodgins DJ (2011) What about those boys? [Presentation materials]. http://www.rhpl.org/images/stories/Youth/EarlyLit/presentation_materials/Daniel_Hodgins_What_About_Those_Boys.pdf. Accessed 15 Jun 2013
Hollingsworth H, Bonner JL (2012) Why single-sex education is spreading across the U.S. Christ Sci Monit 8:2012
Humphries KH, Izadnegahdar M, Mackay MH (2012) Sex differences in presentation of myocardial infarction. JAMA 307:2486–2487
Hyde JS (2005) The gender similarities hypothesis. Am Psychol 60:581–592
Hyde JS (2014) Gender similarities and differences. Annu Rev Psychol 65:373–398
Ingalhalikar M, Smith A, Parker D, Satterthwaite TD, Elliott MA, Ruparel K, Hakonarson H, Gur RE, Gur RC, Verma R (2014) Sex differences in the structural connectome of the human brain. Proc Natl Acad Sci 111:823–828
Jordan-Young RM (2010) Brainstorm: the flaws in the science of sex differences. Harvard University Press, Cambridge
Jordan-Young R, Rumiati RI (2012) Hardwired for sexism? Approaches to sex/gender in neuroscience. Neuroethics 5:305–315
Kam K (2009) Her guide to a heart attack: recognising female heart attack symptoms. http://www.webmd.boots.com/heart-disease/features/her-guide-to-a-heart-attack. Accessed 28 Dec 2013

Kaufmann C (n.d.) How boys and girls learn differently. http://www.rd.com/advice/how-boys-and-girls-learn-differently/. Accessed 1 Jul 2013

Khadaroo ST (2012) Judge stops W. Va. single-sex classes: were they a success or pseudoscience? Christ Sci Monit 31:2012

Lenroot RK, Gogtay N, Greenstein DK, Wells EM, Wallace GL, Clasen LS, Blumenthal JD, Lerch J, Zijdenbos AP, Evans AC, Thompson PM, Giedd JN (2007) Sexual dimorphism of brain developmental trajectories during childhood and adolescence. Neuroimage 36:1065–1073

Leonard CM, Towler S, Welcome S, Halderman LK, Otto R, Eckert MA, Chiarello C (2008) Size matters: cerebral volume influences sex differences in neuroanatomy. Cereb Cortex 18:2920–2931

Liberman M (2007) High crockalatorum. http://itre.cis.upenn.edu/~myl/languagelog/archives/004926.html. Accessed 15 Jun 2013

Lindeløv JK (2013) This PNAS paper could be the go-to example of how not to interpret statistics. http://lindeloev.net/?p=64. Accessed 19 Jan 2014

Longley R (2013) Women's heart attack symptoms are different from men's. http://usgovinfo.about.com/cs/healthmedical/a/womensami.htm. Accessed 28 Dec 2013

Mäntylä T (2013) Gender differences in multitasking reflect spatial ability. Psych Sci 24:514–520

Maas AH, van der Schouw YT, Regitz-Zagrosek V, Swahn E, Appelman YE, Pasterkamp G, Ten Cate H, Nilsson PM, Huisman MV, Stam HC, Eizema K, Stramba-Badiale M (2011) Red alert for women's heart: the urgent need for more research and knowledge on cardiovascular disease in women. Eur Heart J 32:1362–1368

Mackay MH, Ratner PA, Buller CE, Johnson JL, Humphries KH (2009) Gender differences in reported symptoms of acute coronary syndromes. Can J Cardiol 25(Suppl B):294

Mackay MH, Ratner PA, Johnson JL, Humphries KH, Buller CE (2011) Gender differences in symptoms of myocardial ischaemia. Eur Heart J 32:3107–3114

McBride W (2008) Girls will be girls and boys will be boys. [Presentation materials]. http://crr.math.arizona.edu/GenderKeynote.pdf. Accessed 1 Jul 2013

McCarthy MM, Arnold AP, Ball GF, Blaustein JD, De Vries GJ (2012) Sex differences in the brain: the not so inconvenient truth. J Neurosci 32:2241–2247

Meana M (1998) Depression: comorbidity in women. Can J Psychiatry 43:893–899

Meder E (2012) Schools battling courts, cost for single-gender education. SCNow Morning News Online. http://www.scnow.com/news/local/article_bb4dbc75-5465-5a28-94be-91fb75a0ca17.html. Accessed 13 Oct 2013

Melcangi RC, Garcia-Segura LM (2010) Sex-specific therapeutic strategies based on neuroactive steroids: In search for innovative tools for neuroprotection. Horm Behav 57:2–11

Mobbs D, Lau HC, Jones OD, Frith CD (2007) Law, responsibility, and the brain. PLoS Biol 5(4): e103

Mogil JS (2012) Sex differences in pain and pain inhibition: Multiple explanations of a controversial phenomenon. Nat Rev Neurosci 13:859–866

Multiplying Connections (2012). Male and female brains are not the same. http://www.multiplyingconnections.org/become-trauma-informed/male-and-female-brains-are-not-same. Accessed 1 Jul 2013

Neufang S, Specht K, Hausmann M, Güntürkün O, Herpertz-Dahlmann B, Fink GR, Konrad K (2009) Sex differences and the impact of steroid hormones on the developing human brain. Cereb Cortex 19:464–473

Penn Medicine (2013) Brain connectivity study reveals striking differences between men and women. http://www.uphs.upenn.edu/news/News_Releases/2013/12/verma/. Accessed 21 Dec 2013

Poldrack RA (2006) Can cognitive processes be inferred from neuroimaging data? Trends Cogn Sci 10:59–63

Ridgeway G (2013) Illustrative effect sizes for sex differences. Figshare. http://figshare.com/articles/Illustrative_effect_sizes_for_sex_differences/866802. Accessed 21 Dec 2013

Rivers C, Barnett RC (2011) The truth about boys and girls: challenging toxic stereotypes about our children. Columbia University Press, New York

Roy D (2012) Neuroethics, gender and the response to difference. Neuroethics 5:217–230

Sample I (2013) Male and female brains wired differently, scans reveal. The Guardian, December 2 2013. http://www.theguardian.com/science/2013/dec/02/men-women-brains-wired-differently. Accessed 21 Dec 2013

Sax L (2005) Why gender matters: what parents and teachers need to know about the emerging science of sex differences. Doubleday, New York

Sax L (2006) Six degrees of separation: what teachers need to know about the emerging science of sex differences. Edu Horizons 84:190–200

Schott B (2010) Neurosexism. http://schott.blogs.nytimes.com/2010/12/13/neurosexism/. Accessed 21 Dec 2013

Scott SK (2013) Asking questions about men and women by looking at teenagers. https://sites.google.com/site/speechskscott/SpeakingOut/askingquestionsaboutmenandwomenbylooking-atteenagers. Accessed 20 Jan 2014

Smith K (2011) Neuroscience vs. philosophy: taking aim at free will. Nature 477:23–25

Sorge RE, LaCroix-Fralish ML, Tuttle AH, Sotocinal SG, Austin JS, Ritchie J, Chanda ML, Graham AC, Topham L, Beggs S, Salter MW, Mogil JS (2011) Spinal cord Toll-like receptor 4 mediates inflammatory and neuropathic hypersensitivity in male but not female mice. J Neurosci 31:15450–15454

Today@UCI (2005, January) Intelligence in men and women is a gray and white matter. http://archive.today.uci.edu/news/release_detail.asp?key=1261. Accessed 15 Jun 2013

Tobin AM (2009) Women have same heart attack symptoms as men, including chest pain. http://www.inspire.com/groups/womenheart/discussion/study-women-have-same-heart-attack-symptoms-as-men-including-chest-pain/. Accessed 28 Dec 2013

Tytus RH (2010) Heart attack symptoms in women. http://www.drtytus.com/heart-attack-symptoms-in-women.html. Accessed 28 Dec 2013

U.S. Department of Education (2006) 34 Code of Federal Regulations, Part 106, 71. www2.ed.gov/legislation/FedRegister/finrule/2006-4/102506a.html. Accessed 15 Jun 2013

Weisberg DS, Keil FC, Goodstein J, Rawson E, Gray JR (2008) The seductive allure of neuroscience explanations. J Cogn Neurosci 20:470–477

Whitmire R (2011) Chadwell on the science article debate. http://blogs.edweek.org/edweek/whyboysfail/2011/09/chadwell_on_the_science_article_debate.html. Accessed 21 Dec 2013

Money and Morals

Ending Clinical Trials for Financial Reasons

Margaret L. Eaton, Brian K. Kwon and Christopher Thomas Scott

Abstract Too often, biopharmaceutical companies stop their clinical trials solely for financial reasons. In this chapter, we discuss this phenomenon against the backdrop of a 2011 decision by Geron Corporation to abandon its stem cell clinical trial for spinal cord injury (SCI), the preliminary results of which were released in May 2014. We argue that the resultant harms are widespread and are different in nature from the consequences of stopping trials for scientific or medical reasons. We examine the ethical and social effects that arise from such decisions and discuss them in light of ethical frameworks, including duties of individual stakeholders and corporate sponsors. We offer ways that sponsors and clinical sites can ensure that trials are responsibly started, and once started adequately protect the interests of participants. We conclude with recommendations that industry sponsors of clinical trials should adopt in order to advance a collective and patient-centered research ethic.

Keywords Neuroscience · Neurons · Stem cells · Human embryonic stem cells · Ethics · Informed consent · Clinical trials · Spinal cord injury · Biotechnology · Geron Corporation

Contents

M. L. Eaton · C. T. Scott (✉)
Center for Biomedical Ethics, Stanford University, Stanford, CA, USA
e-mail: cscott@stanford.edu

B. K. Kwon
Department of Orthopedics, Faculty of Medicine, University of British Columbia, Vancouver, Canada

C. T. Scott
University of British Columbia, Vancouver, Canada

Curr Topics Behav Neurosci (2015) 19: 297–315
DOI: 10.1007/7854_2014_337

Published Online: 26 July 2014

1 Introduction

We believe, along with other commentators, that any company that stops a clinical trial for financial reasons creates a unique set of negative and potentially harmful consequences that raise ethical concerns (Malmqvist et al. 2011; Iltis 2004; Lièvre et al. 2001; Evans and Pocock 2001). What is missing from the literature is a discussion of this two-part question: What are companies expected to do to minimize the risk of stopping trials for financial reasons, and what are their obligations after they abandon such trials? Here, we discuss halting trials for financial reasons against the backdrop of the 2011 decision by Geron Corporation to abandon its stem cell clinical trial for spinal cord injury (SCI). We examine the social impact and harms that arise from such decisions and discuss them in light of the duties that corporate sponsors assume when initiating such trials. We further discuss ways that sponsors and clinical sites should ensure that trials are responsibly started, and once started adequately protect the interests of human participants. We conclude with recommendations that sponsors of clinical trials and their collaborators should adopt in order to advance a collective and patient-centered ethic.

2 Definition

This topic requires an explanation of what we mean by *stopping a corporate-sponsored clinical trial for financial reasons*. We address situations where trials are abandoned solely or primarily for economic and business reasons unrelated to safety, efficacy, or feasibility. We exclude from discussion trials that are stopped for other reasons, such as the emergence of unacceptable rates of toxicity, discoveries that alter the understanding of the therapeutic intervention, or problems with research execution. Though trials stopped in this fashion may have financial consequences, we do not include them in our definition. We recognize that there may be mixed reasons to abandon a trial but address instances where money is the primary, driving motivation for a sponsor to stop a trial.

3 Examples of Corporations that Stopped Clinical Trials for Financial Reasons

There is little information about the prevalence of this phenomenon, most probably because corporations are not required to reveal the details of their research failures. While it was not possible to discover prevalence data, we did locate cases that illustrate the range of circumstances that exist when clinical trials are stopped by corporations for financial reasons.

In 1996, Hoechst AG shut down the European trials of Pimagedine (under study to slow the progression of renal disease in diabetics) 2 years after recruitment had begun. The clinical investigators strongly objected to the decision on ethical grounds since the decision was based primarily on financial considerations (Keen et al. 1997). In 1997, Hoechst Marion Roussel drew fire when the company stopped a trial after treating 500 subjects with Cardizem, a drug being tested to prevent myocardial re-infarction. The reason given for the decision was that Cardizem faced competition from a generic product. That same year, the Liposome Company halted a study of doxorubicin in metastatic breast cancer, citing strategic reasons (Langer 1997; Hopf 1997). In 2000, after enrolling nearly 1,500 patients, Novartis stopped a placebo-controlled trial of fluvastatin intended to prevent hypercholesterolemia in individuals aged 70–85 years. Novartis feared that a competitor's clinical trial of a similar drug would end sooner and stated that this decision was necessary "to reallocate resources…to the newer growth assets" and cited "the competition entering the elderly segment" (Lièvre et al. 2001).

In another case, Pharmacia stopped a large-scale trial for hypertension in 2003 for financial reasons and criticisms about study design (Black et al. 2003; Psaty and Rennie 2003). The trial aimed to enroll over 15,000 patients to compare the ability of three drugs to reduce the incidence of myocardial infarction, stoke, and cardiovascular death. In 2006, Antigenics lacked sufficient funds to conduct a confirmatory trial to verify preliminary data showing that its vaccine was safe and effective in preventing recurrence of intermediate-stage renal cell carcinoma (Goldman and DeFrancesco 2009). Renal cell cancer patients spoke out about their disappointment that a potentially effective and demonstrably safe cancer vaccine might never become available (Anand 1986).

More recently, ReVision Therapeutics, Inc., stopped the development of its drug fenretinide because it lacked the funds to complete its clinical trials. The drug had been under clinical study for over four years to treat dry age-related macular degeneration, a leading cause of blindness in the elderly for which there are no FDA-approved treatments (PR Newswire 2011; Roberts 2012). Most recently, in October 2012, Aveo Pharmaceuticals, Inc., stopped two clinical trials midstream and announced a cost-cutting layoff and restructuring plan to focus the business on its more promising renal cell carcinoma drug (Aveo 2012; Bonanos 2012).

To better illustrate the consequences and concerns about corporate abandonment of clinical research for financial reasons, we take an in-depth look at Geron Corporation's SCI stem cell clinical trial.

4 The Geron SCI Clinical Trial

Geron Corporation (Menlo Park, CA) is a publicly traded company that garnered international headlines for initiating the first ever phase 1 clinical trial of a human embryonic stem cell (hESC)-based therapy for SCI. The highly publicized trial began with the enrollment of their first patient in October 2010 and came to an abrupt halt on November 14, 2011 after enrolling only 4 out of the planned 10 patients. The CEO, John Scarlett, cited the cost of the research and "capital scarcity" as the reasons (Loftus 2011). The company was abandoning its stem cell programs to focus on its cancer programs, which would produce quicker profits and issued the statement:

> we anticipate having sufficient financial resources to reach these important near-term value inflection points for shareholders without the necessity of raising additional capital. This would not be possible if we continue to fund the stem cell programs at the current levels (Geron 2011a).

Geron then laid off 66 workers, representing 38 % of its workforce.

Public expectations in this trial had been high and Geron was central in maintaining that expectation.[1] In advance of the phase 1 trial, the company had raised significant amounts of venture capital funding and, after going public, had returned to the capital market 24 times to support its programs. Geron had also obtained a $25 million loan from the state's granting agency, the California Institute for Regenerative Medicine (CIRM), to fund hESC research (Scott and Huggett 2012). Geron spent $45 million alone submitting its 22,500 page Investigational New Drug Application to the FDA, the largest application the FDA had ever received (Gawrylewski 2008). All told, Geron had spent $250 million and had taken 12 years to get to the start of this first-in-human study of hESCs (Scott and Huggett 2012). This study involved seven research centers and investigator teams that Geron trained to perform the treatment. Informed consents were long and involved, adding to worries that patients would not fully understand the risks and conflate an experimental

[1] Geron funded both the initial derivation of hESCs in 1998 and the research that produced videos in 2002 of spinal cord injured rats walking after being transplanted with cells made from hESC-derived oligodendrocyte precursors.

procedure with a treatment or cure (Kimmelman et al. 2006). After the trial commenced, some patients contacted the investigators, offering a million dollars and more to receive the cells. Another patient, a jockey paralyzed in a fall, recruited his doctor to move "heaven and earth" to get him into a trial and commenced a letter writing campaign by other paralyzed patients on his behalf (Regalado 2011).

One year after the study started, only four out of an anticipated ten subjects had been recruited and transplanted; and this was only the first of a series of studies required to merely assess safety. The fifth patient had been enrolled, but not transplanted, when the company announced its termination of the trial. After discussions with clinical staff and family, an agreement was reached to add her to the cohort and proceed with the transplant (Conger 2011).

Geron's announcement that it was discontinuing all of its cell therapy research programs—neural, cardiac, and pancreatic—was seen as a blow to the nascent field of regenerative medicine (Salahi 2011). Geron's president, David Greenwood, justified the decision by stating that the change would save the company at least $25 million per year over the next few years (Moran 2011). It announced that it would commit $8 million to wind down the SCI study and follow the patients with periodic assessments for 15 years (Geron 2011b). It refunded $6.5 million it had used from state coffers. All told, these moves allowed Geron to retain about $151 million in cash reserves.

In January 2013, BioTime, a blood plasma company, acquired Geron's stem cell assets including its stem cell intellectual property. BioTime's subsidiary, Asterias Biotherapeutics, now owns multiple lots of the hESC-derived oligodendrocytes used in the Geron trial, which were starting materials to manufacture additional lots of the cells for cancer immunotherapy, chondrocytes for cartilage and disc repair, and cardiomyocytes for heart disease. The deal transferred all of the clinical and regulatory documents pertaining to the SCI clinical trials (Brown 2013; Businesswire 2013). After two–three years of clinical follow-up with the five subjects, Asterias announced the trial was successful and that no serious adverse events associated with the cells or the associated immunosuppression had been identified (Asterias 2014). This report was followed just weeks later with news that CIRM had approved a $14.3 million award to Asterias, which would support the company's planned Phase 1/2 dose escalation trial in cervical spinal cord injury (CIRM 2014). Two and half years since the Geron trail ended, the clinical research is still waiting to restart.

Using the examples cited above, we will examine the consequences of trials prematurely stopped for financial reasons.

5 Harms to Enrolled Subjects

Stopping a trial for financial reasons may cause physical and emotional harm to human subjects, especially for those in medical need and where decision-making is compromised. In the examples cited above, the human subjects were elderly

(70–85 years of age) or were enrolling in studies of drugs to treat severe illness (cancer, diabetic renal disease, heart attack, and impending blindness). Stopping trials mid-stream on these patients most likely resulted in some emotional (if not physical) harm. In the Geron trial, the human subjects were particularly vulnerable going in since, to qualify for the study, they had to have recently experienced complete paralysis from a life-altering traumatic injury. This vulnerability is underscored by (1) the severity of injury (complete and likely permanent paraplegia); (2) post injury and surgical pain; (3) the charged emotional atmosphere of concerned family members; and importantly, (4) the brief window of time (7–14 days) in which the patient had to decide to undergo the transplant (Scott 2008; Bretzner et al. 2012; Illes et al. 2011).

The last and fifth patient to receive the stem cell transplant was particularly complex. At age 23, she was suddenly paraplegic and the decision to enroll in the study was emotionally trying. Eventually, she signed the consent but then learned that Geron had stopped the study. She did not know if the transplant would take place. She also recalls worrying about getting proper care and monitoring after the procedure. Would another company step forward and continue the research? Eventually, she was re-consented and elected to undergo the procedure after being informed of the status of the trial (Conger 2011). By her own account, she believes her decision was the right one. But she admits to being disappointed upon learning that Geron was stopping the study and she remains concerned about whether another company will take the research forward (personal communication).

Geron's subjects were warned in the consent form of the risk that the transplanted cells might cause tumor growth within the spinal cord, the consequences of which are unknown. They were also warned of the risk of developing neuropathic pain (Siddall et al. 2003).[2] The development of these adverse events would require medical care, and the subjects were not assured that Geron would cover the costs of care or other costs associated with these adverse effects after termination of the trial. Finally, patients would likely be precluded from participating in future research studies of novel SCI treatments because of the potentially confounding effects of the transplants.

Whether or not physical harm results, trials that end in this fashion can make patients feel like that they have been treated merely as means to an end, and denied opportunities to fulfill an altruistic act (Murdoch and Scott 2010). One scholar commented that Geron's human subjects had been left stranded "in a kind of twilight zone between patient and research participant" (Baylis 2011). One volunteer, Ryan Neslund, highlighted this complicated dynamic when he told a reporter about his thoughts going into the study, "Whatever the dangers are, I don't care. I just want to do something rather than nothing," adding that he was glad he participated because he hoped the cell transplant would eventually lead to something positive. But after learning that the trial had been stopped, Neslund

---

[2] Since this pain is quite common in SCI patients, it is unclear in such cases how the role of transplanted cells would be adjudicated.

said, "You have these things shot in your back and then they tell you that they ran out of money. It just doesn't seem right to me" Dizikes (2011).

Before the Geron trial ended, another patient, TJ Atchison, wrote that though he feared the development of tumors at the injection site, he had acted altruistically:

> I realized that I had a great responsibility to fulfill. I'd be the one to help doctors and researchers learn how these cells actually work in humans. I'd be able to encourage continued research in this controversial field from the perspective of someone who had been through the type of injury the researchers hope to treat (Atchison and Minus 2011).

6 Harms to Patient Communities

Patients and patient advocates follow the progress of clinical trials to learn about their failures and successes, and relay this information through websites, meetings, and advocacy efforts. Prior to Geron's SCI clinical trial, patients had been expressing their frustration about the legal, funding, and religious roadblocks that had hindered progress in the development of hESC treatments. One SCI patient expressed this frustration to a reporter:

> Imagine being paralyzed by a SCI in your teens, watching for decades as medical treatment progresses but not quite fast enough, and knowing that it could have been faster (Kinsley 2000).

Expectations for the Geron trial were especially high. The paralyzed movie actor Christopher Reeve lobbied for aggressive approaches to SCI, and spinal injured patients testified ardently in support of California's stem cell research bond initiative. Sabrina Cohen, who was paralyzed in a car accident and runs a stem cell research foundation based in Florida, summed up her dismay at the news that Geron had terminated its trial: "It was like someone ripped my heart out" (Brown 2011). Daniel Heumann, who is paralyzed and is a board member of the Christopher and Dana Reeve Foundation, said, "To get people's hopes up and then do this for financial reasons is despicable. They're treating us like lab rats" (Stein 2011). These comments illustrate the tenuous trust that exists between patient volunteers and sponsors of clinical research. Patients can likewise lose trust in the physician who recruited them for a study that was terminated. Newspaper and blog reports of these dissatisfactions can dissuade others from volunteering as human subjects.

7 Harm to Researchers and Their Institutions

When trials stop for financial reasons, researchers will certainly lose the money they would have been paid if the trial continued and may have invested time, money, and personnel in the trial that cannot be recouped. Clinicians and their trainees may be disappointed to have lost opportunities to help their patients and publish the results. The lead clinical investigator of the Geron trial at Northwestern University, Richard Fessler, said this about the Geron trial's premature end: "It is both disturbing and annoying and atypical when compared to other areas of research" (Dizikes 2011).

However, residual benefits may result. Fessler points out the advantages of learning how to purify, store, and administer the stem cell derived products. He added that the trial "keeps us thinking about (paralysis) and trying to figure out ways to treat it effectively, and it advances our knowledge of stem cell biology" (Dizikes 2011). And, the opportunity to conduct a high-profile clinical research may give an institution needed expertise and exposure required to raise funds and further develop its clinical programs.

8 Loss of Knowledge and Delay

One of the primary benefits of a clinical trial is its ability to add to generalizable knowledge. When a trial is terminated early, important scientific information often remains concealed. For industry-sponsored trials, preclinical research and information that led to federal approvals is confidential and protected to preserve corporate trade secrets (Code of Federal of Regulations[3]). If a company abandons a trial or an area of research altogether and does nothing to publish, sell or otherwise transfer the technology, then the data may be lost to the scientific community and thus to society. Also lost are opportunities to learn from past mistakes. This failed social obligation undermines trust between sponsors of research, human volunteers, medical scientists, and future stakeholders that stand to benefit. To its credit, Geron did announce that, as part of its commitment to follow the five human subjects for 15 years, it would report the results to the FDA and medical community (Geron 2011a). Asterias, the current owner of Geron's hESC technologies, did present the results of following the five patients at the 2014 American Society for Gene and Cell Therapy (ASGCT) Annual Meeting in Washington, DC (Asterias 2014).

If a trial is stopped before a reasonable judgment can be made about the safety or effectiveness of an intervention, then opportunities for continued research and the inertia required to complete the study are also lost. If a company does attempt to sell the technology, a lag will occur during efforts to find a buyer, and another lag in know-how will ensue once the transfer is made. The acquiring party may buy the technology defensively and do nothing with it, protecting its own competing products.

As noted above, a company, BioTime, did buy Geron's stem cell assets and its subsidiary, Asterias, has plans to develop the hESC technology. However, in the face of this, raising cash for a new clinical trial and filing approvals with the federal government would remain enormous obstacles for any company attempting to

[3] 21 Code of Federal Regulations, Subchapter F, Part 601.

resume Geron's research. Geron was able to overcome these tall hurdles but whether Asterias can do the same is uncertain.

9 Relevant Ethical and Social Considerations

When analyzing the ethical ramifications of these events, the magnitude of the problem matters primarily within a utilitarian construct. The fewer instances of research abandonment and the fewer people harmed, the more inconsequential the problem becomes. However, within the ethical frameworks of rights and justice, the prevalence of harm is less relevant. Violations of rights and occurrences of injustice are legitimate concerns no matter how often they occur. Below, we discuss the major ethical and social consequences that arise from these decisions.

9.1 Corporate Duty

For decades a debate has existed about the social responsibilities of business beyond maximizing shareholder value. Regardless of opinions on this question in general, many believe that bio-pharmaceutical companies should be held to a higher social responsibility standard given how fundamentally their products affect people's lives (Dresser 2006). Others have agreed that the social importance of medical products requires companies to adopt ethical obligations more in line with the medical professions (Relman and Angell 2000; Psaty et al. 2004). Especially when these companies engage in human research, their activity spills outside of the corporate realm and comes within the purview of the Declaration of Helsinki and other ethical standards that require primary emphasis on the well being of the human subject, informed and voluntary consent, and recognition that human subject research is justified because of its usefulness to society. When companies sponsor and control so many aspects of this kind of research, they have responsibilities along with the actual investigators to abide by these principal duties, to respect these rights, and to preserve the social utility of their research. The industry trade associations acknowledge these corporate duties (PhRMA 2011). These duties are derived from several ethical concepts. The stakeholder theory of business ethics requires companies to consider the consequences to the many research partners and participants impacted by corporate actions. Principles of distributive justice require that the burdens born by human subjects impose duties on sponsors that benefit from such research (The Belmont Report 1978). Fairness requires that a company mitigate the harm caused by its research, since the company initiates and controls the research and is often better positioned, financially and otherwise, to mitigate any resulting harm (National Bioethics Advisory Commission 2001).

9.2 Compromising the Risk-Benefit Contract

Researcher Steven N. Goodman, a physician and biostatistician at Johns Hopkins University, has said that when a research subject's sacrifice and altruism are for naught,

> In the ethical world, two things need to be considered—harms and wrongs. People in unnecessary trials are sometimes harmed, but I would say they are always wronged. And in the world of clinical research, wrongs are almost worse than harms (Brown 2006).

We believe that the lack of disclosure about the risk that the company will shut down the research for business reasons alone constitutes one such wrong.

In consent documents, sponsors typically state that they reserve the right to discontinue trials at any time. We call this the "reservation clause." Malmqvist and colleagues have argued that this disclosure is sufficient to fully inform subjects about this risk:

> If subjects consent to participation knowing that a trial may be stopped and why, there is no commitment, and no violation [of the consent agreement] occurs. This is so regardless of whether the trial is terminated for financial or other reasons (Malmqvist et al. 2011)

The use of the reservation clause, at least as it is usually written, does not meet Malmqvist's requirement since the typical reservation clause does not say "why" a trial may stop. Also, we believe that more is required. Without revealing the risk that the trial may be stopped for financial reasons and the subsequent consequences to the subjects, the requirement for fully informed consent will not be met and a violation of the research contract with subjects will occur.

To understand our opinion, we can compare the typical disclosure of the reservation clause to the disclosure of other risks in the consent document. For instance, subjects are informed of the *specific* risks of bodily harm and the risk of death from participation in a study. It is reasonable to assume that this kind of disclosure is sufficient to apprise the potential subject of some of the immediate downstream consequences of these risks, such as additional morbidity, the need for further medical care, or even death. However, when the consent document states that the sponsor reserves the right to terminate the trial at any time for any reason (the typical disclosure), the possibility and consequences of this risk are not so readily apparent.

The clause is likely to be discounted as boilerplate legal cover and is so nonspecific that it conveys no information. Without information, subjects have no reason to inquire whether, for example, the sponsor is sufficiently funded to finish the trial so that the subject's participation can matter in determining the safety or efficacy of the technology. The ability of patients to delve this deeply into the circumstances of any trial is not as far-fetched as it once was, now that patients can research companies and their trials online. Additionally, unlike the risk of bodily harm disclosures, only the savviest of subjects can envision personal consequences if a trial ends prematurely. From our collective experiences with clinical research, we believe that, despite reading the typical reservation clause, most subjects assume that the study will be completed and are naive to any of the harms that may flow from the sponsor abandoning the study.

Further, the contention by Malmqvist et al. that disclosing that a trial may stop means that the sponsor has made “no commitment” to finish the trial requires comment. Recruiting patients with the representation that the clinical trial can advance medical progress and lead to an approved therapeutic implies that the company has made a commitment to complete the trial. We are not arguing that a human subject who recognizes the potential for abandonment of a trial for financial reasons will necessarily weigh this heavily in her decision to participate. But, the requirement that subjects be informed so that they can provide a knowing consent means that sponsors should do more to convey information about this specific risk (if it exists) and its consequences. If the sponsor is aware of this risk, it seems only fair that the human subjects should be as well.

Finally, to avoid the therapeutic misconception, researchers make diligent efforts to convey the fact that subjects may receive no personal medical benefit from enrolling in studies. Subjects are encouraged to believe in the value of participating for the possible benefit only to medical knowledge or to future patients. When commerce and not science stops a trial midstream, the possibility of these benefits diminishes significantly, since the incomplete data set is typically not as instructive as compared to what would have emerged from the finished trial or even a trial stopped for medical reasons. The corporate sponsor thereby nullifies this basis upon which consent was given. Once the potential benefits disappear, so too do the grounds on which human subjects have given their consent (Boyd 2001).

9.3 Compromising the Social Utility of Clinical Research

Unlike when trials are stopped for scientific or clinical reasons, stopping a trial for financial reasons compromises the calculus of risk and benefit that makes such research justifiable. Typically, when trial data show that the risk-benefit ratio of continuing is no longer justifiable, it makes sense to stop in order to prevent harm to human subjects and future patients who are spared further exposure to ineffective or harmful products. Preventing this harm can thus be seen as a benefit. Harm is usually confined to those directly involved in the original research, making medically based decisions to stop generally understandable and acceptable since the net result produces more overall benefit than harm.

When the reasons for stopping are purely financial, the net effect is likely to be just the opposite—the harms outweigh the benefits. If the only reason to stop is financial, the positive risk-benefit ratio of the investigational product that initially justified the trial may still hold and that product may retain the ability to improve future medical care. Yet, it is abandoned. Therefore, when there is a significant risk at the outset that a sponsor will not be able to complete the research, the social requirements that the research has potential utility and be able to contribute to generalizable knowledge are only tenuous at best. It would obviously be preferable to devise a study plan with a higher probability of success. Otherwise, proceeding with a trial with a significant risk of financial failure undermines the product's potential, exposes subjects to avoidable risk, and wastes resources.

Loss of trust in the research endeavor may also result when companies stop trials for financial reasons. Subjects are typically informed that their welfare was the primary concern of the researcher and that IRBs exist to ensure that this is the case. When companies stop trials for financial reasons, subjects can conclude that the corporate bottom line was the real concern, as illustrated by Ryan Nusland's comments. These factors can result in an erosion of trust in the research, the consequences of which include reluctance of subjects to volunteer and increased difficulty in performing clinical research.

This does not mean that no knowledge accrues from a trial stopped in this fashion. As we explain in the Geron example, some researchers and clinical sites learned a sophisticated methodology for cell delivery into the spinal cord and techniques for expanding and manipulating cell populations. Investigators likely learned a great deal about recruitment challenges and the complications of performing transplants after patients had been stabilized with spinal fusion hardware. Regulatory agencies, too, plowed new ground with the approval process and will surely apply this knowledge to future applications. Institutions and their IRBs gained from reasoned discussions that took place before trials commenced. And though the cohort is very small, we can hope that forthcoming papers will provide the field with information about technique, outcomes (both positive and negative), and challenges for this promising area of regenerative medicine.

But this upside should not sway us from considering the duty to minimize the harms and wrongs described above. We conclude that, as regrettable as it is that clinical trials fail for scientific or medical reasons, prematurely stopping clinical trials solely for financial reasons, from an ethical standpoint, is worse. Therefore, we propose that companies take a number steps to minimize the possibility of having to stop trials for financial reasons and that all engaged in the process of corporate-sponsored trials warn about the risk and minimize the harm from such an event as much as possible.

10 Recommendations

Some commentators suggest that it is unacceptable to terminate a trial early for financial reasons if there is not yet sufficient benefit to be gained from the study to offset the risks to which participants have been exposed (Iltis 2004). Simply stated, if it is unacceptable to stop a trial, the trial should not be stopped. We understand the impetus behind this conviction but the recommendation is not practical. Even the best plans of competent, well-intentioned companies can go awry. We believe that companies have the duty to do as much as reasonably possible to prevent having to stop trials prematurely for financial reasons and, if this is not possible, to mitigate harm. The range of concern should extend beyond the company to patients, researchers, collaborating institutions, and the public at large. We encapsulate this

Table 1 Recommendations summary

Pre-trial obligations	(1) Convene an independent ethics advisory board (2) Confirm individual trials are properly funded and devise contingency plans if needed (3) Assess how potential financial failure might harm stakeholders outside the company and devise plans to insulate them from the identified harms
Intra-trial obligations	(4) Be vigilant for signs of impending financial problems (5) Refrain from hype as an investigational product enters a clinical trial
Obligations if the research has been abandoned for financial reasons	(6) Fulfill obligations to patients and researchers. Transfer data and disseminate results
Obligations of non-corporate stakeholders	(7) Review and approve protocols based on a collective, patient-centered ethic

proposition in seven recommendations that rest on corporate social responsibility with the patient at the center (Table 1).

10.1 Pre-trial Obligations

(Recommendation 1)*Convene an independent ethics advisory board.*

Given the novelty and ethical sensitivity surrounding the use of hESC's, it is interesting to note that, while Geron began its hESC program with ethics advice, it is unclear whether such advice was sought when the human research was started or when the research was abandoned (Eaton 2004).[4] Nonetheless, many biomedical companies, understanding the ethical complexity of their work, have incorporated ethics into their decision-making. Geron Bio-Med's Chief Executive Officer Simon Best recognized the need for ethics advice when he said, "We in the industry are not experts in ethics. Forming an ethics advisory board to deal with both scientific discoveries and the conduct of business is therefore a strategic and moral necessity" (Brower 2002). This statement is a recognition that ethics advisory boards (EABs) can play an important role in assisting companies to ensure that safeguards for stakeholders are in place before human trials begin (Eaton 2007). In addition, EABs can expand their usefulness by assisting companies in executing their obligations in the event that a trial ends prematurely.

[4] According to Geron, the groups that provided advice on the SCI human subjects protocol were as follows: Geron's clinical steering committee, an independent data monitoring committee, an embryonic stem cell research oversight committee, investigators, the FDA, 7 independent IRBs, and numerous other committees at the clinical trial sites. If Geron did use ethics advice, the company must have done so confidentially.

Specific to our main concern, EABs should examine whether the financial commitment, resources, and track record exist for a company to complete a planned trial. Financial experts can be appointed to the EAB to assist this deliberation. The language of informed consent should also be clear about the risk of stopping a trial for all reasonably foreseeable causes, including economic. Also, the consent should describe the consequences, plans, and funding for follow-up care if the trial ends prematurely. Any clause signaling the right of the sponsor to discontinue any study for any reasons at any time should be eliminated.

(*Recommendation* 2) *Ensure individual trials are properly funded.*

An objective, good faith analysis of the ability to fund a trial to completion is required. Sponsors certainly plan for financial success and strive to guard against losses from unsuccessful clinical programs, but they should also design trials within the company's means and plan for any reasonably foreseeable financial contingencies that may require abandoning the trial. These include avoiding overly optimistic cost and time projections, lack of sufficient funds, unrealistic faith in the ability to raise money during the trial, emerging competing products that would make this product obsolete or inadequate, and inability to meet milestone requirements of a major funder of the research. If any of these factors pose a significant risk of financial failure, the company should re-consider proceeding and devise contingency plans to better assure successful study completion. We note the likelihood that clinical research will cost more and/or take longer to complete when the therapeutic in question is novel, unique, or associated with ethical or political controversy, thereby increasing the difficulty of making these determinations. We also note that novel therapeutics associated with significant scientific concern that first-in-human research is premature also have a higher likelihood of failure.

(*Recommendation* 3) *Assess how potential financial failure might harm stakeholders outside the company and devise plans to insulate them from the identified harms.*

After an investigation of the likely consequences, informed consents should reveal to subjects and stakeholders any significant risk of abandonment for financial reasons. Patients and researchers should be told not just that the company reserves the right to abandon the trial, but that there is a risk of this particular kind of failure and the consequences that may result. If it is likely that the subjects will have ongoing medical needs after the research has been stopped, the company could consider funding a trust to pay for the costs of that future medical treatment. If these medical needs stem from exposure to a unique or first-in-human therapeutic, the company may want to identify and/or train those physicians best placed to provide competent care for the subjects. These measures can promote confidence in human volunteers and researchers and improve the willingness to participate in clinical trials.

Sponsors should also decide in advance what it would do to preserve the data and the technology if the company steps away, providing that both retain their value to society. Investigating opportunities for sale, transfer, and/or license of the intellectual property rights should be a component of this obligation. A commitment to publish the results should be made if the trial ends prematurely or if the program is abandoned.

Once the company has assured itself that it has a good faith belief that it can fund the trial to completion and that protective contingencies are in place, managers need to make a commitment to finish the trial in order to prevent the harms discussed here. If the company cannot assure itself of these factors, it should not initiate the trial.

10.2 Intra-trial Obligations

(*Recommendation* 4) *Be vigilant for signs of impending financial problems.*

It is by no means a given that markets have the same optimism in the research progress as does the company conducting it. Neither should the company assume that there is an unending appetite to fund the company's continuing research. Early detection of signs that the funding will run out should lead the company to execute contingency plans in the event that market forces impede trial progress. Such plans may take the form of the identification of better-funded research partners, co-licensing technology, or merger with another company in the same financial sector.

(*Recommendation* 5) *Refrain from hype as a product enters a clinical trial.*

Companies are always motivated to project an optimistic view of their technology, since shareholders, investors, and the marketplace value the start of an approved clinical trial. During product development, companies use optimism to drive shareholder value and raise money. They reason, correctly, that any indication that a research program is in trouble destroys value.

However, overly optimistic statements are unfairly misleading and unethical. Inflating the promise of the investigational product can also induce patients to unwisely volunteer for the research and as such violate moral norms of protecting the vulnerable (Goodin 1998; Hawkins and Emanuel 2008). Even if optimistic statements result in the infusion of capital investments, values will plunge when unembellished research data emerge or studies are abandoned.

10.3 Obligations if the Research Has Been Abandoned for Financial Reasons

(*Recommendation* 6) *Fulfill obligations to patients and researchers. Transfer data and disseminate results.*

If the company has taken reasonable steps in advance to address this possibility, mitigating the resultant harm will be much less problematic. Prior commitments to subjects and researchers can be fulfilled, the protocols can be transferred if possible, the data disclosed, and the intellectual property made available to others who have the capability of making use of it.

10.4 Obligations of Non-corporate Stakeholders

(*Recommendation* 7) *Review and approve protocols based on a collective, patient-centered ethic.*

Additionally, we propose that institutions and researchers insist that a corporate sponsor commits to completing its clinical trial before a trial begins. If there is reasonable uncertainty, sponsors should be asked by researchers and IRBs to provide evidence that proper resources exist to fund trials to completion and ensure that the findings become available to society. Furthermore, we recommend that the usual reservation clause where the sponsor "reserves the right to discontinue any study for any reasons at any time" (or similar vague language) should no longer be present in clinical trial protocols or informed consents and that IRBs should refuse to approve trials with this language. The language should make clear that the risk of stopping a trial for various reasons, including business reasons, exists and then explains what will happen in these cases. IRBs, on behalf of their clinical researchers and institutions, should ask companies what commitments they are making to take care of abandoned research subjects. Researchers should also ask the company whether they have plans to preserve the utility of the technology if the trial is stopped. This is in many ways similar to the assurances researchers seek from corporate sponsors that they will be eventually free to publish the data regardless of the outcome. Finally, institutions and their investigators should consider carefully whether it is in their and their patients' best interests to participate in future clinical trials from industry sponsors who have previously abandoned clinical trials for financial reasons.

11 Conclusions

Stopping a clinical trial prematurely for reasons related to safety, efficacy, medical unknowns, or feasibility problems is an unavoidable aspect of the endeavor that can be minimized by careful planning but not entirely eliminated. Stopping a trial because the company sponsor has run out of funds or has decided to spend its money more profitably elsewhere is different–both in the harms created and in its social and ethical acceptability. Companies should take steps to minimize the risk of trial abandonment and to protect the interests of all major stakeholders affected by the research decisions of the company but most especially the interests of the human subjects. This ethical stance obligates companies to refrain from starting clinical trials that they cannot reasonably finish, to commit to finish clinical trials that they start, to implement every reasonable strategy to prevent trial cessation for financial reasons, and to mitigate the harm caused when they cannot abide by these commitments. Research institutions, and clinical sites and their investigators have a reciprocal obligation to engage in this collective ethic and encourage the undertaking of an approach whose successful ethical structure matches a financial one.

Acknowledgments C. T. S. thanks the Stanford University Center for Biomedical Ethics and the Stanford Institute for Stem Cell Biology and Regenerative Medicine for financial support. The authors thank Judy Illes for her valuable assistance with the manuscript and Emily Borgelt and Nora Nguyen for help during the preparatory phase of the submission. We also thank Marleen Eijkholt and members of the University of British Columbia Core for Neuroethics for discussions during conceptual phases of the project.

References

Anand G (1986) Burden of proof, cancer drug fails, so maker tries new pitch. Wall Street Journal at A1

Asterias (2014) http://asteriasbiotherapeutics.com/asterias-biotherapeutics-inc-announces-new-results-from-first-in-man-clinical-trial-of-a-cell-therapy-derived-from-embryonic-stem-cells/. Accessed 25 May 2014

Atchison AT, Minus T (2011) The right decision. MIT Technol Rev. http://www.technology review.com/article/424413/the-right-decision/. Accessed 18 Jan 2013

Aveo (2012) Aveo announces strategic restructuring and provides third quarter financial results. http://investor.aveooncology.com/phoenix.zhtml?c=219651&p=irol-newsArticle&ID=1751465&highlight. Accessed 18 Jan 2013

Baylis F (2011) Geron's discontinued stem cell trial: what about the research participants? The Hastings Center Bioethics Forum. http://www.thehastingscenter.org/Bioethicsforum/Post.aspx?id=5640&blogid=140&. Accessed 30 Jan 2013

Black HR, Elliot WJ, Grandits G et al for the CONVINCE Research Group (2003) Principal results of the controlled onset Verapamil investigation of cardiovascular end points (CONVINCE) trial. JAMA 289:2073–2082

Bonanos P (2012). Preparing for Tivonzanib approval, Aveo restructures with staff cuts. The pink sheet daily. http://www.elsevierbi.com/publications/the-pink-sheet-daily?issue=Oct-30-2012. Accessed 18 Jan 2013

Boyd (2001) Early discontinuation violates Helsinki Principles. BMJ 322:605–606

Bretzner F, Gilbert F, Baylis F, Brownstone RM (2012) Target populations for first-in-human embryonic stem cell research in spinal cord injury. Cell Stem Cell 8:468–474

Brower V (2002) Biotech embraces bioethics; but are they being exemplary or expedient? And who are the bioethicists? http://www.biospace.com/b2/Articles/061499_bioethics.cfm. Accessed 22 June 2002

Brown D (2006) Superfluous medical studies called into question. Washington Post (2 Jan): A06. http://www.washingtonpost.com/wp-dyn/content/article/2006/01/01/AR2006010100749.html. Accessed 18 Jan 2013

Brown E (2011) Economics, not science, thwarts embryonic stem cell therapy, Los Angeles Times. http://articles.latimes.com/2011/nov/21/health/la-he-geron-stem-cell-20111121. Accessed 2 Feb 2013

Brown S (2013) BioTime acquires stem cell assets from Geron, raises $10 million. San Francisco Business Times. http://www.bizjournals.com/sanfrancisco/blog/biotech/2013/01/biotime-acquires-stem-cell-assets-geron.html. Accessed 18 Jan 2013

Businesswire (2013) Asterias Biotherapeutics, Inc., a subsidiary of BioTime, Inc., acquires Geron's embryonic stem cell assets. http://www.businesswire.com/news/home/20131001006076/en/Asterias-Biotherapeutics-Subsidiary-BioTime-Acquires-Geron%E2%80%99s-Embryonic. Accessed 5 Dec 2013
CIRM (2014) Asterias biotherapeutics. http://www.cirm.ca.gov/our-progress/institutions/asterias-biotherapeutics. Accessed 1 June 2014
Conger K (2011) Stanford/SCVMC team treats final patient in landmark stem cell therapy trial. http://med.stanford.edu/ism/2011/november/geron.html. Accessed 18 Jan 2013
Dizikes C (2011) Stem cell trial's cancellation disappoints paraplegic patient and a Northwestern researcher. Chicago Tribune. http://articles.chicagotribune.com/2012-01-10/news/ct-met-stem-cell-trial-20120110_1_cell-research-geron-cell-transplantation-program. Accessed 18 Jan 2013
Dresser R (2006) Private-sector research ethics: marketing or good conflicts management? Theor Med and Bioeth 27:115–139
Eaton M (2004) Geron Corporation and the role of ethics advice, in ethics and the business of bioscience. Stanford University Press, Stanford, pp 490–517
Eaton M (2007) Managing the risks associated with using biomedical ethics advice. J Bus Ethics 77:99–109
Evans S, Pocock S (2001) Societal responsibilities of clinical trial sponsors. BMJ 322:569–570
Gawrylewski A (2008) Stem cell trial nearly a go? The Scientist. http://www.the-scientist.com/?articles.view/articleNo/26829/title/Stem-cell-trial-nearly-a-go-/. Accessed 21 Jan 2013
Geron (2011a) Geron to focus on novel cancer programs http://ir.geron.com/phoenix.zhtml?c=67323&p=irol-newsArticle&ID=1635764&highlight. Accessed 22 Jan 2013
Geron (2011b) Spinal cord injury trials program. http://cell-therapies.geron.com/grnopc1. Accessed 18 Jan 2013
Goldman B, DeFrancesco L (2009) The cancer vaccine roller coaster. Nat Biotechnol 27:129–139
Goodin RE (1998) Reasons for welfare: the political theory of the welfare state. Princeton University Press, Princeton, p 33
Hawkins JS, Emanuel EJ (2008) Exploitation and developing countries: the ethics of clinical research. Princeton University Press, Princeton, p 9
Illes J, Reimer JC, Kwon BK (2011) Stem cell clinical trials for spinal cord injury: readiness, reluctance, redefinition. Stem Cell Rev and Rep 7:997–1005
Iltis AS (2004) Stopping trials early for financial reasons: the risk-benefit relationship as a moral compass. J Med Ethics 31:410–414
Keen A, Fukker JH, Menzinger G (1997) Early closure of European Pimagedine trial. Lancet 350:214–215
Kimmelman J, Baylis F, Glass KC (2006) Stem cells: lessons from gene transfer research. Hastings Cent Rep 36(1):23–26
Kinsley M (2000) Reason, faith, and stem cells. Washington Post at A17
Langer A (1997) Early stopping of trials. Lancet 350:890–891
Hopf G (1997) Early stopping of trials. Lancet 350:890–891
Lièvre M, Ménard J, Bruckert E et al (2001) Premature discontinuation of clinical trial for reasons not related to efficacy, safety, or feasibility. BMJ 322:603–605
Loftus P (2011) Geron's stem-cell exit arose from financial uncertainty. Wall Street J. http://online.wsj.com/news/articles/SB10001424052970204323904577040462527322408. Accessed 19 Nov 2013
Malmqvist M, Juth N. Lynoe N et al (2011) Early stopping of clinical trials: charting the ethical terrain. Kennedy Inst Ethics J 21: 51–78
Moran N (2011) hESC therapy pioneer Geron abandons stem cell R&D, BioWorld. http://www.bioworld.com/content/hesc-therapy-pioneer-geron-abandons-stem-cell-rd-1. Accessed 18 Jan 2013
Murdoch CE, Scott CT (2010) Stem cell tourism and the power of hope. Am J of Bioeth 10:16–23
National Bioethics Advisory Commission (2001) Ethical and policy issues in research involving human participants: report and recommendations of the National Bioethics Advisory Commission, vol 1: at 123

Psaty BM, Furberg C, Ray CW et al (2004) Potential for conflict of interest in the evaluation of suspected adverse drug reactions: use of cerivastatin and risk of Rhabdomyolysis. JAMA 292:2629

PhRMA (2011) Principles for the conduct of clinical trials and communication of clinical trial results. http://phrma.org/sites/default/files/pdf/042009_clinical_trial_principles_final_0.pdf. Accessed 19 Nov 2013

Psaty BM, Rennie D (2003) Stopping medical research to save money. JAMA 289:2128–2131

PR Newswire (2011) Phase 2b data suggest ReVision's oral fenretinide (RT-101) slows lesion growth, preserves vision and reduces incidence of neovascularization in geographic atrophy. http://www.prnewswire.com/news-releases/phase-2b-data-suggest-revisions-oral-fenretinide-rt-101-slows-lesion-growth-preserves-vision-and-reduces-incidence-of-neovascularization-in-geographic-atrophy-patients-105166174.html. Accessed 18 Jan 2013

Regalado (2011) A stem-cell gamble. MIT Technol Rev. http://www.technologyreview.com/featuredstory/424392/stem-cell-gamble/. Accessed 10 Mar 2013

Relman AS, Angell M (2000) America's other drug problem: how the drug industry distorts medicine and politics. The New Republic, 16 Dec 2000, pp 27–41

Roberts D (2012) Fenretinide may slow vision loss in patients with dry MD and Stargardt's Disease. Macular Degeneration Support. http://www.mdsupport.org/library/fenretinide.html. Accessed 18 Jan 2013

Salahi L (2011) Geron announcement throws stem cell research into question ABC News. http://abcnews.go.com/Health/Wellness/scientists-rethink-embryonic-stem-cell-research-geron-announcement/story?id=14966735. Accessed 19 Nov 2013

Siddall PJ, McClelland JM, Rutkowski SB et al (2003) A longitudinal study of the prevalence and characteristics of pain in the first 5 years following spinal cord injury. Pain 103:249–257

Scott CT (2008) Stem cells: new frontiers of ethics, law and policy. Neurosurg Focus 3–4:E24

Scott CT, Huggett B (2012) Geron's quixotic fate. Nat Biotechnol 30:497

Stein R (2011) First test of human embryonic stem cell therapy in people discontinued. The Washington. http://articles.washingtonpost.com/2011-11-14/national/35280838_1_cell-therapy-cell-research-george-q-daley. Accessed 2 Dec 2013

The Belmont Report (1978) Ethical principles and guidelines for the protection of human subjects of research, report of the National Commission for the Protection of Human Subjects of Biomedical and Behavioral Research, Sect. B.3. Available at: http://www.hhs.gov/ohrp/humansubjects/guidance/belmont.html. Accessed 19 Nov 2013

Index

Curr Topics Behav Neurosci (2015) 19: 317–320
DOI: 10.1007/978-3-662-44866-3

CPSIA information can be obtained
at www.ICGtesting.com
Printed in the USA
LVOW02*1925110316
478808LV00001B/6/P

9 783662 448656